STATISTICAL METHODS IN E-COMMERCE RESEARCH

STATISTICS IN PRACTICE

Founding Editor

Statistics in Practice is an important international series of texts which provide detailed coverage of statistical concepts, methods and worked case studies in specific fields of investigation and study.

With sound motivation and many worked practical examples, the books show in down-to-earth terms how to select and use an appropriate range of statistical techniques in a particular practical field within each title's special topic area.

The books provide statistical support for professionals and research workers across a range of employment fields and research environments. Subject areas covered include medicine and pharmaceutics; industry, finance and commerce; public services; the earth and environmental sciences, and so on.

The books also provide support to students studying statistical courses applied to the above areas. The demand for graduates to be equipped for the work environment has led to such courses becoming increasingly prevalent at universities and colleges.

It is our aim to present judiciously chosen and well-written workbooks to meet everyday practical needs. Feedback of views from readers will be most valuable to monitor the success of this aim.

A complete list of titles in this series appears at the end of the volume.

STATISTICAL METHODS IN E-COMMERCE RESEARCH

WOLFGANG JANK AND GALIT SHMUELI
Department of Decision, Operations and Information Technologies, R.H. Smith School of Business, University of Maryland, College Park, Maryland

WILEY
A JOHN WILEY & SONS, INC., PUBLICATION

This book is printed on acid-free paper.

For ordering and customer service, call 1-800-CALL-WILEY.

Wiley also publishes its books in variety of electronic formats. Some content that appears in print may not be available in electronic format. For more information about Wiley products, visit our web site at www.wiley.com.

Library of Congress Cataloging-in-Publication Data:

Jank, Wolfgang, 1970-
Statistical methods in ecommerce research/Wolfgang Jank, Galit Shmueli.
p. cm. -- (Statistics in practice)
Includes bibliographical references and index.
ISBN 978-0-470-12012-5 (cloth)
1. Electronic commerce -- Statistical methods. I. Shmueli, Galit, 1971- II. Title.
HF5548.32.J368 2008
381′.142015195--dc22

2007050394

Printed in the United States of America

10 9 8 7 6 5 4 3 2 1

CONTENTS

PREFACE

Electronic commerce (e-commerce) is part of our everyday lives. Whether we purchase a book on Amazon.com, sell a DVD on eBay.com, or click on a sponsored link on Google.com, e-commerce surrounds us. E-commerce also produces a large amount of data: When we click, bid, rate, or pay, our digital "footprints" are recorded and stored. Yet, despite this abundance of available data, the field of statistics has, at least to date, played a rather minor role in contributing to the development of methods for empirical research related to e-commerce. The goal of this book is to change that situation by highlighting the many statistical challenges that e-commerce data pose, by describing some of the methods currently being used and developed, and by engaging researchers in this exciting interdisciplinary area. The chapters are written by researchers and practitioners from the fields of statistics, data mining, computer science, information systems, and marketing.

The idea for this book originated at a conference that we organized in May 2005 at the University of Maryland. The theme of this workshop was rather unique: "Statistical Challenges and Opportunities in Electronic Commerce Research." We organized this workshop because, during our collaboration with nonstatistician researchers in the area of e-commerce, we found that there was a disconnect between the available data (and its challenges) and the methods used to analyze those data. In particular, there was a strong disconnect between statistics (which, as a discipline, is based upon the science of data) and the domain research, where statistical methods were used for analyzing e-commerce data. The conference was a great success: We were able to secure a National Science Foundation (NSF) grant; over 100 participants attended from academia, industry, and government; and finally, the conference resulted in a special issue of the widely read statistics journal *Statistical Science*. Moreover, the conference has become an annual event and is currently in its third year (2006 at the University of Minnesota, 2007 at the University of Connecticut; 2008 at New York University, and 2009 at Carnegie Mellon University). All in all, this inaugural conference has created a growing community of researchers from statistics, information systems, marketing, computer science, and related fields. This book is yet another fruitful outcome of the efforts of this community.

E-commerce has surged popularity in recent years. By *e-commerce*, we mean any transaction using the Internet, like buying or selling goods or exchanging information related to goods. E-commerce has had a huge impact on the way we live today compared to a decade or so ago: It has transformed the economy, eliminated borders, opened the door to many innovations, and created new ways in which consumers and businesses interact. Although many predicted the death of e-commerce with the burst of the Internet bubble in the late 1990s, e-commerce is thriving more than ever.

There are many, examples of e-commerce. These include electronic transactions (e.g., online purchases); selling or investing; electronic marketplaces like Amazon.com and online auctions like eBay.com; Internet advertising (e.g., sponsored ads by Google, Yahoo! and Microsoft); clickstream data and cookie-tracking; e-bookstores and e-grocers; Web-based reservation systems and ticket purchasing; marketing email and message postings on web logs; downloads of music, video, and other information; user groups and electronic communities; online discussion boards and learning facilities; open source projects; and many, many more. All of these e-commerce components have had a large impact on the economy in general, and they have transformed consumers' and businesses' life.

The public nature of many Internet transactions has allowed empirical researchers new opportunities to gather and analyze data in order to learn about individuals, companies, and societies. Theoretical results, founded in economics and psychology and derived for the offline brick-and-mortar world, have often proved not to hold in the online environment. Possible reasons are the worldwide reach of the Internet and the related anonymity of users, its unlimited resources, constant availability, and continuous change. For this reason, and due to the availability of massive amounts of freely available high-quality web data, empirical research is thriving.

The fast-growing area of empirical e-commerce research has been concentrated in the fields of information systems, economics, computer science, and marketing. However, the availability of this new type of data also comes with many new statistical challenges in the different stages of data collection, preparation, and exploration, as well as in the modeling and analysis stages. These challenges have been widely overlooked in many of these research efforts. The absence of statisticians from this field is surprising. Two possible explanations are the physical distance between researchers from the fields of information systems and statistics and a technological gap. In the academic world, it is rare to find the two groups or departments located within the same school or college. Information systems departments tend to be located within business schools, whereas statistics departments are typically found within the social sciences, engineering, or the liberal arts and sciences. The same disconnect often occurs in industry, where it appears that only now are statisticians slowly being integrated into e-commerce companies. This physical disconnect has kept many statisticians unaware of the exciting empirical work done in information systems departments. The second explanation for the disconnect is the format in which e-commerce data often arrive. E-commerce data, although in many cases publicly available on the Web, arrive in the form of HTML pages. This means that putting together a standard database requires the collection and extraction of HTML pages to obtain the desired information. These skills are not common components of the statistics education. Thus, the discipline is often unaware of web crawling and related data collection technologies which open the door to e-commerce empirical research.

Our collaboration with information systems and marketing colleagues has shown just how much the two sides can benefit from crossing the road. E-commerce data are different than other types of data in many ways, and they pose real statistical challenges. Using off-the-shelf statistical methods can lead to incorrect or inaccurate results; furthermore, important real effects can be missed. The integration of statistical thinking into the entire process of collecting, cleaning, displaying, and analyzing e-commerce data can lead to more sound science and to new research advances. We therefore see this as an opportunity to establish a new interdisciplinary area: empirical research in e-commerce.

This book is driven by two components: methods and applications. Some chapters offer methodological contributions or innovative statistical models that are needed for e-commerce empirical research. In other chapters, the emphasis is on applications, which tend to challenge existing statistical methods and thus motivate the need for new statistical thought. And finally, some chapters offer introductions or surveys of application areas and the statistical methods that have been used in those contexts. The chapters span a wide spectrum in terms of the types of methods (from probabilistic models for event arrivals, to data-mining methods for classification, to spatial models, functional models, or differential equation models), the e-commerce applications (from online auctions, to search engines, to Wikipedia), and the topics surveyed (from economic impact to privacy issues). We hope that this diversity will stir further research and draw more researchers into the field of empirical e-commerce research.

ACKNOWLEDGMENTS

We'd like to thank the many people whose help has led to the creation of this book.

To Ravi Bapna, Rob Kauffman, and Paulo Goes, who introduced us to the area of e-commerce research and have pushed for collaborations between information systems researchers and statisticians.

To Ed George, Steve Fienberg, Don Rubin, and David Banks, who have been involved in and very supportive of our efforts.

To our colleagues from the Department of Decision, Operations and Information Technologies at the Smith School of Business, with whom we informally discussed many of these ideas.

To the authors of the chapters, who have contributed their knowledge and time in support of the book.

To the many reviewers who helped improving the content of this book.

And to our families, Angel Novikov-Jank, Waltraud, Gerhard and Sabina Jank, and Boaz and Noa Shmueli, for their endless support and encouragement.

CONTRIBUTOR LIST

Deepak Agarwal, Yahoo! Research, Santa Clara, CA, USA

Edoardo M. Airoldi, Computer Science Department and Lewis-Sigler Institute for Integrative Genomics, Princeton, University, Princeton, New Jersey

David L. Banks, Department of Statistics, Duke University, Durham, North Carolina

Ulf Böckenholt, Faculty of Management, McGill University, Montreal, Canada

Mayukh Dass, Area of Marketing, Rawls College of Business, Texas Tech University, Lubbock, TX

Stephen E. Fienberg, Department of Statistics, Machine Learning Department, and Cylab Carnegie Mellon, University, Pittsburgh, Pennsylvania

Chris Forman, College of Management, Georgia Institute of Technology, 800 West Peachtree Street NW, Atlanta, Georgia

Anindya Ghose, Information, Operations and Management Sciences Department, Leonard Stern School of Business, New York University, New York, New York

Avi Goldfarb, Rotman School of Management, University of Toronto, 105 St George St, Toronto, Ontario, Canada

Avi Harel, Ergolight Ltd., Haifa, Israel

Valerie Hyde, Applied Mathematics and Scientific Computation Program, University of Maryland, College Park, Maryland

Wolfgang Jank, Department of Decision and Information Technologies, R.H. Smith School of Business, University of Maryland, College Park, Maryland

P.K. Kannan, Department of Marketing, R.H. Smith School of Business, University of Maryland, College Park, Maryland

Robert J. Kauffman, W.P. Carey Chair in Information Systems, W.P. Carey School of Business, Arizona State University, Tempe, AZ 85287

Ron S. Kenett, KPA Ltd., Raanana, Israel, and Department of Applied Mathematics and Statistics, University of Torino, Torino, Italy

Benn Konsynski, Emory University, Goizueta Business School, Atlanta, GA

Mahesh Kumar, Department of Decision and Information Technologies, R.H. Smith School of Business,University of Maryland, College Park, Maryland

Bitao Liu, Department of Statistics, University of California, Davis, California

Hans-Georg Müller, Department of Statistics, University of California, Davis, California

Alon Matas, Media Boost Ltd., Ohr Yehuda, Israel

Eric Overby, Georgia Institute of Technology, College of Management, Atlanta, GA

Nitin R. Patel, Massachusetts Institute of Technology and Cytel Software, Cambridge, Massachusetts

Claudia Perlich, IBM T.J. Watson Research Center, Yorktown Heights, New York

Srinivas K. Reddy, Department of Marketing and Distribution, Terry College of Business, University of Georgia, Athens, Georgia

Saharon Rosset, IBM T.J. Watson Research Center, Yorktown Heights, New York

Fabrizio Ruggeri, CNR IMATI, Milano, Italy

Ralph P. Russo, Department of Statistics and Actuarial Science, The University of Iowa, Iowa City, Iowa

Yoni Schamroth, Media Boost Ltd., Ohr Yehuda, Israel

Galit Shmueli, Department of Decision and Information Technologies, R.H. Smith School of Business, University of Maryland, College Park, Maryland

Nariankadu D. Shyamalkumar, Department of Statistics and Actuarial Science, The University of Iowa, Iowa City, Iowa

Peter G.M. Van Der Heijden, Department of Methodology and Statistics, Utrecht, The Netherlands

Bin Wang, Assistant Professor, College of Business Administration, University of Texas—Pan American, Edinburg, TX 78539

Shanshan Wang, Modeling and Analytical Services, DemandTec Inc., San Carlos, California

Robert H. Warren, School of Computer Science, University of Waterloo, Waterloo, Ontario, Canada

SECTION I

OVERVIEW OF E-COMMERCE RESEARCH CHALLENGES

1

STATISTICAL CHALLENGES IN INTERNET ADVERTISING

DEEPAK AGARWAL

Yahoo! Research, Santa Clara, CA, USA

1.1 INTRODUCTION

Internet advertising is a multi-billion-dollar industry, as is evident from the phenomenal success of companies like Google, Yahoo, Microsoft, and continues to grow at a rapid rate. With broadband access becoming ubiquitous, Internet traffic continues to grow in both volume and diversity, providing a rich supply of inventory to be monetized. Fortunately, the surge in supply has been accompanied by an increase in demand, with more dollars being diverted to Internet advertising relative to traditional advertising media like television, radio, and newspapers.

Marketplace designs that maximize revenue by exploiting billions of advertising opportunities through efficient allocation of available inventory are the key to success in this scenario. Due to the massive scale of the problem, an attractive way to accomplish this is by learning the statistical behavior of the environment through the huge amounts of data constantly flowing into the system. Furthermore, automated learning reduces overhead and has a low marginal cost per transaction, making Internet advertising a lucrative business. However, learning in these scenarios is highly nontrivial and gives rise to a series of challenging statistical problems, including prediction of rare events from massive amounts of high-dimensional data, experimental designs to learn emerging trends, and protecting advertisers by constantly monitoring traffic quality. In this chapter, I provide a perspective on some of the statistical challenges through illustrative examples.

Statistical Methods in e-Commerce Research. Edited by W. Jank and G. Shmueli

1.2 BACKGROUND

Web advertising supports a broad swath of today's Internet ecosystem, with an estimated $15.7 billion in revenues for 2005 (www.cnnmoney.com). Traffic and content on the Web continue to grow at a rapid rate, with users spending a larger fraction of their time on the Internet. This trend has caught the eye of the advertising industry, which has been diverting more advertising dollars to the Internet. Thus, revenue continues to grow, both in the United States and in international markets.

Currently, two main forms of advertising account for a large fraction of the total Internet revenue. The first, called Sponsored Search advertising, places ads on result pages from a Web search engine like Google, Yahoo!, or MSN, where the ads are driven by the originating query. In contrast to these search-related ads, the second, more recent advertising mechanism, called Contextual Advertising or Content Match, refers to the placement of commercial text ads within the content of a generic Web page. In both Sponsored Search and Content Match, usually there is a commercial intermediary, called an *ad network*, in charge of optimizing the ad selection, with the twin goals of increasing revenue (shared by the publisher and the ad network) and improving the user's experience. Typically, the ad network and the publisher are paid only when the user visiting the Web page or entering keywords in a query box *clicks* on an advertisement (often referred to as the *pay-per click* (*PPC*) model. For instance, both Google and Yahoo! have such ad networks in the context of Content Match which cater to both large Web publishers (e.g., AOL, CNN) and small Web publishers (e.g., owners of blog pages). Introduced by Google, Content Match provides an effective way to reward publishers who are creators of popular content. In Sponsored Search, most major search engines often play the twin roles of publisher and ad network; hence, they receive the entire proceeds obtained from clicks on advertisements.

Yet another form of Internet advertising that still has a lucrative market is the display of graphical or banner ads on content pages. For instance, this advertising model is used extensively by Yahoo! on its properties like Mail, Autos, Finance, and Shopping. One business model charges advertisers by the number of displays or *impressions* of advertisements instead of clicks. In general, this is a rapidly evolving area and there is scope for new revenue models.

Of the three forms of Internet advertising just discussed, Sponsored Search typically display ads that are more relevant since the keywords typed by the user in the query box are often better indicators of user intent. In Content Match, user intent is inferred indirectly from the context and content of the page being visited; hence, the ads being shown typically tend to be less relevant than those on Sponsored Search. For banner ads, intent information is typically weaker compared to both Sponsored Search and Content Match; it is generally used by advertisers as a brand awareness tool. In both Sponsored Search and Content Match, since advertisers are charged only when ads are clicked (the amount paid is often called *cost per click* or *CPC*), the clicks provide a meterable way to measure user feedback. Also, advertisers can monitor the effectiveness of their Sponsored Search or Content Match advertising campaigns by tracking *conversions* (sales, subscriptions, etc.)

that accrue from user visits routed to their Websites through clicks on ads in Sponsored Search and Content Match. For banner ads, advertisers are typically charged per display (also called *cost per milli* (thousand) or *CPM*). As expected, CPC for Sponsored Search is typically higher than that for Content Match, and the CPM model in banner ads typically yields lower revenue than Sponsored Search and Content Match per impression. Finally, all three advertising mechanisms are automated procedures with algorithms deciding what ads to display in which context. Automation enables the system to work at scale, with low marginal cost, and leads to a profitable business.

The rest of the chapter is organized as follows. We begin by providing a brief high-level overview of search engines in Section 1.3. In Sections 1.4 and 1.5, we provide a brief description of ad placement in the context of Sponsored Search and Content Match, followed by a detailed description of the important statistical problem of estimating click-through rates. Section 1.6 describes the problem of measuring the quality of clicks received in Sponsored Search and Content Match, also known as *click fraud* in popular media. In Section 1.7, we discuss next-generation search engines and the challenges that arise thereof. We conclude in Section 1.8.

1.3 SEARCH ENGINES

This section provides a brief high-level overview of how search engines work. This is useful in understanding some of the statistical challenges we discuss later in the chapter.

Before delving into the details of search engine technology, we provide a brief description of how the World Wide Web (WWW) works. In the most common scenario, a user requests a webpage by typing in an appropriate URL on the web browser. The page is fetched via an http (protocol to transmit data on the WWW) request issued by the user's web server (typically, a machine running a software program called Apache) to the destination web server. The transmission of data takes place through a complex mechanism whereby another server, called the *Domain Names Server* (DNS), translates the URL, which is in human-understandable language, into an IP address. The IP address is used to communicate to the destination web server through special-purpose computers called *routers*. With the availability of broadband technology, this entire mechanism is amazingly fast, typically taking only a few milliseconds. Once the destination server receives the request, it transmits the requested page back to the user's web server via the routers.[1] The files requested are mostly written in Hypertext Markup Language (HTML) (files in other formats, like ppt and pdf, can also be requested), in which tags are used to mark up the text. The tags enable the browser to display the text content on the requested HTML page. The HTML page contains a wealth of information about the webpage and is extremely useful in extracting features that can be used for various statistical modeling tasks. Among other things, it contains *hyperlinks* that are

[1] A complete description of how this transfer takes place is beyond the scope of this chapter.

typically URLs providing links to other pages. The hyperlinks are extremely useful and have been used for various modeling tasks, including computation of the popular PageRank algorithm (Page et al. 1998). Each hyperlink is annotated with text called *anchor text*. Anchor text provides a brief, concise description of pages and hence serves as a useful source from which important features can be extracted. For instance, if anchor text from several hyperlinks pointing to a page agree closely on the content, we get a fairly good idea of page content.

We now provide a brief overview of how search engines work. There are three main steps: (a) continuously getting updated information on the WWW by running automatic programs called *crawlers* or *spiders*; (b) organizing content in retreived pages efficiently, with the goal of quick retrieval during query time (called *indexing*); (c) at query time, retrieving relevant pages and displaying them in rank order, with the more relevant ones being ranked higher. This has to be done extremely fast (typically in less than a few milliseconds).

1.3.1 Crawler

The Web is huge and diverse, and storage space and network bandwidth are finite. Hence, it is not feasible for search engines to keep a current copy of the entire WWW. Thus, the crawler has to be smart in selecting the pages to crawl. Typically, the search engine starts with a seed of domain names, crawls their home pages, crawls the hyperlinks on these pages, and recurses. The problem is compounded since the arrival rate of new content on the Web is high, and it is important to keep up with fresh content that might be of interest to users. There are several other technical issues that make crawling difficult in practice. Servers are often down or slow, hyperlinks can put the crawler into cycles, requests per second on an individual website are limited due to politeness rules, some websites are extremely large and cannot be crawled in a small amount of time while obeying the politeness rules, and many pages have dynamic content (also referred as the *hidden web*) which can be only retrieved by running a query on the page. Prioritizing page crawls is a challenging sequential design problem. Note that sampling of pages here is more involved than traditional sequential design due to the graph structure induced by hyperlinks. The sequential design should be able to discover new content efficiently under all the constraints mentioned above to keep the index fresh. Also, we want to minimize the number of recrawls for pages that do not change much. In other words, we may want to recrawl pages based on their estimated change frequency (see Cho and Ntoulas 2002; Cho and Garcia-Molina 2003 for details). However, crawling high-frequency pages may not be an optimal strategy to discover new content. A naive strategy of crawling all new pages may also be suboptimal. This is so because old low-change-frequency pages may contain links to other old but high-change-frequency pages which, in turn, provide links to a large number of new pages. What is the best trade-off between recrawling old pages and crawling new pages? Detailed discussion of this and some other issues mentioned above can be found in Dasgupta et al. (2007), along with an initial formulation using the *multi–armed bandit* framework, perhaps one of the oldest formulations of sequential design popularized in statistics by seminal works of Gittins (1979) and Lai and Robbins (1985). The main idea in multi–armed bandit problems is to devise an adaptive sampling procedure which will identify the best of k given hypothesis using a small number of samples. The sampling procedure at any given time point is a rule which depends on the outcomes that have been observed so far. Adaptive sequential designs are routinely used in statistics (Rosenberger and Lachin 2002) in the context of clinical trials, but in this context the problem is high-dimensional, with constraints imposed by the structure of the hyperlink graph. Moreover, the objective function here is quite different from the ones used in clinical trials literature. This is a promising new area of research for statisticians with expertise in experimental design and sampling theory.

Once we crawl a page, the next question is, what information should we store about the page? The typical information we store includes words in the title, body, inlinks, outlinks, anchor text, etc. Some pages might be long, and storing every word might not add much value. Thus, the statistical problem here is to characterize a set of sufficient statistics that capture most of what the page is about. This is also referred to as *feature extraction* in machine learning and data mining. Of course, computing such sufficient statistics would require a statistical model, which may be driven by editorial judgments on a small set of pages and click feedback obtained on a continuous basis when pages are shown by the search engine in response to queries. Several such models based on ideas from machine learning, data mining, and statistics are currently used by search engines, the details of which are often closely guarded secrets. However, there is substantial scope for improvement. The abstract statistical problem in this context can be stated as follows: Given an extremely large number of features and two response variables, the first one being more informative but subjective and costly to obtain and the second one being less informative but inexpensive to obtain, how does one devise statistical procedures that can do effective variable selection? The problem gets even more complex since a nonignorable fraction of pages are affected by spam, which is perpetuated mainly to manipulate the ranking of pages by search engines.

1.3.2 Indexing

Once content from crawled pages is extracted, one needs to organize it to facilitate fast lookup at query time. This is done by creating an *inverted index*. In general, this is done by first forming a dictionary of features and, for each feature, associating all document identities that contain the feature. In reality, the index is huge and has to be spread across several machines, with clever optimization tricks used to make the lookup faster.

1.3.3 Information Retrieval

The last step consists of procuring documents from the index in response to a query and displaying them in a rank-ordered fashion. This is an active area of research in computer science, SIGIR being a major conference in the area. We refer the reader to Manning et al. (2008) for an introduction. At a high level, the search engine looks up words contained in the query in the inverted index and retrieves all relevant pages. It then rank orders the documents and displays them to the user. The entire process has to be fast, typically taking a few milliseconds. The ranking is based on a number of criteria, including the number of words on the page that match the query, the location of matches on the page, frequency of terms, page rank (which provides a measure of the influence the page has on the hyperlink graph), the click rate on the page in the context of a query, editorial judgments, etc. Again, creating algorithms to combine information from such disparate sources to provide a single global ranking is a major statistical challenge that determines the quality of the search engine to a large extent. In general, algorithms are trade secrets and are not revealed. Also, making changes to algorithms is routine but evaluating the effect of these changes on quality is of paramount importance. One effective way to solve this problem is through classical experimental design techniques (e.g., factorial designs).

1.4 ESTIMATING CLICK-THROUGH RATES

We now provide a brief high-level description of procedures that are used to place ads both in Sponsored Search and Content Match. We then introduce an important problem of estimating click-through rates (CTR) in both Sponsored Search and Content Match and discuss some statistical challenges.

In Sponsored Search, placement of ads in response to a query depends on three factors: (a) relevance of the ad content to the query, (b) the amount of money an advertiser is willing to pay per click on the ad, and (c) the click feedback received for the ad. Relevance is determined by keywords that are associated with ads and decided by advertisers a priori when planning their advertising campaigns. Along with the keyword(s), the advertiser also places a *bid* on each ad, which is the maximum amount he or she is willing to pay if the ad is clicked once. Typically, advertisers also specify a *budget*, i.e., an upper bound on the amount of money they can spend. As with search results, candidate ads to be shown for each query are obtained by matching keywords on ads with the query. The exact forms of matching functions are trade secrets that are not revealed by ad networks. In general, there is an algorithm that determines if the keyword(s) match(es) the query exactly (after normalization procedures like removing stop words, stemming, etc.) and a series of algorithms which determine if there is a close conceptual match between query and keyword(s). The candidate ads are then ranked according to revenue ordering, that is, according to a product of the bid and a relevance factor related to the expected CTR of the ad. Thus, an ad can be highly ranked if it is highly relevant (i.e., CTR is high), and/or the advertiser is willing to pay a high price per click. The rankings determine the placement of ads on the search engine page: the top-ranked ad is placed in the top slot, and so on. The CTR, of ads placed higher on a page is typically higher than the CTR of ads placed in lower slots. The actual amount paid by an advertiser when a click occurs is determined by an extension of the second price auction (Varian 2007; Edelman et al. 2006) and in general depends on the CTR and bid of the ad that is ranked directly below the given ad. In the most simple form, if all CTRs are equal, an advertiser's payment per click is the bid of the next highest bidder.

In Content Match, every showing of an ad on a webpage (called an *impression*) constitutes an event. Here, among other things, matching of ads is based on the content of the page, which is a less precise indicator of user intent than a query provided in Sponsored Search.

In both Sponsored Search and Content Match, estimating the CTR for a given (query/page, ad) pair in different contexts is a challenging statistical problem. The context may include the position on the page where the ad is placed, user geography derived from the ip address, other user features inferred from browsing behavior, time-of-day, day-of-week, day-of-year information, etc. A rich class of features is also available from the query/page and the ad. The estimation problem is challenging for several reasons, some of which are as follows:

- *Data sparsity*: The feature spaces are extremely large (billions of query/pages, millions of ads, with great diversity and heterogeneity in both query/pages and ads) and the data are extremely *sparse*, since we observe only a few interactions for a majority of query/page-ad feature pairs.
- *Rarity of clicks*: The CTR, defined as the number of clicks per impression (number of displays) for a majority of page-ad feature pairs, is small.
- *Massive scale*: The number of observations available to train models is huge (several billions), but one generally has access to a grid computing environment. This provides a statistical computing challenge of scaling up computations to fit sophisticated statistical models by harnessing the computing power available.
- *Ranking*: Although we have formulated the problem as estimating CTRs, in reality what is needed is a method that can rank the ads. Thus, transforming the problem to predict a monotone function of CTR to produce a rank-ordered list is a good approach and opens up new opportunities to obtain *clever* approximations.

To provide an idea of the sparsity inherent in the data, Figure 1.1a shows the frequency of (page, ad) pairs and Figure 1.1b shows the same distribution for a subset of impressions where a user clicks on the ad being shown on the page from a Content Match application. Clearly, an overwhelming majority of (page, ad) pairs are extremely rare, and a small fraction account for a large fraction of total impressions and clicks. Naive statistical estimators based on frequencies of event occurrences incur high statistical variance and fail to provide satisfactory predictions, especially for rare events. The usual procedure involves either removing or aggregating rare events to focus on the frequent ones. While this might help estimation at the "head" of the curve, the loss in information leads to poor performance at the "tail." In Internet advertising, the tail accounts for several billion dollars annually, making reliable CTR estimation for tail events an important problem.

Replacing pages and ads with their features and fitting a machine learning model is an attractive and perhaps the most natural approach here. In general, a machine

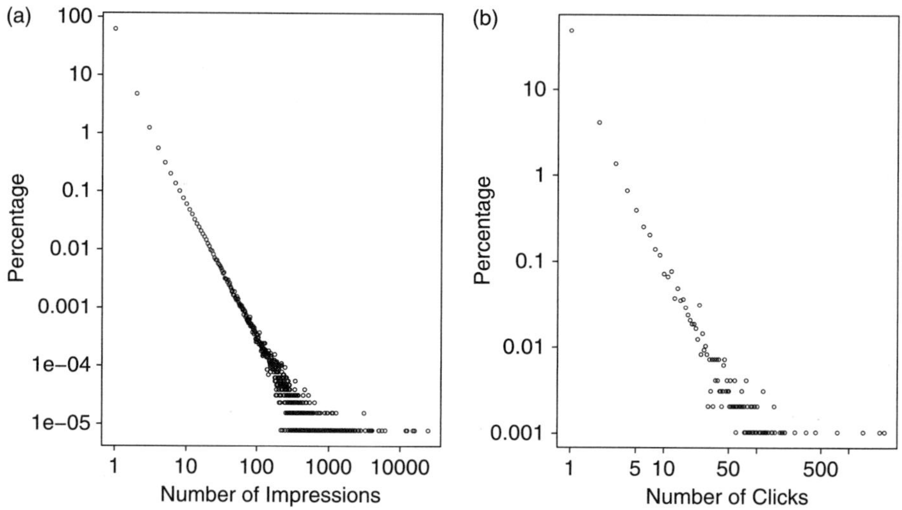

Figure 1.1 (a) Distribution of impression events and (b) click events. Plots are on log-log scale but ticks are on the original scale; 99.7% of impression events had no clicks.

learning model with a reasonable number of features performs well at the head; the problem begins when one starts fitting features to "chase" the tail. A large fraction of features tend to be sparse, and we may end up overfitting the data. One solution to this problem is to train machine learning models on huge amounts of data (which are available in our context), but that opens up the problem of scaling computations. Typically, one has access to a grid computing environment which is generally a cluster of several thousand computers that are optimized to perform efficient distributed computing. However, algorithms for fitting machine learning and statistical models were not developed to perform distributed computing, and hence the subject needs more research.

The rarity of clicks with sparseness of features makes the problem even more challenging. There is substantial literature on machine learning for predicting imbalanced or rare response variables (Japcowicz 2000; Chawla et al. 2003, 2004). Most of the approaches rely on sampling the majority class to reduce the imbalance. In statistics, the paper by King and Zeng (2001) discusses logistic regression with rare response. The authors note that with extreme imbalance, the logistic regression coefficients can be sharply underestimated, and suggest sampling and bias correction as a remedy. Recently, an interesting paper by Owen (2007) derived the limiting behavior of logistic regression coefficients as the amount of imbalance tends to infinity. The authors provides a $O(p^3)$ (p is the number of features) algorithm to compute the regression coefficients. However, the method requires estimation of feature distribution for cases in the majority class. This is a daunting task in our scenario. Further research on methods to predict rare events in the presence of large and sparse features is required. Methods based on "shrinkage" estimation may prove useful here. However, the challenge is to scale them to massive datasets. Some recent work that may be relevant includes techniques described in Ridgeway and Madigan (2002) and Huang and Gelman (2005). Yet another approach that has been pursued in the data mining community is that of scaling down the data using an approach called *data squashing* (Du Mouchel 2002; Du Mouchel and Agarwal 2003). Another approach that could be useful to reduce the dimension of feature space is clustering. However, the clustering here is done to maximize predictive accuracy of the model as opposed to a classical clustering approach that finds homogeneous sets in the feature space. It is also possible that clustering using an unsupervised approach may provide a good set of features for the prediction task and simplify the problem. The actual algorithms that are currently used by search engines are a complex combination of a number of methods.

1.5 ONLINE LEARNING

The discussion in the previous section pertains to estimating CTRs using retrospective data. Theoretically, if a model can predict CTR for all query/page, ad combination in different contexts, we are done. However, the number of queries/pages and ads is *astronomical*, making this infeasible in practice. Hence, one only ranks a subset of ads for a given query/page. The subset is decided based on some relevance criteria (e.g., consider only sports ads if the page is about sports). Thus, a large portion of query/page, ad space remains unexplored and may contain combinations that can lead to a significant increase in revenue. Also, the system is nonstationary and may change over time. Thus, ads that have been ruled out completely today in a given context might become lucrative after a month, but the retrospective estimation procedure would fail to discover them since it does not collect any data on such events. Designing efficient experiments to recover some of the lost opportunities is an important research problem that may lead to significant gains. Online learning or sequential design provides an attractive framework whereby a small fraction of traffic gets routed to the online learning system to conduct live experiments on a continuous basis. Although several online learning procedures exist, we will discuss the complexity of the problem and propose some potential solutions using a multi–armed bandit formulation.

We begin by providing a high-level overview of the multi–armed bandit problem and establish the connection to the CTR estimation problem in our context. In particular, we illustrate ideas using Content Match. The *multi–armed bandit problem* derives its name from an imagined slot machine with $k(\geq 2)$ arms. The ith arm has a payoff probability p_i which is unknown. When arm i is pulled, the player wins a unit reward with payoff probability p_i. The objective is to construct N successive pulls of the slot machines to maximize the total expected reward. This gives rise to the familiar explore/exploit dilemma where, on the one hand, one would like to gather information on the unknown payoff probabilities, while on the other hand, one would like to sample arms with the best payoff probabilities empirically estimated so far. A bandit policy or allocation rule is an adaptive sampling process that provides a mechanism to select an arm at any given time instant based on all previous pulls and their outcomes. Readers lacking a background in statistics may ignore the technical details in the next two paragraphs, but it will be insightful to understand the essential idea of the sampling process; the sampling scheme selects an arm that seems to have the potentail of getting the highest payoff at a given time instant. Thus, an arm with a worse empirical mean but high variance might be preferred to an arm with a better mean but low variance (exploration); after the sampling is continued for a while, we should learn enough to sample the arm that will provide the highest payoff (exploitation). A good sampling scheme should reach this point quickly. For instance, treating the ads that could be shown on a fixed webpage as arms of a bandit, an ad that has been shown on the page only twice and has received 1 click might be placed again on the page compared to an ad that had been shown 100 times and received 55 clicks.

A popular metric used to measure the performance of a policy is called *regret*, which is the difference between the expected reward obtained by playing the best arm and the expected reward given by the policy under consideration. A large body of bandit literature has considered the problem of constructing policies that achieve tight upper bounds on regret as a function of the time horizon N (total number of pulls) for all possible values of the payoff probabilities. The seminal work of Lai and Robbins (1985) showed how to construct policies for which the regret is $O(\log N)$ asymptotically for all values of payoff probabilities. The authors further proved that the asymptotic lower bounds for the regret are also $\Omega(\log N)$ and constructed policies that actually attain them. Subsequent work has constructed policies that are simpler and achieve the logarithmic bound uniformly rather than asymptotically (see Auer et al. 2002 and references therein). The main idea in all these policies is to associate with each arm a priority function which is the sum of the current empirical payoff probability estimate plus a factor that depends on the estimated variability. Sampling the arm with the highest priority at any point in time, one explores arms with little information and exploits arms which are known to be good based on accumulated empirical evidence. With increasing N, the sampling variability is reduced and one ends up converging to the optimal arm. This clearly shows the importance of the result proved by Lai and Robbins (1985), which proves that one cannot construct the variance adjustment factor to make the regret better than $\Omega(\log N)$, thereby providing a benchmark for evaluating policies.

Two policies that both have $O(\log N)$ regret might involve different constants in the bounds (the constant depends on the *margin* of the bandit, which is the difference between the payoffs of the best two arms; the smaller the margin, the higher the constant) and may behave differently in real applications, especially when considering short-term behavior. One way of comparing the short-term behavior of policies that are otherwise optimal in the asymptotic sense is by using simulation experiments. One can also evaluate short-term behavior by proving the finite sample properties of policies, but this may become extremely hard to derive except in simple situations. The main difficulty is caused by the presence of dependencies in the sampling paths.

Focusing on Content Match for the sake of illustration, we can consider the online learning problem of matching ads to pages as a set of bandit processes. Thus, for each page, we have a bandit where ads are the arms and CTRs are the payoff probabilities. However, high dimensionality makes the bandit convergence slow and involves a significant amount of exploration leading to revenue loss. In fact, asymptotic guarantees are not good enough in our situation, and we need procedures that can guarantee good short-term performance. Also, we need to learn the CTRs of the top few arms instead of the best arm, since we may run out of best ads due to budget constraints imposed by advertisers. Hence, given two policies that have similar revenue profiles, we would prefer the one whose CTR estimates have lower mean squared error.

To deal with the difficulties mentioned above, reducing dimensionality is of paramount importance. One approach is to assume that CTRs are simple functions of both page and ad features (Abe et al. 2003). Another approach is to cluster the pages and ads and conduct learning at coarser resolutions. Panoly et al. (2007) discuss such an approach where CTRs are learned at multiple resolutions, from coarser to finer, by using an online multistage sampling approach coupled with a Bayesian model. The authors report significant gains compared to a bandit policy that uses single-stage sampling. Further, they show that use of a Bayesian model leads to substantial reduction in mean square error without incurring loss in revenue. We note that sequential designs have been mainly considered in statistics in the context of clinical trials (see Rosenberger and Lachin 2002 for an overview). However, the problems in Internet advertising are large and require further research before sequential designs become an integral part of every ad network.

1.6 DISCOUNTING ADVERTISER TRAFFIC

The pay-per-click (PPC) revenue model used in Sponsored Search and Content Match is prone to abuse by unscrupulous sources. For instance, in Content Match, publishers who share the revenue proceeds from advertisers with the ad network might be tempted to use a service which uses sophisticated methods to produce false clicks for ads shown on the publisher's webpage. Although ad networks may benefit in the short term, collusion between publishers and ad networks is ruled out since such false clicks dilute the traffic quality received by advertisers through clicks on ads and lead to substantial losses to the ad network in the long run. Hence, monitoring traffic quality on the publisher's webpage is extremely important and, to a large extent, determines the feasibility of the PPC model in the long run. Ad networks have built sophisticated systems to detect such false clicks in order to protect their advertisers. In Sponsored Search, a competitor might use a similar behavior to drain a competitors' advertising budget. The problem, popularly known as *click fraud*, has received a lot of attention in recent times, including lead articles in *Business Week* and the *New York Times*. Another fraudulent behavior used in Sponsored Search is known as *impression fraud*. Here, an advertiser may use a robot to artificially inflate the impression volume and hence substantially deflate the CTR of competitors' ads. This, in turn, increases the rank of the advertiser's ads (ads are ranked using relevance measured by both CTR and bid) and increases his or her CTR. Thus, the advertiser gets better conversion rates at a lower cost.

The problems described above are difficult, and a complete solution seems to be elusive at this time. Simple frauds[2] intiated by a single individual manually (e.g., relatives of a blog owner clicking on ads, a person hired to click on ads of a competitor) are fairly obvious. Those that are initiated by more sophisticated means (e.g., randomizing false clicks over a large set of ips) are difficult to detect. An indirect approach is to use labels on good clicks to determine overall quality of clicks that are received on a publisher's website in Content Match and for an advertiser in Sponsored Search. Such labels can be obtained by tracking the behavior of users once they get to the landing page (the website of the advertiser) of the clicked ad. However, such data might be hard to obtain since advertisers are reluctant to allow the ad network to track the revenue generated through advertisements. Fortunately, some advertisers (not representative of the entire population) have agreed to share such data with the ad network, providing a valuable resource to validate automated algorithms built to detect false clicks. As more advertisers agree to provide such data, the situation will improve. The ideal approach here would be to have algorithms which can score every click as valid or invalid in an online fashion. However, this may be too ambitious, and an alternative approach which provides a global measure of click quality separately for advertisers and publishers based on a large pool of click data retrospectively may be a more feasible approach. Hybrid approaches that combine online and offline scoring may also be attractive.

Statistics has an important role to play here. One helpful approach is to detect abnormal click behavior in the highly multidimensional feature space that includes ip addresses, queries, advertisers, users (tracked by their browser cookie), ads and their associated features, and webpages and their associated features through time. This problem, known as *anomaly detection*, has received considerable attention in recent times in biosurveillance (Fienberg and Shmeuli 2005; Agarwal et al. 2006), telecommunications (Hill et al. 2006), monitoring help lines (Agarwal 2005), and numerous other areas. However, the percentage of anomalies in all the applications cited above is rare, which is typically not the case for click fraud. Popular press articles cite numbers ranging from 10% to 15% (although the distribution across several segments can vary widely). Time series methods to monitor the system over time (e.g., West and Harrison 1997) are germane in this context. Semisupervised learning approaches (sequentially labeling data to learn a classifier with a small set of labels but a large set of unlabeled examples) (Chapelle et al. 2006) are also important in this context. Not much research has been done in the statistics literature on semisupervised learning.

1.7 SOCIAL SEARCH

Internet advertising and search engines are a recent phenomenon, but they have had a profound impact on our lives. However, the current technology is constantly changing, and statisticians, computer scientists, machine learners, economists, and

[2]The term *fraud* is used loosely here; it means "unethical" in this context.

social scientists have an important role in shaping the next generation of search engines and Internet advertising. One important direction is social search. The popularity of Web-based tagging systems like Del.icio.us, Technocrati, and Flickr, which allow users to annotate resources like blogs, photographs, web pages, etc with freely chosen keywords (*tags*) (see Marlow et al. 2005 for an overview) has provided a rich source of data that can potentially be exploited to improve and broaden search quality, which will in turn increase ad revenue. These tagging systems also allow users to share their tags among friends. How does one exploit this rich source of information and the corresponding social network among users to enhance search quality? Let us consider the social bookmarking site Del.icio.us, for example. In Del.icio.us, users can *post* the URLs (called *artifacts*) of their favorite webpages into their Del.icio.us account and annotate these artifacts with informative tags. Users can also include their friends and other like-minded people in their social network. When searching for artifacts relevant to a particular keyword, it seems intuitive that apart from the relevance of content in artifacts to the keyword, one could further improve the relevance of search results by incorporating the tagging behavior of the user and others in his or her social network. For instance, a search for the keyword *conference* by the author should rank all statistics conference higher for the author, since most of his friends have bookmarked recent statistics conferences. A matching based on content alone might provide high rank to a conference on chemistry, which is perhaps not that interesting to the author. Incorporating user tags and the social network of users to personalize the search is a promising new area.

Currently, the search engine and publisher network monetizes its services through an ad network. Is it possible to build a network where individuals in a social network actively participate in providing answers to queries and enhancing the search? How would one design incentives to create a reasonable probability of extracting answers out of the network? Kleinberg and Raghavan (2005) explore theoretical properties of this fascinating idea.

1.8 CONCLUSION

In this chapter, we have provided an overview of Internet advertising and emphasized the important role statisticians can play as technology creators (as opposed to technology aiders) through a set of examples. The challenges discussed in this chapter are by no means exhaustive and provide a perspective based on the author's experience at a major search engine company for a period of one year. As a disclaimer, the views expressed are solely the author's own and are in no way representative of the official views of his employer. Although several statisticians have made a transition to this exciting area, more will be needed in the coming years. Internet advertising provides a unique opportunity to shape the future of the Web and invent technology that can affect the lives of millions of people. The author would like to urge statisticians to consider this area when making a career decision. One important component in conducting research in the area is the availability of data. Several search engine companies are trying their best to provide sanitized data for academic research.

However, the recent AOL debacle wherein search logs containing private information about users were released to the public demonstrates the difficulty of the problem. Hence, for research that depends critically on real data, the best method at the moment seems to involve working in close collaboration with companies that collect such data on an ongoing basis.

ACKNOWLEDGMENTS

I thank Chris Olston and Arpita Ghosh for discussions and pointers to related work in crawling and auction theory. I also benefited from discussions with Srujana Merugu, Michael Benedikt, and Sihem Amer-Yahia on social search. I would also like to thank an anonymous referee and the editors, whose insightful comments improved the presentation of the chapter.

REFERENCES

Abe, N., Biermann, A.W., and Long, P.M. (2003). Reinforcement learning with immediate rewards and linear hypotheses. *Algorithmica*, 37(4): 263–293.

Agarwal, D. (2005). An empirical bayes approach to detect anomalies in dynamic multidimensional arrays. *International Conference on Data Mining*.

Agarwal, D., McGregor, A., Phillips, J.M., Venkatasubramanian, S., and Zhu, Z. (2006). Spatial scan statistics: Approximations and performance study. *SIGKDD. Knowledge Discovery and Data Mining*.

Auer, P., Cesa-Bianchi, N., and Fischer, P. (2002). Finite-time analysis of the multiarmed bandit problem. *Machine Learning*, 47: 235–256.

Chapelle, O., Schlkopf, B., and Zien, A. (eds.). (2006). *Semi-supervised Learning*. Cambridge, MA: MIT Press.

Chawla, N., Japkowicz, N., and Kolcz, A. (eds.). (2003). Learning from Imbalanced Datasets. *Proceedings of the icml2003 Workshop*.

Chawla, N., Japkowicz, N., and Kolcz, A. (2004). Editorial: special issue on learning from imbalanced data sets. *ACM SIGKDD Explorations Newsletter*, 6(1): 1–6.

Cho, J. and Garcia-Molina, H. (2003). Estimating frequency of change. *ACM Transactions on Internet Technology*, 3(3): 256–290.

Cho, J. and Ntoulas, A. (2002). Effective change detection using sampling. *Very Large Databases*.

Dasgupta, A., Ghosh, A., Kumar, R., Olston, C., Pandey, S., and Tomkins, A. (2007). Discoverability of the web. *World Wide Web*.

DuMouchel, W. (2002). *Data Squashing: Constructing Summary Data Sets*. Norwell, MA: Kluwer Academic Publishers.

DuMouchel, W. and Agarwal, D. (2003). Applications of sampling and fractional factorial designs to model-free data squashing. *Knowledge Discovery and Data Mining*.

Edelman, B., Ostrovsky, M., and Schwarz, M. (2006). Internet advertising and the generalizaed second price auction: Selling billions of dollars worth of keywords. Second Workshop on Sponsored Search Auctions, Ann Arbor, Michigan, June.

Fienberg, S.E. and Shmueli, G. (2005). Statistical issues and challenges associated with rapid detection of bio-terrorist attacks. *Statistics in Medicine*, 24(4): 513–529.

Gittins, J.C. (1979). Bandit processes and dynamic allocation indices. *Journal of the Royal Statistical Society, Series B*, 41: 148–177.

Hill, S., Agarwal, D., Bell, R., and Volinsky, C. (2006). Building an effective representation for dynamic graphs. *Journal of Computational and Graphical Statistics*, 15: 584–608.

Huang, Z. and Gelman, A. (2005). Sampling for bayesian computation with large datasets. Technical Report, Columbia University.

Japkowicz, N. (2000). Learning from imbalanced data sets: Papers from the aaai workshop. aaai, 2000. Technical Report WS-00-05,

King, G. and Zeng, L. (2001). Logistic regression in rare events data. *Political Analysis*, 9(2): 137–163.

Kleinberg, J.M. and Raghavan, P. (2005). Query incentive networks. In *FOCS '05: 46th Annual IEEE Symposium on Foundations of Computer Science*.

Lai, T. and Robbins, H. (1985). Asymptotically efficient adaptive allocation rules. *Advances in Applied Mathematics*, 6: 4–22.

Manning, C.D., Raghavan, P., and Schutze, H. (2008). *Introduction to Information Retrieval*. Cambridge University Press.

Marlow, C., Naaman, M., Boyd, D., and Davis, M. (2005). Position paper, tagging, taxonomy, flickr, article, toread. *WWW, Collaborative Web Tagging Workshop*.

Owen, A. (2007). Infinitely imbalanced logistic regression. *Journal of Machine Learning Research*, 8: 761–773.

Page, L., Brin, S., Motwani, R., and Winograd, T. (1998). Pagerank citation ranking: Bringing order to the web. Technical Report, Stanford University.

Pandey, S., Agarwal, D., Chakrabarti, D., and Josifovski, V. (2007). Bandits for taxonomies: a model based approach. *Proceedings of the Siam Data Mining Conference*.

Ridgeway, G. and Madigan, D. (2002). A sequential monte carlo method for bayesian analysis of massive datasets. *Journal of Data Mining and Knowledge Discovery*, 7: 301–319.

Rosenberger, W.F. and Lachin, J.M. (2002). *Randomization in Clinical Trials: Theory and Practice*. New York: Wiley.

Varian, H.R. (2007). Position auctions. *International Journal of Industrial Organization*, 25: 1163–1178.

West, M. and Harrison, J. (1997). *Bayesian Forecasting and Dynamic Models*. Springer-Verlag.

2

HOW HAS E-COMMERCE RESEARCH ADVANCED UNDERSTANDING OF THE OFFLINE WORLD?

CHRIS FORMAN

College of Management, Georgia Institute of Technology, 800 West Peachtree Street NW, Atlanta, Georgia

AVI GOLDFARB

Rotman School of Management, University of Toronto, 105 St George St, Toronto, Ontario, Canada

2.1 INTRODUCTION

Statistical research in electronic commerce has made great advances in recent years. Researchers have gained an understanding of how use of consumer electronic markets leads to lower prices (Brynjolfsson and Smith 2000), greater selection for consumers (Brynjolfsson et al. 2003; Ghose et al. 2006), and a great many other benefits. While early statistical work on electronic commerce often examined behavior in the online world while holding constant characteristics of the offline world, increasingly researchers are examining similarities, differences, and interactions between these two settings. Such studies help researchers and practitioners arrive at a more nuanced understanding of online behavior. Moreover, electronic commerce data allows researchers to apply new identification strategies to study traditional (offline) questions in economics, marketing, and information systems.

Statistical Methods in e-Commerce Research. Edited by W. Jank and G. Shmueli

In this chapter, we review recent statistical electronic commerce work that takes this latter approach of studying the interaction of the online and offline worlds. It is not our goal to provide a comprehensive review of work in this area; rather, we highlight recent work in economics, marketing, and information systems that follows this approach and offers (in our opinion) particularly good examples of how electronic commerce research can improve our understanding of traditional questions.

Section 2.2 describes how electronic commerce research has contributed to traditional questions in marketing, including word-of-mouth marketing, stockouts, and brand loyalty. Section 2.3 discusses how electronic commerce research has informed our understanding of the role of geography in the economy through research in international trade and in the economics of cities. Section 2.4 demonstrates how the relationship between online and offline electronic markets informs our knowledge of channel substitution, search costs, discrimination, vertical structure, and tax distortions. Section 2.5 concludes the chapter.

2.2 MARKETING: WORD-OF-MOUTH, STOCKOUTS, BRANDING, AND CONSIDERATION SETS

Electronic commerce data have allowed marketers to answer a number of questions that had previously been difficult to address due to data limitations. In this section, we discuss four areas of the marketing literature that have benefited from electronic commerce data: word-of-mouth measurement, stockouts, brand loyalty, and consideration sets.

2.2.1 Word-of-Mouth Marketing

Prior to the arrival of online data, word-of-mouth marketing was difficult to measure. Much of the literature was limited by the challenge of the Reflection Problem (Manski 1993), which suggests that similar people live near each other, communicate, and use the same technologies and products. Therefore, without either observing conversations or finding an effective instrument, it is not possible to measure word-of-mouth effects. Internet research has overcome this difficulty by allowing researchers to observe conversations. Godes and Mayzlin (2004) argue that online conversations provide an opportunity to measure word-of-mouth. In particular, online postings are publicly observable and can easily be converted into data for analysis. They show that online conversations help predict the success of new television shows. Dellarocas and Narayan (2006) have developed further methods for converting online postings into usable word-of-mouth metrics.

A rich literature has followed that examines how word-of-mouth affects behavior, a subject previously impossible to measure properly outside a laboratory. The works of Chevalier and Mayzlin (2006), Chen et al. (2006), Forman et al. (2007), and Li and Hitt (2007) are prominent examples. Chevalier and Mayzlin examine reviews at Barnesandnoble.com and Amazon.com to show that better reviews at one site increase sales at that site relative to the other site. In other words, they find strong evidence that word-of-mouth (in the form of reviews) drives sales and that negative word-of-mouth has a bigger impact than positive word-of-mouth. Chen et al. show that online product reviews have a larger impact when they are written by reviewers with good reputations and when more people report that they "found the review helpful." Forman et al. show that in addition to the product information available in reviews, social information about reviewers has a significant impact on product sales. Finally, Li and Hitt show that because preferences of early buyers may differ from those of later buyers, there may be systematic trends in product reviews that may influence the relationship between reviews and product sales. Overall, online data has allowed a much deeper understanding-of-how word of mouth works.

2.2.2 Stockouts

Stockouts (or unexpected product unavailability) are a substantial problem for marketers. A rich literature discusses short-run consumer choices in response to a stockout (e.g., Campo et al. 2000). While there have been limited attempts to assess the longer-run impact of stockouts (Bell and Fitzsimons 1999), previous research had been unable to assess how and why stockouts affect future choices for two reasons. First, stockouts are endogenous. Stores run out of a product due to unexpectedly high sales. This may impact future outcomes. Second, it is not always possible to track purchases by specific households in the aftermath of a stockout. Goldfarb (2006a) uses online data to resolve both of these difficulties to understand why unavailability affects long-run behavior. To do so, he combines clickstream data on the online behavior of 2,651 households with public information on denial-of-service attacks on Yahoo, CNN, and Amazon. The denial-of-service attacks help overcome the endogeneity problem. The attacks (and their timing) can be reasonably viewed as exogenous from the point of view of the websites and their visitors. The clickstream data allow those households that were unable to visit an attacked site to be observed for several weeks before and after the attacks. The results show that customers who attempted and bailed to visit the attacked website during the attack were less likely to return in the future. For example, Yahoo lost an estimated 7.56 million visits in the 53 days following the attacks. Goldfarb argues that if the impact is solely due to changing preferences, then all competitors of the attacked website gain in proportion to their share; if, however, there is lock-in, then the competitor that is chosen instead of the unavailable product should gain disproportionately more. The results show that lock-in drives 51% of the effect on Yahoo, but it dissipates much more quickly than the effect of changing preferences.

2.2.3 Branding

Internet data have also helped increase our understanding of the function of branding and brand loyalty. The Internet provides a "natural experiment" in which search costs and switching costs appear to be very low (Bakos 1997). If people still choose brands (and pay more for them) online, this suggests that a brand's value extends beyond reducing search and switching costs. Using data on every website visited by a panel of households, Goldfarb (2006b) finds that switching costs (as opposed to underlying individual-level preferences) generate 11% to 15% of market share for Internet portals. Since switching costs should be very low online, he argues that a likely source of these switching costs is brand loyalty. Danaher et al. (2003) also find that brands matter online. They compare purchase behavior at the online and offline stores of a large grocery retailer and find that better-known brands display especially an especially high degree of loyalty online. Overall, these papers branding on have reinforced the idea that brands serve to provide information on experience goods rather than simply reducing search costs (Goldfarb et al. 2006).

2.2.4 Consideration Sets

There is a rich literature in marketing that asserts a two-stage choice process in purchase decisions. In the first stage, consumers select a subset of all products available in order to examine them more thoroughly. In the second stage, consumers choose a single product from this smaller *consideration set*. Shocker et al. (1991) review much of the theoretical literature on the subject. There have been several attempts to capture empirically the existence of these consideration sets using scanner panel data, including the work of Siddarth et al. (1995), Andrews and Srinivasan (1995), and Mehta et al. (2003). However, data limitations mean that stage one is never observed. Instead, it is captured by combining sophisticated statistical techniques with models of consumer behavior. Moe (2006) shows that Internet clickstream data allow researchers to observe consideration sets directly because the data include each product viewed by the consumer before the final purchase. By using clickstream data to observe consideration sets, Moe showed that the determinants of consideration set inclusion and final purchase are different: Consumers use simpler decision rules in the first stage. Clickstream data enabled Moe to provide a deep understanding of consideration set formation based on observed, rather than inferred, consideration sets. Her results will be especially useful to future researchers when modeling each stage in the purchase process without direct information on the first stage.

The four examples of word-of-mouth, stockouts, branding, and consideration sets show that online data and the online environment have enabled researchers in marketing to gather evidence on questions that were previously difficult to answer.

2.3 LOCATION AND THE ECONOMY

Electronic commerce data have allowed economists to better understand the role of location in economic transactions. In this section, we discuss how the reduction in communication costs due to the Internet has provided researchers with an experiment to observe how social networks and local preferences influence international trade and the economics of cities.

2.3.1 International Economics

Data on Internet usage patterns have informed our understanding of the geographic patterns of trade. *Gravity* is a well-established empirical regularity in international trade: A given country will trade more with large and nearby countries than with small and distant countries (see Disdier and Head (2008) for a meta-analysis of this literature). Internet data have enabled researchers to examine trade when transportation costs approach zero. They have also provided researchers with customer-level information for estimating spatial correlation in preferences and for understanding the importance of trust. Using Internet data has therefore enabled researchers to determine if spatial correlation in preferences and trust are important factors in the distance effect in trade without having transportation costs confound the analysis.

Next, we describe several papers that use Internet data to address this question. First, Blum and Goldfarb (2006) examine the website visiting behavior of 2654 Americans. They show that these Americans are much more likely to visit websites from nearby countries (e.g., Mexico and the United Kingdom) than from countries farther away (e.g., Spain and Australia), even controlling for language, income, immigrant stock, and a number of other factors. Since they only look at websites that do not ship items to consumers, transportation costs cannot account for the distance effect observed in the data. To understand the reason distance matters for digital goods, they further split the data into taste-based categories like music, games, and pornography and non-taste-based categories like software. The distance effect holds only in the taste-based categories, suggesting that spatial correlations in taste (or cultural factors) may be an important reason for the distance effect in trade. A second paper by Hortacsu et al. (2006) finds that distance matters in transactions at the online marketplaces eBay (in the United States) and MercadoLibre (in Latin America), even after controlling for shipping costs and time. Their results suggest that both culture and trust play an important role in explaining the distance effect. Finally, while it is not the primary objective of their work, Jank and Kannan (2005, this volume) provide further confirmation of the role of tastes by showing that consumer preferences are strongly geographically correlated.

2.3.2 The Economics of Cities

Internet research has also informed our knowledge of the role of cities in the economy. By examining the impact of a substantial drop in long-distance communication costs, Internet research has allowed us to identify some of the ways in which cities facilitate communication and to better understand constraints on social interaction not related to communication costs.

Gaspar and Glaeser (1998) introduced the question of how Internet communications technologies (ICTs) affect personal interactions to the literature on the economics of cities. They argue that ICTs may be a substitute for or a complement to face-to-face communication. They may be a substitute because instead of face-to-face interaction, it is possible to communicate by electronic means. On the other hand, ICTs may be a complement to face-to-face interactions because they may make such interactions more efficient. Subsequent empirical literature has found support for both arguments. Sinai and Waldfogel (2004) find that isolated individuals are more likely to connect to the Internet than others, suggesting that ICTs act as a substitute for face-to-face communication. For example, blacks in mainly white communities are more likely to connect. However, Sinai and Waldfogel also find that people in larger communities have more online content of interest to them and therefore are more likely to connect overall. Forman et al. (2005) show that while rural businesses are especially likely to connect to the Internet for basic communication services, urban businesses are most likely to adopt sophisticated ICTs because of the lower cost of implementation in cities. Agrawal and Goldfarb (2006) also address this question. They primarily find support for the Internet as a complement to face-to-face communication. In particular, they examine the effect of Bitnet (a 1980s academic version of the Internet) on collaboration between electrical engineering professors and find that the reduction in communications costs associated with the technology led to an overall increase in collaboration. Interestingly, this increase was strongest between researchers at top-tier and second-tier schools in the same city. Rather than facilitating collaboration between Harvard and Stanford researchers, Bitnet had its biggest impact on collaboration between Harvard and Northeastern engineering professors. This suggests that, at least initially, electronic communication especially strengthened the value of local social networks that existed within cities. This group of papers shows that, while a reduction in communication costs does facilitate communication across large distances, social networks often are local. As a result, cities will continue to play a role in facilitating social network formation even after a reduction in communications costs.

In summary, the Internet reduced communications costs. Researchers in both international trade and the economies of cities used this change to identify the role of local networks and local preferences on economic transactions.

2.4 RELATIONSHIP BETWEEN ONLINE AND OFFLINE CONSUMER ELECTRONIC MARKETS

Electronic commerce data often enable the researcher to observe both online and offline variables. This ability enables the researcher to examine how differences between these environments affect behavior. Researchers have used this identification strategy to answer standard marketing, economics, and information systems questions on channel substitution, search costs, discrimination, vertical integration, and tax distortions.

2.4.1 Substitution Between Online and Offline Channels

Channel choice is an important marketing decision. Channel substitution and channel management have a rich history of research in marketing and economics (e.g., Balasubramanian 1998; Fox et al. 2004), and electronic commerce research has the opportunity to understand these phenomena in a new setting with very different channel properties.[1] When consumers have a choice, when do they use the online channel and when do they use the offline channel? Why? For example, online channels can simultaneously provide consumers with better convenience, selection, and

[1]For a review of the early theoretical literature on this decision, see Chapter 9 of Lilien et al. (1992).

price (Forman et al. 2007); however, despite these advantages, many consumers have been slow to adopt electronic channels (Langer et al. 2007).

As Goolsbee (2001) notes, one difficulty of conducting research in this area is the difficulty of finding data that include transactions and prices from both markets. For example, this line of research is generally less conducive to the use of data scraped from websites, often a source of data in other areas of electronic commerce research. Researchers in this area have sometimes implemented a strategy of observing how changes in proxies for prices—such as retail competition and tax rates—in one market influence transaction behavior in another (e.g., Goolsbee 2001; Prince 2007; Avery et al. 2007; Forman et al. 2007; all discussed below).

Several papers have investigated the cross-price elasticity across online and offline markets. Goolsbee (2001) estimates a price index for local retail computers using a hedonic regression, and then uses it to show how local prices influence where a consumer will purchase a computer (online or offline). He shows that conditional on purchasing a computer, the elasticity of buying online with respect to local prices is roughly 1.5. Using a similar methodology, Prince (2007) measures the cross-price elasticity for PCs purchased online and offline for several years and finds that 1998 is the first year for which there is significant cross-price elasticity. He explores several candidate demand-side and supply-side explanations for the increase in cross-price elasticity, and finds that expansion of multichannel sales models and increasing opportunities for consumers to physically inspect and customize PCs are the reason for the increase. Chiou (2005) uses data on consumers' choice of retailer in the market for DVDs to measure consumers' cross-price elasticity between online and offline retailers and how it is moderated by demographic characteristics such as consumer income. She finds evidence of significant cross-price elasticities across stores and shows that these elasticities vary with income. Overall, this literature has shown that price is a key determinant of channel choice.

Of course, consumers may substitute between channels for reasons other than price. Balasubramanian (1998) develops an analytical model that allows a consumer's decision between traditional retail stores and the direct channel to depend upon such factors as the distance to the closest retail store, the transportation cost, and the disutility of using the direct channel, as well as prices across the two channels. Increases in the distance to an offline retailer will increase the attractiveness of purchasing from the direct retailer. Though Balasubramanian's model is framed as one for a generic direct retailer, it can easily be applied to competition between online and offline markets. Forman et al. (2007) examine how changes in consumers' offline options—that is, the local supply of stores—influence online purchaes. In particular, they examine how local store entry influences the composition of the most popular products purchased online, identified through Amazon.com's "Purchase Circles" data. They find strong evidence that offline transportation costs matter and that there is a substantial disutility of buying online. In sum, they find evidence of substitution between online and offline channels for reasons other than price, providing empirical support for the Balasubramanian (1998) model.

Avery et al. (2007) also examine the relationship between local store openings and consumer behavior online. In particular, they investigate whether a retailer's new-store opening will cannibalize or complement the retailer's direct-channel sales. They find that while direct-channel sales do fall temporarily in markets experiencing retail store entry, this cannibalization effect dissipates over time. This may be due either to a dissipation of the cannibalization effect over time or to increasing complementarity effects.

2.4.2 Search Costs

One of the earliest comparisons researchers have made between online and offline markets involved price differences across the two channels, and how the lowering of search costs on the Internet might contribute to lower prices online and offline. This research contributed to a long theoretical literature that had argued that lowering consumer search costs would decrease average prices and eventually make the distribution of prices more concentrated (e.g., Stahl 1989; Bakos 1997). However, until the advent of electronic commerce research, it was unusual to find a natural experiment that had such a dramatic effect on search costs. Brynjolfsson and Smith (2000) address this question by comparing online and offline prices. They find that books and CDs sold on the Internet cost an average of 9–16% less than identical items sold via conventional channels. However, they find evidence of substantial price dispersion across retailers online. Brown and Goolsbee (2002) show that the use of Internet price comparison sites also lowers the prices that consumers pay for life insurance policies offline. Building on prior work on search theory (in particular Stahl 1989), they argue that decreases in search costs can lead to both increases and decreases in price dispersion, and their empirical work supports this theory.[2]

This early work explored how the Internet led to lower prices and changes in price dispersion of commodity products (such as books, CDs, and term life insurance policies) due to lower search costs. For some products, such as term life insurance, awareness of prices online can influence the prices that consumers pay offline by lowering search costs. However, another set of work has examined how Internet comparison sites can lower the prices paid by consumers offline even when products are very heterogeneous and search costs are high, as they are for automobiles. Scott Morton et al. (2001) examine the relationship between use of Internet car referral services and prices paid and find that consumers of an online service pay on average 2% less for their car. Later work, discussed below, has examined the mechanisms through which Internet use is associated with lower purchase prices.

Therefore, lower search costs may lead to lower average prices paid online and offline. However, the ability to search for and find a larger selection of products online may also create significant benefits; Brynjolfsson et al. (2003) argue that these benefits are five to seven times more important than lower prices. Brynjolfsson et al. (2007) examine whether the lower search costs of online channels

[2]A large literature has investigated the extent of price dispersion online and has examined how lower Internet prices have led to better consumer welfare. This work, though important, is outside the scope of our review. See Baye et al. (2006) for a recent review.

shift the distribution of products purchased. Like prior work on prices and price dispersion, this work is motivated by a substantial theory literature on search costs that predates electronic commerce (e.g., Diamond 1971; Wolinsky 1986; Stahl 1989). To identify the effects of lower search costs separately from those of greater product variety, they examine the distribution of sales in two different channels—an Internet channel and a non-Internet (catalog) channel—of the same retailer. By examining sales in two different channels of the same retailer, they hold selection and supply-side drivers constant and are able to isolate the impact of search costs. They find that consumers in the online channel have a significantly less concentrated sales distribution than consumers who buy through the catalog channel; this is true even when one examines the subset of consumers who buy from both channels. Thus, the Internet lowers product search costs, and these lower search costs lead consumers to purchase less popular products.

As noted above, Internet use is associated with lower prices in online (e.g., Brynjolfsson and Smith 2000) and offline (e.g., Scott Morton et al. 2001; Brown and Goolsbee 2002) channels. Zettelmeyer et al. (2006) examine the mechanisms through which Internet use leads to lower prices for consumers. They investigate how use of online buying services for automobiles is associated with lower prices for consumers. They find that the use of such services lowers prices for two reasons. First, the Internet provides information about dealers' invoice prices, which improves consumers' bargaining position and enables them to negotiate a lower price. Second, online buying services' dealer contracts help consumers to obtain lower prices through the referral process. This research required detailed demographic data about consumers, the kind of data that are often not available in statistical electronic commerce research. To address their question, the authors supplemented their data with a survey mailed to 5250 consumers. Thus, use of the Internet is associated with lower prices through lower search costs. This works partially by improving consumers' bargaining position.

2.4.3 Discrimination

Electronic commerce research has provided a deeper understanding of the mechanisms behind observed discrimination. In particular, another way in which use of the Internet can benefit consumers is by concealing who they are from (potentially discriminating) sellers. Prior work has found that women and minorities pay significantly more for automobiles: In earlier work, Ayres and Siegelman (1995) find that black male and black female testers pay prices that are significantly higher ($1100 and $410, respectively) than those paid by white men. However, this literature leaves unresolved the question of whether price discrimination in car buying has a "disparate impact" on minorities (dealer practices that are applied to all groups but that have a greater impact on minority groups) or affects them because of "disparate treatment" (dealers explicitly treat minority groups differently). Scott Morton et al. (2003) suggest that disparate impact is the reason for the higher prices paid by minorities. In particular, they utilize a unique feature of electronic commerce to identify the effects of disparate impact separately from those of disparate treatment: While dealers can easily identify the racial characteristics of Internet buyers from their names, they are unable to use common cues that are often used to determine consumers' willingness to pay. They find that online minority consumers pay almost the same prices us white consumers once demographic characteristics are controlled. In addition to examining an important issue for electronic commerce marketers, this research helped us to further understand discrimination in the broader economy. Again, data from an electronic commerce setting offer a new identification strategy for an important phenomenon.

2.4.4 Vertical Organization and Adoption of an Online Distribution Channel

Online research has also informed the literature on how the organization of a company affects its decision to adopt new technology and new distribution channels. In particular, the arrival of the Internet provided a natural experiment to compare the decisions of companies that owned their retail channels (i.e., vertically integrated companies) with those of companies that did not own their retail channels. Gertner and Stillman (2001) examine the adoption of online retailing. They find that less vertically integrated firms (in particular, those for which offline customer contact is managed by an independent firm) will be slower to adopt online retailing. This could be due to a number of reasons: higher coordination costs, greater transaction costs, production complementarities, or channel conflicts. Forman and Gron (2005) find that property and casualty insurers with independent agents are slower to adopt new information technology (IT) that is specific to the distribution relationship than firms that are vertically integrated into distribution. They attribute these differences to the lower transaction costs within vertically integrated firms. These papers have informed a prior debate on the consequences of firm boundaries on decision making (e.g., Williamson 1985). They show that the organizational structure of the firm is an important determinant of technology adoption and channel distribution decisions. The natural experiment using Internet data enabled these researchers to show that firm structure lead to specific choices rather than vice versa.[3]

2.4.5 Measuring Tax Distortions

Research in electronic commerce has also enhanced our understanding of how sales taxes influence consumer behavior. Research prior to the introduction of electronic commerce demonstrated that consumers living along state borders are particularly sensitive to changes in tax rates (e.g., Mikesell 1970; Fox 1986; Walsh and Jones 1988; Holmes 1998). Because consumers effectively pay no sales tax for electronic commerce transactions conducted with retailers located out of state,[4] electronic

[3]For a review of other empirical work on IT adoption, see Forman and Goldfarb (2006).

[4]Consumers are required to pay a use tax for electronic commerce sales transactions with out-of-state retailers. However, noncompliance is very common, so effective tax rates are close to zero (Goolsbee 2000).

commerce allows any consumer to arbitrage tax rates as easily as those living along state borders (Goolsbee 2000). The sensitivity of consumers to such sales tax changes has important implications for state tax policy. As a result, recent research has examined how offline sales taxes influence consumer behavior online (Goolsbee 2000; Brynjolfsson et al. 2004; Anderson et al. 2006; Ellison and Ellison 2006; Forman et al. 2007).

Goolsbee (2000) was the first to examine how offline sales taxes influence consumers' behavior online. Using a large database that includes information on 25,000 people with online access, he shows that consumers in high-sales-tax states are more likely purchase products online than are consumers in low-sales-tax states. To demonstrate that these results are not driven by other differences in consumer propensity to buy online across high and low sales tax states, he also shows that consumers in high-tax locations are no more likely to own a computer, buy other electronic goods, or use the Internet more frequently than are consumers in low-tax locations. As Ellison and Ellison (2006) have pointed out, however, there are two concerns with Goolsbee's analysis: (1) he examines only how sales taxes influence consumers' propensity to conduct at least one transaction on the Internet, and says nothing about how offline sales tax influences the intensive margin of consumers' purchases, and (2) unobservable differences in consumer preferences may be partially influencing his results.

Recent work has attempted to use alternative identification strategies to address the concerns with Goolsbee's pioneering work, as well as to show how sales taxes can influence other aspects of online consumer behavior. Ellison and Ellison (2006) use data from an online market for memory chips (Pricewatch.com) to show how various factors influence consumer propensity to purchase from an online retailer, including prices, sales taxes, shipping times, and retailer location (in particular, whether consumers prefer to purchase from retailers located in the same state). Like Goolsbee (2000), they exploit cross-sectional variation in sales tax rates to examine how sales taxes shape consumer demand. However, they also exploit time series variation in prices of different types of memory chips available online and their offline equivalents to measure consumer sensitivity to prices and taxes. This additional source of identification removes concerns about how cross-sectional differences in preferences may influence their analysis and provides yet another example of how the rich data available from websites can be used to implement alternative identification strategies. While they find that consumers are very sensitive to the prices they pay online, they find that consumers are somewhat less sensitive to differences in taxes than to differences in pretax prices. Like other researchers who have examined the role of geography, they find that location shapes consumer behavior online: They show that consumers are more likely to purchase from in-state retailers and those with faster shipping times, other things equal.

Anderson et al. (2006) use a natural experiment to show how sales taxes influence unit sales, thereby alleviating the concern that cross-sectional differences in preferences are responsible for the correlation between sales tax rates and online sales. In particular, they show that local entry by a physical retail store from a direct retailer will reduce sales more for consumers in the same state than for those in adjacent states. They further show that the effects of store entry will be moderated by consumers' history with the retailer. Last, they demonstrate that retailers who conduct a large fraction of their business on the Internet will be relatively less likely to have stores in high-sales-tax states than similar retailers who conduct a smaller share of business online. Forman et al. (2007) also use offline store entry to show how sales taxes shape online-offline channel substitution: They find that consumers in high-sales-tax locations will find offline retail stores to be a poorer substitute for online stores compared to consumers in low-sales-tax locations.

Thus, a variety of studies have shown that changes in offline sales taxes will shift consumers' propensity to purchase online. While the data exist, our current knowledge of how sales taxes influence the intensive margin of consumer purchases is limited. There are many opportunities for future research using Internet data to assess tax distortions.

2.5 CONCLUSION

In this chapter, we have reviewed recent work that has used data and settings from the online world to understand questions and phenomena from the offline world. One objective has been to review recent research on the interplay between the online and offline worlds. Another objective has been to show that electronic commerce research, by providing new sources of data and new identification strategies (i.e., natural experiments), has allowed researchers to develop a better understanding of many traditional issues in economics, marketing, and information systems. Our review has touched on diverse topics such as word-of-mouth, stockouts, brands, consideration sets, international economics, the economics of cities, channel substitution, search, discrimination, vertical integration, and taxation. This list highlights the diversity of topics that have used electronic commerce data and information to address research questions that preceded the Internet.

This is by no means a complete list of the research areas that Internet data have affected. We have focused on the areas in which we have particular expertise. For example, though we have discussed at some length substitution between online and offline channels for consumer goods, we have not discussed physical and electronic market mechanisms for business-to-business commerce (e.g., Garicano and Kaplan 2001; Banker and Mitra 2005; Overby and Jap 2006). Further, we have not discussed recent literature that examines how online file-sharing has impacted offline record sales (e.g., Hong 2006; Liebowitz 2006; Rob and Waldfogel 2006). A number of other research areas have benefited from Internet data, and some already have rich literature surveys. In particular, Bajari and Hortacsu (2004) review the literature on how online auctions have informed our knowledge of auction theory and of how auctions work in practice. Baye et al. (2006) review the literature on online price dispersion. Understanding online price dispersion helps improve our modeling of search and the connections between search and price dispersion.

In conclusion, this chapter has shown that Internet data and methods have proved extremely useful in answering a number of classic marketing, economics, and information systems questions. We look forward to future research that addresses these and other questions using the online environment as its setting.

REFERENCES

Agrawal, A. and Goldfarb, A. (2006). Restructuring research: Communication costs and the democratization of university innovation. Working Paper No. 12812, National Bureau of Economic Research.

Anderson, E.T., Fong, N.M., Simester, D.I., and Tucker, C.E. (2006). Do internet tax policies place local retailers at a competitive disadvantage? Working Paper, MIT.

Andrews, R.L. and Srinivasan, T.C. (1995). Studying consideration effects in empirical choice models using scanner panel data. *Journal of Marketing Research*, 32(1): 30–41.

Avery, J., Caravella, M., Deighton, J., and Steenburgh, T.J. (2007). Adding bricks to clicks: The effects of store openings on sales through direct channels. Working Paper, Harvard Business School.

Ayres, I. and Siegelman, P. (1995). Race and gender discrimination in bargaining for a new car. *American Economic Review*, 85(3): 304–321.

Bajari, P. and Hortacsu, A. (2004). Economic insights from Internet auctions. *Journal of Economic Literature*, 42(2): 457–486.

Bakos, J.Y. (1997). Reducing buyer search costs: Implications for electronic marketplaces. *Management Science*, 43(12): 1676–1692.

Balasubramanian, S. (1998). Mail versus mall: A strategic analysis of competition between direct marketers and conventional retailers. *Marketing Science*, 17(3): 181–195.

Banker, R. and Mitra, S. (2005). Impact of information technology on agricultural commodity auctions in India. *Proceedings of the 2005 International Conference on Information Systems*.

Baye, M., Morgan, J., and Scholten, P. (2006). Information, search, and price dispersion. In *Handbook in Information Systems, Volume 1: Economics and Information Systems* (T. Hendershott, ed.). Amsterdam: Elsevier.

Bell, D.R. and Fitzsimons, G.J. (1999). An experimental and empirical analysis of consumer response to stockouts. Working Paper, Wharton School of Business, University of Pennsylvania.

Blum, B. and Goldfarb, A. (2006). Does the Internet defy the law of gravity? *Journal of International Economics*, 70(2): 384–405.

Brown, J.R. and Goolsbee, A. (2002). Does the Internet make markets more competitive? Evidence from the life insurance industry. *Journal of Political Economy*, 110(3): 481–507.

Brynjolfsson, E., Hu, Y., and Simester, D. (2007). Goodbye pareto principle, hello long tail: The effect of search costs on the concentration of product sales. Working Paper, Sloan School of Management, MIT.

Brynjolfsson, E., Hu, Y., and Smith, M. (2003). Consumer surplus in the digital economy: Estimating the value of increased product variety. *Management Science*, 49(11): 1580–1596.

Brynjolfsson, E. and Smith, M. (2000). Frictionless commerce? A comparison of Internet and conventional retailers. *Management Science*, 46(4): 563–585.

Brynjolfsson, E., Smith, M.D., and Montgomery, A.L. (2004). The great equilizer? An empirical study of consumer choice at a shopbot. Working Paper, MIT.

Campo, K., Gijsbrechts, E., and Nisol, P. (2000). Towards understanding consumer response to stock-outs. *Journal of Retailing*, 76(2): 219–242.

Chen, P.-Y., Dhanasobhon, S., and Smith, M.D. (2006). All reviews are not created equal. Available at SSRN: http://ssrn.com/abstract=918083.

Chevalier, J. and Mayzlin, D. (2006). The effect of word of mouth on sales: Online book reviews. *Journal of Marketing Research*, 43(3): 345–354.

Chiou, L. (2005). Empirical analysis of retail competition: Spatial differentiation at Wal-Mart, Amazon.com, and their competitors. Working Paper, Occidental College.

Danaher, P.J., Wilson, I.W., and Davis, R.A. (2003). A comparison of online and offline consumer brand loyalty. *Marketing Science*, 22(4): 461–476.

Dellarocas, C. and Narayan, R. (2006). A statistical measure of a population's propensity to engage in post-purchase online word-of-mouth. *Statistical Science*, 21(2): 277–285.

Diamond, P. (1971). A model of price adjustment. *Journal of Economic Theory*, 3: 156–168.

Disdier, A.-C. and Head, K. (2008). The puzzling persistance of the distance effect on bilateral trade. *Review of Economics and Statistics*, 90(1): 37–48.

Ellison, G. and Ellison, S.F. (2006). Internet retail demand: Taxes, geography, and online-offline competition. NBER Working Paper No. 12242.

Forman, C., Ghose, A., and Goldfarb, A. (2007). Geography and electronic commerce: Measuring convenience, selection, and price. Working Paper, University of Toronto.

Forman, C., Ghose, A., and Wiesenfeld, B. (2007). A Multi-level examination of the impact of social identities on economic transactions in electronic markets. Working Paper, New York University.

Forman, C. and Goldfarb, A. (2006). Diffusion of information and communication technologies to businesses. In *Handbook in Information Systems, Volume 1: Economics and Information Systems* (T. Hendershott, ed.). Amsterdam: Elsevier.

Forman, C., Goldfarb, A., and Greenstein, S. (2005). How did location affect adoption of the commercial Internet: Global village vs. urban leadership. *Journal of Urban Economics*, 58(3): 389–420.

Forman, C. and Gron, A. (2005). Vertical integration and information technology adoption: A study of the insurance industry. *Proceedings of the 38th Hawaii International Conference on System Sciences.*

Fox, W. (1986). Tax structure and the location of economic activity along state borders. *National Tax Journal*, 14: 362–374.

Fox, E. J., Montgomery, A. L., and Lodish, L. M. (2004). Consumer shopping and spending across retail formats. *Journal of Business*, 77(2): S25–S60.

Garicano, L. and Kaplan, S.N. (2001). The effects of business-to-business e-commerce on transaction costs. *Journal of Industrial Economics*, 49(4): 463–485.

Gaspar, J. and Glaeser, E. (1998). Information technology and the future of cities. *Journal of Urban Economics*, 43(1): 136–156.

Gertner, R.H. and Stillman, R.S. (2001). Vertical integration and Internet strategies in the apparel industry. *Journal of Industrial Economics*, 44(4): 417–440.

Ghose, A., Smith, M., and Telang, R. (2006). Internet exchanges for used books: An empirical analysis of product cannibalization and welfare implications. *Information Systems Research*, 17(1): 1–19.

Godes, D. and Mayzlin, D. (2004). Using online conversations to study word-of-mouth communication. *Marketing Science*, 23(4): 545–560.

Goldfarb, A. (2006a). The medium-term effects of unavailability. *Quantitative Marketing and Economics*, 4(2): 143–171.

Goldfarb, A. (2006b). State dependence at internet portals. *Journal of Economics and Management Strategy*, 15(2): 317–352.

Goldfarb, A., Lu, Q., and Moorthy, S. (2006). Measuring brand value in an equilibrium framework. Working Paper, University of Toronto.

Goolsbee, A. (2000). In a world without borders: The impact of taxes on Internet commerce. *Quarterly Journal of Economics*, 115(2): 561–576.

Goolsbee, A. (2001). Competition in the computer industry: Online versus retail. *Journal of Industrial Economics*, 49: 487–499.

Holmes, T. (1998). The effect of state policies on the locale of manufacturing: Evidence from state borders. *Journal of Political Economy*, 106: 667–705.

Hong, S.-H. (2006). The effect of digital technology on sales of copyrighted goods: Evidence from Napster. Working Paper, University of Illinois.

Hortacsu, A., Martinez-Jerez, F. de A., and Douglas, J. (2006). The geography of trade on eBay and MercadoLibre. NET Institute Working Paper No. 06–09.

Jank, W. and Kannan, P.K. (2005). Understanding geographical markets of online firms using spatial models of customer choice. *Marketing Science*, 24: 623–634.

Langer, N., Forman, C., Kekre, S., and Sun, B. (2007). Ushering buyers into electronic channels. Working Paper, Tepper School of Business, Carnegie Mellon University.

Li, X. and Hitt, L. (2007). Select selection and information role of product reviews. Working Paper, University of Connecticut.

Liebowitz, S.J. (2006). File sharing: Creative destruction of just plain destruction? *Journal of Law and Economics*, 49(April): 1–28.

Lilien, G., Kotler, P., and Sridhar Moorthy, K. (1992). *Marketing Models*. Upper Saddle River, NJ: Prentice Hall.

Manski, C.F. (1993). Identification of endogenous social effects: The reflection problem. *Review of Economics Studies*, 60(3): 531–542.

Mehta, N., Rajiv, S., and Srinivasan, K. (2003). Price uncertainty and consumer search: A structural model of consideration set formation. *Marketing Science*, 22(1): 58–84.

Mikesell, J. (1970). Central cities and sales tax rate differentials: The border city problem. *National Tax Journal*, 23: 206–213.

Moe, W. (2006). An empirical two-stage choice model with decision rules applied to Internet clickstream data. *Journal of Marketing Research*, 43(4): 680–692.

Overby, E.M. and Jap, S.D. (2006). Electronic vs. physical market mechanisms: Testing multiple theories in the wholesale automotive market. Working Paper, Emory University.

Prince, J.T. (2007). The beginning of online/retail competition and its origins: An application to personal computers. *International Journal of Industrial Organization*, 25(1): 139–156.

Rob, R. and Waldfogel, J. (2006). Piracy and the high C's: Music downloading, sales displacement, and social welfare in a sample of college students. *Journal of Law and Economics*, 49(April): 29–62.

Scott Morton, F., Zettelmeyer, F., and Silva-Risso, J. (2001). Internet car retailing. *Journal of Industrial Economics*, 49(4): 501–519.

Scott Morton, F., Zettelmeyer, F., and Silva-Risso, J. (2003). Consumer information and discrimination: Does the Internet affect the pricing of new cars to women and minorities? *Quantitative Marketing and Economics*, 1: 65–92.

Shocker, A.D., Ben-Akiva, M., Boccara, B., and Nedungadi, P. (1991). Consideration set influences on consumer decision-making and choice: Issues, models and suggestions *Marketing Letters*, 2(3): 181–197.

Siddarth, S., Bucklin, R.E., and Morrison, D.G. (1995). Making the cut: Modeling and analyzing choice set restriction in scanner panel data. *Journal of Marketing Research* 32(3): 255–266.

Sinai, T. and Waldfogel, J. (2004). Geography and the Internet: Is the Internet a substitute or a complement for cities? *Journal of Urban Economics*, 56(1): 1–24.

Stahl, D.O. (1989). Oligopolistic pricing with sequential consumer search. *American Economic Review*, 79: 700–712.

Walsh, M. and Jones, J. (1988). More evidence on the "border tax" effect: The case of West Virginia. *National Tax Journal*, 14, 362–374.

Williamson, O. (1985). *The Economic Institutions of Capitalism*. New York: Free Press.

Wolinksy, A. (1986). True monopolistic competition as a result of imperfect information. *Quarterly Journal of Economics*, 101(3): 493–512.

Zettelmeyer, F., Scott Morton, F., and Silva-Risso, J. (2006). How the Internet lowers prices: Evidence from matched survey and automobile transaction data. *Journal of Marketing Research*, 43: 168–181.

3

THE ECONOMIC IMPACT OF USER-GENERATED AND FIRM-GENERATED ONLINE CONTENT: DIRECTIONS FOR ADVANCING THE FRONTIERS IN ELECTRONIC COMMERCE RESEARCH

Anindya Ghose

Information, Operations and Management Sciences Department, Leonard Stern School of Business, New York University, New York, New York

3.1 INTRODUCTION

An important use of the Internet today is in providing a platform for consumers to disseminate information about products they buy and sell, as well as about themselves. Indeed, through the use of Web 2.0 tools like blogs and opinion forums, a large amount of content is being generated by users in the online world. Consequently, we have seen online markets develop into social shopping channels, and facilitate the creation of online communities and social networks. Firms are also using technology-mediated spaces to reveal information about their buyers. The increasing avenues for online content creation have changed the fundamental nature of information in terms of volume, availability, and importance. A lot of that publicly available content has concrete economic value that is often embedded in it. To date, businesses, government organizations, and customers have not fully incorporated such information in their decision-making and policy formulation processes, either because the value of the intellectual capital or appropriate methods for

Statistical Methods in e-Commerce Research. Edited by W. Jank and G. Shmueli

measuring that value have not been identified. This chapter is a call for research that aims to measure the economic value of various kinds of user-generated and firm-published content on the Web.

As an example, a vital piece of sociocultural information that is increasingly being published on the Web is the *geographical location* of market participants and members of virtual communities. The availability of users' location information, either disclosed by users themselves or published by firms, opens up a plethora of opportunities for research examining how geographical location shapes consumer search and purchase behavior on the Internet. As another example of increasingly ubiquitous user-generated social information, members who use electronic markets as a forum for social interaction reveal a lot of personal information about themselves. The availability of this self-descriptive information generated by users in an online community can enable researchers to examine how economic exchanges in the online world are being influenced by social exchanges between various entities.

While the above examples concern the availability of users' personal information, there are more detailed forums that provide information about users' actual experiences with sellers or with products in a rich textual format. For example, when buyers cannot deterministically assess the quality of a seller's fulfillment characteristics exante in an electronic market, the textual feedback posted by buyers describing their transaction experience can influence other buyers' purchase decisions and thus affect sellers' future performances. Similarly, based on the rich theoretical literature that suggests that consumer-generated word of mouth influences product sales, we can hypothesize that textual content of user-generated reviews is also likely to influence sales. Most studies of reputation systems or online reviews so far have used only numeric information about sellers or products to examine their economic impact. The understanding that "text matters" has not been fully realized in electronic markets or in online communities. Insights derived from text mining of user-generated feedback can thus provide substantial benefits to businesses looking for competitive advantages.

At the same time, excessive content in the online world can cause information overload among individuals, resulting in various cognitive costs incurred by users. These human-computer interaction costs include, for example, search costs incurred by consumers in locating the right information, cognitive costs of processing textual information prior to making purchase decisions, and decision-making or menu costs incurred by managers in adjusting price information. These costs arise due to delays in information diffusion and are brought about by the bounded rationality (limited ability to process complex information) of humans. To date, businesses have generally formulated strategies in the online world without factoring in such costs. Hence, such policies can be suboptimal. This calls attention to the need to identify and measure these costs in order to formulate optimal pricing policies.

The overarching theme across the above phenomena is that much of this user-generated and firm-published online information has an *economic value* that can be measured, monetized, and utilized intelligently in formulating business strategies. Extracting this economic value from publicly available online content and leveraging it has become increasingly important for all participants in a competitive market. To identify the economic value of online content, we need to examine three related questions: (1) How does the Internet influence consumers' information-seeking and purchase behavior by providing newer distribution channels, newer forms of online advertising, and unique community forums for social exchanges? (2) What is the economic value of user-generated content in Internet-mediated spaces such as reputation systems, review forums, and social networking sites? (3) How do users' information search and processing costs affect firms' pricing strategies in offline and online markets? Answering these questions requires an interdisciplinary approach that builds on theories and tools from multiple fields such as computer science, economics, information systems, machine learning, marketing, social psychology, and statistics to measure how various categories of content on the Internet influence exchanges between participants in digital markets and online communities.

Sections 3.2 to 3.5 constitute the main body of this chapter. Section 3.2 discusses the opportunities for measuring the economic value of information on consumers' information-seeking and purchase behavior in electronic markets. This kind of information is embedded in both user-generated and firm-published content. Section 3.3 discusses the economic value of user-generated textual feedback that is ubiquitous on the Internet, such as in reputation systems in electronic markets, product reviews in online communities, product descriptions in used-good markets, and social networking sites. We also discuss some methodologies that could be used to estimate that value. Section 3.4 analyzes the economic cost of information consumption, such as the search costs of finding information and the costs of processing textual information incurred by consumers, as well as the costs of adjusting product information on electronic markets incurred by firms. It also discusses the impact of search costs and menu costs on the emerging Long Tail phenomenon. In each of these sections, we describe some research opportunities that can build on current work. Section 3.5 concludes the chapter.

3.2 CONSUMERS' INFORMATION-SEEKING AND PURCHASE BEHAVIOR

The Internet has been thought of as a technological advance that removes the disparities between underserved communities and the rest of the country. However, we have little understanding of whether the benefits of the online channel (due to increased convenience, wider selection, and lower prices) are influenced by the concentration of offline retailers, which varies across geographical locations. Similarly, knowledge about how different kinds of online advertising affect consumers' search and purchase behavior is still in its infancy. The emergence of natural and sponsored search keyword advertising is intrinsically related to user-generated queries on search engines. By examining how keyword attributes and user-level demographics affect user search and purchases, one can estimate the business value of search engine advertising. Finally, by exploring the behavior of members in online virtual communities, future research can potentially examine the dynamic interplay between social and economic exchanges on the Internet. The research opportunities described in this section are based on the notion that an analysis of content that specifies the social information of users can increase our understanding of the factors that drive consumer usage of online channels.

3.2.1 Geographical Location and Online Purchase Behavior

Despite a wealth of research on electronic commerce, very little work has measured how geographical location shapes consumer buying behavior in electronic markets. Do consumers in different locations derive different benefits from using the Internet in terms of selection, convenience, and price? Prior studies in this domain have examined substitution between online and offline channels analytically (e.g., Balasubramanian 1998; Ghose et al. 2007) and empirically (e.g., Ellison and Ellison 2004; Prince 2006). Since prior work has focused on price differences across channels, future research can examine how changes in offline shopping convenience and product assortments influence online purchasing behavior. By combining online purchase data with offline data on demographics and availability of local retail channels, one can address an important problem that has been inadequately considered in statistical e-commerce research: how and why consumers substitute between online and offline channels.

Recent research on this question includes that of Forman, Ghose, and Goldfarb (2008), who use data from the web pages on "Purchase Circles" on Amazon.com, where Amazon publishes the geographical locations of its buyers. Purchase Circles are specialized best-seller lists of the top-selling books, music, and DVDs across large and small towns throughout the United States. The Purchase Circles are organized in multiple layers—first by state and then within a state, by town, and by county. Forman et al. match these online data on local demand with those on store openings and closings of major offline competitors of Amazon, which include discount stores such as Walmart and Target and large specialty stores such as Barnes & Noble and Borders, in order to study how geographical variations in offline selection, convenience, and price influence online product purchases. Their findings confirm that consumers do derive considerable benefits from convenience and price. Evidence for selection is demonstrated only for university towns and larger cities. Their results provide empirical support for the assumptions of a widely used framework in models of spatial differentiation that include a direct channel (Balasubramanian 1998). They find that variables and parameters in these models such as offline transportation cost, online shopping disutility cost, market coverage, and the prices of online and offline retailers interact to determine consumers' channel choice in a way that is consistent with these models. Moreover, their results are suggestive about the relative magnitudes of some of these parameters, showing that online disutility costs can be significant, even for products such as books, for which nondigital attributes are relatively unimportant. By looking at how consumers use online channels to compensate for offline retail supply deficiencies, their research contributes to the marketing literature that uses spatial data to capture variations in supply-side and demand-side factors—such as local consumer preferences (Jank and Kannan 2006).

Future analytical modeling studies can use the findings of this research. In particular, the results of Forman, Ghose, and Goldfarb (2008) suggest the usefulness of incorporating the effect of varying offline transportation costs in making optimal product assortment decisions for commodity products in local as well as online stores, and for incorporating the effect of product popularity in modeling the impact of product returns on retailers' pricing decisions, since the cost of returns to retailers and to consumers are likely to vary by product popularity and distance to stores, respectively.

Further research is needed in this domain to increase our understanding of how geography influences the benefits consumers derive from the Internet. To do so, one needs to look at more disaggregated data such as how individual consumer transactions vary across locations. This can highlight how offline geographical distance between buyers and sellers affects their propensity to transact with each other in online markets. Data on demographic characteristics can be obtained from the Bureau of Labor Statistics and matched with the transaction level data. For example, data on population size by county or metropolitan statistical area are provided by the annual census, while demographic data are available at various levels of aggregation in the decennial census. This can shed further light on how demographic factors contribute to differences in online purchase behavior (Scott-Morton et al. 2006). Such research will contribute to the literature on the potential of the Internet to reduce the costs associated with distance (Forman et al. 2004; Sinai and Waldfogel 2004) and increase consumer welfare (Brynjolfsson et al. 2003; Bapna et al. 2006; Ghose et al. 2006).

3.2.2 Web Search, Online Search Advertising, and Social Networks

A vast body of literature has used clickstream data to study consumer behavior in the online world (Bucklin et al. 2002). Similarly, an emerging body of literature has studied incentive mechanisms and auction strategies in keyword search auctions using game theory and computational methods. However, there is great potential for research that skates the boundaries of these two streams. Specifically, one can investigate the correlation between Internet search queries, the associated click-through rates and online sales for different time periods, for different categories of products, and, more importantly, for consumers in different locations. A more rigorous analysis would involve building predictive econometric models that will take the research one step toward identifying causality between these phenomena.

The data that are needed for these kinds of studies can be obtained from firms that advertise on search engines such as Google, MSN, or Yahoo. An ideal dataset would consist of search queries sampled over several weeks, bid prices, and per-query search result click-through rates with product and consumer demographic information. In any economic model required for this study, one needs to incorporate the fact that consumers face decisions at two levels. First, when they receive the result of a search engine query, they decide whether or not to click on it. Second, if they click on a result that is displayed, they can do any one of the following: take no action, make a click-through (without making a purchase), or make a purchase. Research in this domain can use a hierarchical Bayesian modeling framework and estimate the model using Markov chain Monte Carlo methods (Rossi and Allenby 2003).

Ghose and Yang (2007) provide a first step in this direction. Using a unique panel dataset of several hundred keywords collected from a large nationwide retailer that advertises on Google, they empirically model the relationship between different metrics such as click-through rates, conversion rates, bid prices, and keyword ranks. They estimate the impact of keyword attributes on consumer search and purchase behavior as well as on firms' decision-making behavior on bid prices and ranks. They find that the presence of retailer-specific information in the keyword increases click-through rates and conversion rates while the presence of brand-specific information decreases click-through and conversion rate. Moreover, their analysis provides some evidence that advertisers are not bidding optimally with respect to maximizing the profits. They also demonstrate that, as suggested by anecdotal evidence, search engines like Google factor in both the auction bid price and prior click-through rates before allotting a final rank to an advertisement. Finally, they conduct a detailed analysis using product-level variables to explore the extent of cross-selling opportunities across different categories from a given keyword advertisement.

Additionally, if one hard access to historical data on online sales for a variety of goods, one could design predictive models by merging the pricing and sales data of electronic retailers with the Internet search behavior data. Estimation can be challenging in this area because of data sparsity issues and rarity of clicks. Machine learning and statistical techniques that predict imbalanced or rare responses by sampling the majority class to reduce imbalances can be useful in this context (King and Zeng 2001; Chawla et al. 2003). This research can be extended to investigate how consumers' physical distance from retailers influences their online search and click-through behavior on search engines. The additional data needed for this analysis are the locations of the originating search, which are available from advertisers. Future research in this domain can incorporate users' geographical locations to help determine how online advertisements should be customized and targeted across locations. In the business world, this kind of research would be of interest not just to advertisers but also to search engine marketing firms.

Another interesting type of research would examine the relationship between the textual content of search queries and economic variables such as bid prices in keyword auctions, bid slots, page numbers, and click-through rates. One can examine the economic value of various modifiers (basically, adjectives or adverbs) used by consumers in online queries and search keywords. This will involve analyzing the price premiums on keyword bid prices by examining how much the price of a bid change with the addition of a specific modifier. For example, this would mean comparing the difference in bid prices, click-through rates, and conversion rates for a generic keyword such as *airline tickets* with branded keywords like *Orbitz airline tickets* or *discounted airline tickets* to determine the premium that online advertisers need to pay for a branded advertisement or for a keyword that contains a modifier like *discounted*. This study will involve a combination of text mining with economics to infer the economic value of text in keywords and search queries. Research in this domain is in its infancy, with great potential to address these open questions.

Finally, an emerging area of research that offers many interesting possibilities is that of *social search*. Firms are increasingly looking for ways to combine information from social networking sites to improve the quality and accuracy of a search with the final aim of providing a personalized search. Microsoft has made some headway in this area. The popularity of Web-based social networks like Delicious, Technocrati, and Flickr, which allow users to tag resources like blog, pictures, and webpages, generates a rich data trail that can potentially be exploited to improve and broaden search quality. A natural consequence of improving search quality can be an increase in revenues from sponsored advertisements. From the academic research point of view, a combination of machine learning and statistical techniques can be blended with techniques in keyword generation to determine the most valuable keywords at the individual level, enabling increased precision in geo-targeting and contextual advertisements during a Web search.

3.2.3 User-Generated Social Information and Economic Exchanges

The Internet has had a profound impact on at least three areas of life—the way people shop, the way they interact socially, and the way they exchange information. All three are relevant to consumer product reviews posted in IT-enabled electronic markets. Online consumer product reviews provide information that can facilitate economic exchange, which is the central function of electronic markets. However, reviews are also sometimes used as a forum for social exchanges, and this, in part, serves to draw people to such websites, promote purchases, and regulate user behavior. Hence, more research is needed on the important role of social communities in geographically dispersed electronic markets. Some participants in electronic markets also use them as forums for social interaction and, in the process, reveal a lot of information about themselves. Drawing on theories from social psychology such as social identity theory and information processing theory, an interesting arena of research is whether online users view consumer reviews as forums to form social identities and if these social exchanges influence economic outcomes in consumer settings.

In the past, product reviewers had limited opportunities to convey community affiliation because product reviews were nominally focused on the product itself rather than the reviewer. Recent design changes in electronic markets now enable members, who identify with the community, to engage in self-disclosure of social information. For example, on sites such as Amazon.com, information about product reviewers is graphically depicted, highly salient, and sometimes more detailed than information about the products they review. Specifically, it allows reviewers to publicly reveal their *real name*, *geographical location*, *professional interests*, and *hobbies*. Such user-generated self-descriptive information can be explained in social psychology as members' attempts to convey the *social identity* they wish others to associate with them. If online self-disclosure is driven in part by the desire for identification with a community and the need for self-verifying feedback from other community members, then reviewer identity expressions should be patterned to follow community norms. Hence, norm conformity is evidence of an investment on the part of an individual contributor signaling that he or she would like to be viewed as a member of the community (Bartel and Dutton 2001).

In particular, if the types or categories of information that reviewers disclose are consistent with the type of information that is typical or normative in the community, this is suggestive evidence that identification processes can be an important antecedent to reviewer disclosure. Such theories can now be empirically tested with the kind of data available on online review and social networking sites. Furthermore, given the extent and salience of social information about product reviewers, it is also possible to inquire whether such information has an influence on the online consumers who are responsible for product sales.

Forman, Ghose, and Wiesenfeld (2008) explore some of these phenomena. They estimate econometric models using a panel dataset consisting of data on chronologically compiled reviews on sets of products, and the various self-descriptive, social, and personal, information of the reviewers. Using research on information processing (Chaiken 1980), they suggest that in an online community, the social information reviewers provide about themselves is used to supplement or replace product information in shaping community members' purchase decisions and the value they attribute to online reviews. Online community members rate reviews containing identity-descriptive information more positively, but the informative value of reviews attenuates the relationship between reviewer disclosure of identity-descriptive information and community ratings relative to more equivocal reviews. Furthermore, Forman, Ghose, and Wiesenfeld (2008) show that the prevalence of reviewer disclosure of identity information in online reviews is associated with increases in subsequent online product sales after controlling for the valence and volume of reviews.

Future research in this domain can extend an emerging stream of academic work on the relationship between reviews and economic outcomes (e.g., Dellarocas et al. 2005; Reinstein and Snyder 2005; Chevalier and Mayzlin 2006), the work exploring motivations for people to post word-of-mouth (e.g., Hennig-Thurau et al. 2004), and the work on belonging to a community (e.g., Cummings et al. 2002). While Forman, Ghose, and Wiesenfeld (2008) do not directly assess the form of information processing that Amazon.com members used, their results are consistent with the notion that people use more heuristic processing of source characteristics when information overload is high. Future research can evaluate whether the number and diversity of messages influence recipients' response to source and message characteristics. While initial research on common bond and common identity suggested that groups could be characterized as *either* common bond *or* common identity (Prentice et al. 1994), subsequent work suggests that the formation of common bonds and common identities may be related to one another (Ren et al. 2007). Thus, future research may consider whether common bonds promote common identity in online communities like Amazon.com.

This research stream can be combined with *sentiment analysis techniques* from computer science to investigate if sentiment patterns in these reviews can be attributed to the disclosure of self-descriptive information by specific reviewers and how such reviews are rated by the community. Text mining techniques similar to those of Kim and Hovy (2004) can be used to automatically find reviewers who hold opinions about a topic. The reviews will be classified as subjective (opinionated) versus objective (neutral) using techniques such as those in Riloff and Wiebe (2003). In theory, subjective reviews may be rated by the community as less helpful than objective reviews because they provide less useful information to guide purchase decisions. However, this may be less true when the reviews contain more disclosure of personal information because helpfulness ratings of those reviews may be partly driven by the desire to grant membership status. Such propositions can now be empirically tested using the rich trail of data.

3.3 ECONOMIC VALUE OF USER-GENERATED TEXTUAL INFORMATION

The research described in this section is based on the conjecture that the qualitative information contained in text-based feedback, online reviews of products, or descriptions of used goods on the Internet plays a substantial role in influencing social and economic outcomes in electronic markets. Such studies can combine empirical tools such as state-of-the-art techniques in hedonic modeling, multivariate consumer choice modeling, and panel data model with automated text-mining techniques to identify and quantify the impact of this information on economic variables such as revenues, price premiums, and resale rates. Text mining usually involves structuring the input text (usually parsing, along with the addition of some derived linguistic features and the removal of others), deriving patterns within the structured data, and, finally, evaluating the output. The main goal is to discover patterns and trends in the information contained in textual documents (Hearst 1999; Grobelnik et al. 2000).

3.3.1 Feedback in Reputation Systems and Economic Impact

When buyers purchase products in an electronic market, they assess and pay not only for the product they wish to purchase, but also for a set of fulfillment characteristics such as packaging, timeliness of delivery, responsiveness of the seller, and reliability of settlement. Most studies of online reputation so far have based a seller's reputation on the numerical rating that characterizes the seller and the level of experience (Dellarocas 2003; Resnick et al. 2006). This implies that there is great potential for future research on reputation systems based on the analysis of text-based feedback posted by buyers. Such feedback describes their transaction experience with the sellers, and it seems natural that it will play an important role in establishing sellers' reputations in electronic markets, since different sellers in these markets derive their reputation from different characteristics. A comment about "super-fast delivery" can enhance a seller's reputation and thus allow it to increase the price of the listed items by a few cents without losing any sales. On the other hand, feedback about "sloppy packaging" can have the opposite effect on a seller's pricing power. To accomplish this study, one needs information on the transaction price at which the used good was sold, the date of sale, all relevant details on competing offers at the time of the sale, the number of such used goods listed, and corresponding price premiums from an electronic market for used-good exchanges. Prior research (Ghose 2006; Ghose et al. 2006) has demonstrated a novel way of extracting such information using data from Amazon's used-good marketplaces.

Broadly speaking, research on the economics of textual content needs to combine established techniques from econometrics with text mining algorithms from computer science to identify the value of text and assign an economic value to each feedback posting, measuring sentiment effectively and without the need for manual labeling of postings. Ghose et al. (2005) is the first known study to demonstrate the value of some of these automated text-mining methods. The text-mining techniques used in their study are based on research on opinion extraction in the computational linguistics community. The first step in the algorithm consists of parsing feedback postings to identify the dimensions across which the buyer evaluates a seller. For this task, a part-of-speech tagger (such as the Stanford Java NLP tagger) is used for each word. The nouns, noun phrases, and verbal phrases are kept as the dimensions of the seller. Then the adjectives and adverbs that refer to the nouns and verbs extracted in the previous step are retrieved, similar to the study of Turney (2002). For instance, Hatzivassiloglou and McKeown (1997) use a supervised learning technique to identify the *semantic orientation* of adjectives. To associate the adjectives and adverbs with the correct dimensions, they use a syntactic parser. The adjective-noun and adverb-verb pairs serve as the basis for further analysis in their paper. Next, using panel data methods, they estimate econometric models that show that textual feedback adds value to the numerical scores and affects the pricing power of sellers. That is, merchants with negative comments charge lower prices, and merchants with positive comments command higher premiums than their competitors. They show the emergence of a number of unique dimensions of seller reputation (e.g., *shipping*, *packaging*, *delivery*, *responsiveness*) and corroborate that substantial pricing power is associated with each dimension of seller reputation.

The research of Hatzivassiloglou and McKeown (1997) also contributes to an emerging stream of work that examines sentiments in online communities and auctions (Pavlou and Dimoka 2006; Gu et al. 2007). There is tremendous potential for future work in this area. For example, using data from eBay, one can compare how textual feedback in a reciprocity-based reputation system (eBay) differs from that in a market where only buyers rate sellers (Amazon.com). This is important because a reciprocity-based reputation system is less likely to elicit honest feedback from buyers due to the potential for retaliation by sellers in the case of a negative comment. This opens up the possibility of feedback manipulation. Dellarocas (2005) provides a nice theoretical model of this practice, but more empirical research is needed to corroborate it. This is an area where the combination of econometrics with text-mining methods can open up myriad research opportunities.

3.3.2 Sentiments in Product Reviews and Economic Impact

Prior work has shown that the volume and valence of online product reviews influence the sales of books and movies (Dellarocas et al. 2005; Chevalier and Mayzlin 2006). However, this research did not account for the impact of textual content in those reviews. Similarly, computer scientists have conducted extensive sentiment analyses of reviews and news articles (Wiebe 2000; Pang et al. 2002;

Turney 2002; Dave et al. 2003; Hu and Liu 2004; Kim and Hovy 2004; Liu et al. 2005; Popescu and Etzioni 2005) but have not examined their economic impact on sales.

This presents an opportunity for future research to bridge the gap between these two streams of literature. One can examine how sentiment embedded in online reviews acts as a predictor of future demand. To do this, one needs to examine consumer sentiment in online markets. Prior studies have measured the predictive power of word-of-mouth on sales but have not used a combination of sentiment analysis and econometric experiments to extract more precise information from the text. Ghose and Ipeirotis (2006) conduct an initial study in this domain. After training the classifiers appropriately using technical product descriptions provided by manufacturers, they automatically classify reviews as subjective (highly opinionated) or objective. Then they measure their impact on revenues at Amazon.com. This enables them to empirically quantify the effect of textual information in online product reviews on sales in various product categories, as well as allowing them to measure how the extent of subjectivity in the opinions affects consumers' perceptions of review informativeness as measured by peer-provided "helpful votes." A limitation of their study is that they have only crude measures for changes in product demand. Future research can use actual data on demand from an online retailer to do a more refined analysis. There is a lot of potential for future researchers to work on ways to score the utility of consumer reviews, with a focus on extracting the various features of a review and measuring their impact on sales or the informativeness of the review (Archak et al. 2007). Moreover, future research can examine the usefulness of data from blogs and social networking sites in predicting sales. Dhar and Chang (2007) use data from MySpace and Amazon.com to track the changes in online chatter for music albums before and after their release dates. Such studies can provide insights to marketing managers interested in assessing the relative importance of the burgeoning number of Web 2.0 information metrics.

3.3.3 Dimensions of Product Descriptions and Economic Impact

Although e-commerce enables an easier search for new products, such standardized searching has not yet been implemented in used-good markets because of the diversity of seller or product characteristics. Whereas attributes such as product features can be communicated easily in electronic markets, nondigital attributes (product condition and seller integrity) are subject to noise and manipulation. This has the potential to create information asymmetry between buyers and sellers stemming from the unobservability of quality signals in electronic markets. This informational asymmetry is associated with both an individual seller's reputation and the product's self-reported quality. Hence, in such used-good markets, asymmetric information can lead to market failure such as adverse selection (Akerlof 1970). This failure is manifested in the fact that sellers with high-quality goods need to wait longer than sellers of low-quality goods in order to complete a trade (Janssen and Karamychev 2002). Sellers try to minimize this information asymmetry by using textual descriptions of the products such as *pristine condition* or *factory sealed to* signal quality *before purchase*. To what extent can standardized textual descriptions of used goods provided by sellers prior to purchase reduce these information asymmetries?

To study this question, one can combine automated text-mining techniques with econometric methods to infer the dimensions of a used-product description that consumers value the most and quantify the price premiums associated with each dimension. Specifically, using text-mining methods similar to those in Ghose et al. (2005), one can identify features of the used-good descriptions that mitigate the information asymmetry problem between buyers and sellers by identifying the *dimensions of the product description* that maximize price premiums, resale rates, and revenues in used-good markets. Future work on the economics of textual descriptions in used-good markets can build on methods deployed in prior work (Ghose et al. 2006) to advance our understanding of the impact of Internet-based exchanges on social welfare. They can also shed light on how such used-good descriptions can alleviate the information asymmetry in electronic markets found in Dewan and Hsu (2004) and in Ghose (2006). By quantifying the economic value of sentiment in online reviewer communities, such studies will provide methods that can be extended to quantify the economic impact of textual content on social networks, blogs, and social shopping sites, as well as in the emerging practice of tagging products.

3.4 ECONOMIC COST OF INFORMATION SEARCH, PROCESSING, AND MODIFICATION

To monetize the economic impact of firm-published or user-generated content, a key analysis that needs to be conducted is to estimate the cost of processing and disseminating information during various human-computer interactions. One of the main advantages that Internet channels posit over physical markets is a reduction in *friction* costs. These costs include search costs incurred by consumers in locating product-related and price-related information (Bakos 1997), decision-making costs incurred by managers in adjusting prices in response to market conditions (Levy et al. 1997), and cognitive costs of processing textual content in product descriptions incurred by users. There is a lot of potential for research that estimates the cognitive costs incurred by consumers when searching for information or processing textual information online and the decision-making cost incurred by managers while adjusting product-level information. This section describes some opportunities in estimating different kinds of cognitive costs incurred by users in the online world.

3.4.1 Consumer Search Costs

The literature on *rational inattention* argues that observing, processing, and reacting to price change information is not a costless activity. An important implication of rational inattention is that consumers may rationally choose to ignore—and thus not to respond to—small price changes, creating a *range of inattention* along the demand curve (Levy et al. 2006). This applies to online markets too, where consumers face cognitive search costs of processing the vast amount of information published on the Internet (Johnson et al. 2004). This cost includes visiting multiple retailers who differ on various attributes, comparing a diverse set of offerings, and assessing the overall quality of their offerings. Moreover, opportunities for obfuscation (Ellison and Ellison 2004) and the unobserved lack of inventories (Arnold and Saliba 2002) can also create such costs. Indeed, the existence of price dispersion in online markets (Brynjolfsson and Smith 2000; Baye et al. 2004) has often been attributed to the presence of search costs and retailer menu costs. Unless retailers factor these behavioral factors into their strategies, market transparency and prices can be suboptimal.

While the presence of search costs and their impact on consumer demand and retailer strategies has been well established theoretically, very few empirical studies exist in this domain. The main focus of existing work has been to quantify these costs at the consumer level. Brynjolfsson et al. (2004) found that the cost of an exhaustive search is about \$6.45 per consumer on a shopping bot. Hong and Shum (2006) developed a methodology for recovering search cost estimates using only observed price data. Bajari and Hortacsu (2003) quantified the cost of entering into an eBay auction to be \$3.20, which includes search costs and other costs related to auction participation. In a related stream of work, Hann and Terweisch (2003) discussed how search costs and other related frictional costs in electronic markets could be substantial, and found that the median value of these costs ranged from EUR 3.54 for an MP3 portable digital music player to EUR 6.08 for a personal digital assistant.

However, these studies have largely focused on measuring consumer search costs, with less attention being paid to how search costs affect consumer demand structure or how they can affect retailers' business strategies. Few studies looks at the consumer demand structure in electronic markets and whether search costs differ across online retailers. It is generally believed that it is difficult to measure price search costs directly since it requires individual observations of consumer expectations of price distributions. However, by inferring the nature of the consumer demand structure and by analyzing consumer consumers' sensitivity to price changes, it is feasible to verify the presence of price search costs.

The presence of price search costs also implies that it takes time for price information to dissipate among consumers and demand to adjust slowly (Radner 2003). Future research can quantify the impact of this information delay on firms' optimal pricing policies. To do so, one needs to estimate the time it takes for pricing information to percolate into the market and show that the distribution of search cost evolves dynamically over time. Till date, empirical studies on retailing have implicitly assumed that retailers face a demand function with constant price elasticity for any kind of price change (Chevalier and Goolsbee 2003; Ellison and Ellison 2004; Ghose et al. 2006) and have used the estimated price elasticity to make inferences about competition and welfare in online marketplaces. This approach, however, does have a limitation. The assumption of constant price elasticity could lead to biased estimates if the difference in price elasticity for price changes in opposite directions is large. For example, if a retailer faces low price elasticity for price increases but high price elasticity for price reductions, the constant elasticity assumption will average the price elasticity, thereby underestimating the real competitiveness and sensitivity in the market. One exception to the constant elasticity assumption is a recent study by Baye et al. (2005) which shows that price elasticity on shopbots increases dramatically after a retailer reduces its price to become the lowest-priced firm on the market. Future research can explore the impact of having a constant elasticity assumption in much greater depth.

3.4.2 The Cost of Processing Textual Content

The above-described stream of research can be extended to explicitly estimate the cognitive cost of evaluating retail offers in online markets. The Internet has facilitated the creation of markets with a large number of retailers. When consumers search for a unique product, they obtain a list of several offers with varying attributes (price, seller reputation, delivery schedules, etc.) and extended product descriptions. All else equal, in order to choose among the offers displayed on the computer screen, consumers read these description, process the information, and choose the best offer for purchase. In the presence of such information overload, it is natural to expect individuals to bear some cognitive cost of scrolling down the screens and processing the textual description of each offer. Even if users do not cognitively process all the information displayed on the computer screen, a "scan and discard" versus "scan and read more" decision leads to a cognitive cost of thinking (Shugan 1980).

Hence, rather than leaving it up to the consumer to cope with this distracting overload, providers often try to present first the most salient items in their inventory while taking into account the visual real estate available on a given device. Search engines such as Google or Yahoo do not exhibit all their search results on one webpage, but rather prioritize and display them on consecutive pages whose value is assumed to be decreasingly lower to the user (Huberman and Wu 2006). Hence, a natural direction for future research would be to examine the effect of summarizing product descriptions on product sales: Short descriptions reduce the cognitive load of consumers but increase their uncertainty about the product's characteristics, whereas longer descriptions have the opposite effect.

By drawing methods from the literature on summarization in computational linguistics and natural language processing (NLP), one can suggest efficient ways of presenting product information in electronic markets by determining the optimal number of words and sentences in a given product description that minimize the consumer cognitive cost. This requires an empirical analysis of how economic variables like product sales, price premiums, and revenues are affected by the textual content in seller-contributed descriptions. Based on readability metrics such as the FOG, SMOG, and Flesch-Kinciad metrics that estimate the reading difficulty of webpages, techniques such as support vector machine-based approaches that automatically recognize reading levels from user queries (Liu et al. 2004) and advances in statistical language models that incorporate both content and surface linguistic features (Collins-Thompson and Callan 2005), the cost of consumer information processing can be quantified. This will involve building text classifiers, incorporating features based on reading level and nontext features such as average line length.

Using discrete choice models, one can then estimate the likelihood of a consumers' selecting a certain offer conditional on price and other seller attributes.

3.4.3 Managerial Decision-Making Costs

An important requirement for establishing a frictionless market on the Web is the need for retailers to act upon information and adjust prices without incurring substantial costs in the process. Prior academic work has used a variety of techniques and methods to study pricing processes and the menu costs incurred by firms in order to improve our understanding of how firms set and adjust prices. In this literature, menu costs include both the physical and the managerial costs of changing prices.

Lessons from physical markets show that price adjustment is a difficult, costly, time-consuming, and "complex process, requiring dozens of steps and a non-trivial amount of resources" (Levy et al. 1997, p. 792). Thus, retailers to incur substantial costs in implementing price changes. These costs include printing price labels, affixing them to retail products, updating point-of-sale systems for price changes, correcting errors, and supervising these price change activities (see, e.g., Levy et al. 1997). Besides these factors, there are substantial managerial costs as well. To set and adjust prices effectively, firms need to gather and analyze data that may be located in different parts of the organization. They need the ability to assess a wide variety of market factors, from costs to customer segments, to estimating market demand, to understanding customer psychology, to anticipating competitors' reactions, and so on. In addition to this are the complexities of firms selling multiple products through multiple distribution channels and to multiple customers, often internationally (Zbaracki et al. 2004). Consistent with these aspects, recent empirical studies of price adjustment processes by Levy et al. (1997) and Slade (1998), conclude that the price adjustment cost associated with these processes may be significant across industries in many offline markets.

Online retailers face a different environment. The increased market transparency due to the availability of information (Granados et al. 2005a, 2005b) helps firms monitor rivals' pricing activities. With no price labels to print, there is little physical labor cost involved, and hence, in theory, firms can adjust prices costlessly. This can have significant implications for consumer cognitive search costs. If menu costs are negligible and retailers can adjust prices costlessly, then we are likely to see more frequent price changes. Such frequent price changes would impose a higher cognitive search cost on consumers. However, online retailers carry SKUs with much higher value than those in typical offline stores and often achieve a much higher sales volume. In this environment, each price experiment requires careful analysis. As a result, online retailers may incur substantial managerial decision-making costs leading to price rigidity. These costs are similar to data engineering costs, misclassification costs, and active learning costs in the literature on human-computer interactions and machine learning (Turney 2000). Indeed, Bergen et al. (2005) find that Amazon and BN change prices less than once every 90 days, a surprisingly low frequency for online firms. Understanding the exact cause of price stickiness and identifying the magnitude of managerial costs of online retailers would thus be a natural next step in improving our understanding of the nature of consumer search costs.

Despite the extensive literature on menu costs (Levy et al. 1997) in offline markets, no prior work has estimated the actual magnitude of such decision-making costs in electronic markets. Ghose and Gu (2007) take one step in this direction by econometrically estimating and quantifying the magnitude of managerial costs of price adjustment faced by firms in electronic markets. They use the well-known random walk theory model (Dixit 1991) to show that if the optimal price follows a random walk, the optimal solution for the retailer is to change the price if and only if the absolute difference between the actual price and the optimal price is larger than a constant. This enables them to empirically infer menu costs from online retailers' actual price change decisions given their demand fluctuations. A key contribution of their research is the development of statistical methods to separate long-term demand changes that affect retailers' price decisions from transitory demand changes that do not affect these decisions. Future research could extend that work by adopting dynamic programming techniques to explicitly model the online retailer's decision-making process for price changes, such as those used in Aguirregabiria (1999) and Kano (2006). This kind of research will contribute to recent research that measures the costs of online transactions (Hann and Terweisch 2003; Hann et al. 2005), switching costs online (Chen and Hitt 2002), and the welfare changes that online channels provide to consumers (Brynjolfsson et al. 2003; Ghose et al. 2006), as well as the emerging work in computer science that measures the readability and display of various categories of online content (Kumaran et al. 2005).

3.4.4 Impact of Search Costs and Menu Costs on the Long Tail

User-level search costs and firm-level menu costs are likely to influence pricing and product assortment strategies. The Internet is known to facilitate the discovery of lesser-known and obscure products. It has been argued that collectively, these relatively less popular products could make up a significant portion of sales for online retailers (Anderson 2006). This phenomenon has been termed the *Long Tail* and is often presented as a strategic opportunity to reach consumers in niche markets who were previously too costly to serve. While existing articles on the Long Tail highlight the strategic role played by lower *product search costs* in promoting the emergence of the Long Tail (Bryjolffson et al. 2006), more rigorous research is needed to demonstrate the strategic role of *price search costs* in necessitating the emergence of the Long Tail. The impact of search costs on retailers' operations and marketing strategies has been studied both theoretically and empirically. It is now known that a failure to incorporate consumer search costs in the assortment planning process may lead the retailer to erroneously narrow product assortments (Cachon et al. 2005). Prior theoretical studies have also shown that the presence of such search costs could affect the shape and dynamics of consumer demand, resulting in significant differences in retailers' optimal product assortments. For example, Cachon et al. (2006) demonstrate that while lower search costs can intensify price competition, they can also expand the pool of customers visiting a retailer. The market expansion effect can justify a broader assortment, which in turn can lead to higher prices and profits.

While prior studies of the Long Tail phenomenon focus on consumer demand for less popular products, they have not formally analyzed retailers' incentives to provide such products. This presents an exciting opportunity for future research. Rather than focusing on promoting popular products that face increased price competition, should online retailers shift their product assortments toward niche products to take advantage of milder price competition in the emerging Long Tail? It also seems natural that such incentives will be related to the menu costs faced by retailers in making price changes for popular versus unpopular products. This implies that research on the differences in magnitude of menu costs across different categories of products and different kinds of price changes will add to our understanding of firm-level incentives in promoting the Long Tail.

A related research question concerns the composition of product returns for Internet retailers. Is there evidence that popular products are returned more often than obscure products? Or is the opposite pattern more likely? From a managerial perspective, this is important because it highlights whether growth in variety is driven by demand-side or supply-side factors. Thus, any research on the impact of the Long Tail on the Internet also needs to account for the frequency with which online products are returned.

3.5 CONCLUSION

With ongoing technological progress, the amount of user-generated and firm-published content in Web-based systems is growing exponentially. A significant portion of this content has concrete embedded economic value. Prior research has partially studied the impact of content—for instance, by analyzing the impact of numeric information. This chapter is a call for research that expands our understanding of that economic value by measuring the benefits and costs of analyzing other forms of content. Such value can be extracted, for example, from the content that specifies the social information of participants in online exchanges, the feedback and product description information that is captured in conversations between individuals, and finally from estimating the costs incurred by individuals during Internet usage, such as the costs of locating and processing information and the decision-making costs of adjusting information. There are several interesting research opportunities in this domain.

Such research can have many managerial and research implications. First, by studying the economic value of various kinds of textual information on the Internet, one can go well beyond analyzing the impact of numeric information, such as the valence and volume of reviews and reputation scores, that has been done in prior work. This will enable researchers to recommend actionable changes in the design of feedback systems, electronic market–based communities, and social networks. By producing novel ways of measuring the value of user-generated online content, the research activities will make actionable recommendations to practitioners to improve the design of feedback systems and the display of information in online markets. In particular, firms are constantly trying to come up with incentive systems that prevent various kinds of spamming, gaming, and content manipulation. For example, analysis of the information in qualitative textual feedback can lead to the design of more robust reputation systems. Similarly, by mining the product descriptions in social shopping sites and used-good markets, future research in design science can come up with tools that will help businesses display information efficiently by minimizing the information overload on consumers (e.g., Ghose et al. 2005) and maximizing profits.

There are several other ways in which user-generated content can affect e-commerce firms. For example, Delicious, a social bookmarking website, exploits users' social network to produce more relevant search results and hence provide better monetization of content. Since user-generated content has led to the emergence of communities in many online markets (Forman, Ghose, and Wiesenfeld 2008), the availability of such data provides a plethora of opportunities for researchers interested in studying how online network based word-of-mouth marketing affects transactions. A pioneering study in this domain is that of Hill et al. (2006, 2007), who find that consumers linked to a prior customer adopt a telecommunication service much faster than baseline groups selected by the best practices of the firm's marketing team.

Future research on how offline geographical locations affect online purchase and search behavior can provide important insights into the continuing debate over whether there exists a geographical *digital divide* in the United States by providing concrete evidence or lack thereof. This becomes important in light of recent anecdotal evidence that despite the ubiquity of the Internet, many low-income, rural, and small-town communities are being left out of this information revolution and deprived of the economic opportunities it offers. Moreover, research in this arena has the potential to demonstrate that a user's geographical community continues to play a role in his or her behavior in electronic communities. Recent research suggests that users in electronic communities may prefer to communicate with users who have similar socioeconomic and geographical neighborhoods (Van Alstyne and Brynjolfsson 2005). Further, it will help businesses understand the role of social information in driving consumer behavior in online communities, and will provide prescriptive insights to businesses such that advertising and marketing strategies in online channels can be customized by users' geographical locations. It will help create incentives for underserved communities to participate in Internet commerce by understanding the mindsets of users and their need for certain products in environments that are culturally mediated. By exploring virtual communities that emerge based on individuals' relationships to specific economic products, future research can advance our understanding of the dynamic interplay between social and economic exchanges on the Internet.

Finally, a key component in measuring value from online content is an understanding of user costs in searching for, processing, and modifying information from an economic perspective. At present, we have little insight into how to measure various cognitive costs incurred by humans while interacting with computers to process information. An explicit understanding of the magnitude of costs incurred by users during Internet usage will contribute to optimal policies that lead to social and economic benefits for all participants. Such search and menu costs are also related to the emerging phenomenon of the Long Tail by influencing the incentives firms have for stocking niche and obscure products. It would be useful to see more academic research in these areas.

ACKNOWLEDGMENT

I thank Foster Provost for many helpful comments.

REFERENCES

Aguirregabiria, V. (1999). The dynamics of markups and inventories in retailing firms. *Review of Economic Studies*, 66: 275–308.

Akerlof, G.A. (1970). The market for "lemons": Quality uncertainty and the market mechanism. *Quarterly Journal of Economics*, 84: 488–500.

Anderson, C. (2006). The long tail: Why the future of business is selling less of more. Hyperion Press, New York, NY.

Archak, N., Ghose, A., and Ipeirotis, P. (2007). Show me the money! Deriving the pricing power of product features by mining consumer reviews. *Proceedings of the Thirteenth ACM SIGKDD International Conference on Knowledge Discovery and Data Mining (KDD)*, San Jose.

Arnold, M. and Saliba, C. (2002). Price dispersion in online markets: The case of college textbooks. Working Paper.

Bajari, P. and Hortacsu, A. (2003). The winner's curse, reserve prices, and endogenous entry: Empirical insights from eBay auctions. *Rand Journal of Economics*. 34 (Summer): 329–355.

Bakos, Y. (1997). Reducing buyer search costs: Implications for electronic marketplaces. *Management Science*, 43(12): 1676–1692.

Balasubramanian, S. (1998). Mail versus mall: A strategic analysis of competition between direct marketers and conventional retailers. *Marketing Science*, 17(3): 181–195.

Bapna, R., Jank, W., and Shmueli, G. (2006). Consumer surplus in online auctions. Working Paper, Indian School of Business.

Bartel, C.A. and Dutton, J.E. (2001). Ambiguous organizational memberships: Constructing social identities in interactions with others. In *Social Identity Processes in Organizational Contexts* (M.A. Hogg and D. Terry, eds.). Philadelphia: Psychology Press.

Baye, M., Gatti, R., Kattuman, P., and Morgan, J. (2005). Estimating firm-level demand at a price comparison site: Accounting for shoppers and the number of competitors. Working Paper, Indiana University.

Baye, M.R., Morgan, J., and Scholten, P. (2004). Price dispersion in the small and in the large: Evidence from an Internet price comparison site. *Journal of Industrial Economics*, 52(4): 463–496.

Bergen, M.E., Kauffman, R., and Lee, D. (2005). Beyond the hype of frictionless markets: Evidence of heterogeneity in price rigidity on the internet. *Journal of Management Information Systems*, 22(2): 57–89.

Brynjolfsson, E., Dick, A., and Smith, M. (2004). Search and product differentiation at an Internet shopbot. Working Paper, MIT.

Brynjolfsson, E., Hu, J., and Smith, M. (2003). Consumer surplus in the digital economy: Estimating the value of increased product variety. *Management Science*, 49(11): 1580–1596.

Brynjolfsson, E., Hu, Y., and Smith, M. (2006). From niches to riches: The anatomy of the long tail. *Sloan Management Review*, 47(4): 67–71.

Brynjolfsson, E. and Smith, M. (2000). Frictionless commerce? A comparison of Internet and conventional retailers. *Management Science*, 46(4): 563–585.

Bucklin, R., Lattin, J., Ansari, A., Bell, D., Coupey, E., Gupta, S., Little, J., Mela, C., Montgomery, A., and Steckel, J. (2002). Choice and the Internet: From clickstream to research stream. *Marketing Letters*, 13(3): 245–258.

Cachon, G., Terwiesch, C., and Xu, Y. (2005). Retail assortment planning in the presence of consumer search. *Manufacturing and Service Operations Management*, 7(4): 330–346.

Cachon, G., Terwiesch, C., and Xu, Y. (2006). On the effects of consumer search and firm entry on multiproduct competition. *Marketing Science*, forthcoming.

Chaiken, S. (1980). Heuristic versus systematic information processing and the use of source versus message cues In persuasion. *Journal of Personality and Social Psychology*, 39(5): 752–766.

Chawla, N., Japkowicz, N., and Kolcz, A. (2003). Editorial. *Proceedings of the icml2003 Workshop on Learning from Imbalanced Data Sets.*

Chen, P. and Hitt, L. (2002). Measuring switching costs and the determinants of customer retention in Internet-enabled businesses: A study of the online brokerage industry. *Information Systems Research*, 13(3): 255–274.

Chevalier, J. and Goolsbee, A. (2003). Measuring prices and price competition online: Amazon and Barnes and Noble. *Quantitative Marketing and Economics*, 1(2): 203–222.

Chevalier, J. and Mayzlin, D. (2006). The effect of word of mouth online: Online book reviews. *Journal of Marketing Research*, 43(3): 345–354.

Collins-Thompson, K. and Callan, J. (2005). Predicting reading difficulty with statistical language models. *Journal of the American Society for Information Science and Technology*, 56(13): 1448–1462.

Cummings, J, Butler, B., and Kraut, R. (2002). The quality of online social relationships. *Communications of the ACM*, 45(7): 103–108.

Dave, K., Lawrence, S., and Pennock, D. (2003). Mining the peanut gallery: Opinion extraction and semantic classification of product reviews. *World Wide Web Conference.*

Dellarocas, C. (2003). The digitization of word-of-mouth: Promise and challenges of online reputation systems. *Management Science*, 49(10): 1407–1424.

Dellarocas, C. (2005). Strategic manipulation of Internet opinion forums: Implications for consumers and firms. *Management Science*, 52(10): 1577–1593.

Dellarocas, C., Awad, N., and Zhang, M. (2005). Using online ratings as a proxy of word-of-mouth in motion picture revenue forecasting. Working Paper, University of Maryland.

Dewan, S. and Hsu, V. (2004). Adverse selection in electronic markets: Evidence from online stamp auctions. *Journal of Industrial Economics*, 17(4): 497–516.

Dhar, V. and Chang, E. (2007). Does chatter matter? The impact of user-generated content on music sales. Working Paper, New York University–CeDER.

Dixit, A. (1991). Analytical approximations in models of hysteresis. *Review of Economic Studies*, 58: 141–151.

Ellison, G. and Ellison, S. (2004). Search, obfuscation, and price elasticities on the Internet. Working Paper, Massachusetts Institute of Technology.

Forman, C., Ghose, A., and Goldfarb, A. (2008). Competition between local and electronic markets: How the benefit of buying online depends on where you live. Working Paper, New York University.

Forman, C., Ghose, A., and Wiesenfeld, B. (2008). Examining the relationship between reviews and sales: The role of reviewer identity disclosure in electronic markets. *Information Systems Research*, forthcoming.

Forman, C., Goldfarb, A., and Greenstein, S. (2004). How did location affect adoption of the commercial Internet? Global village, Urban leadership, and industry composition. Working Paper, Tepper School of Business, Carnegie Mellon University.

Ghose, A. (2006). Trade patterns and adverse selection in electronic secondary markets. Working Paper, New York University.

Ghose, A. and Gu, B. (2007). Estimating menu costs in electronic markets. Working Paper, New York University.

Ghose, A. and Ipeirotis, P. (2006). Towards an understanding of the impact of customer sentiment on product sales and review quality. *Proceedings of the 2006 Workshop on Information Technology and Systems (WITS 2006).*

Ghose, A., Ipeirotis, P., and Sundararajan, A. (2005). The dimensions of reputation in electronic markets. Working Paper, New York University.

Ghose, A., Smith, M., and Telang, R. (2006). Internet exchanges for used books: An empirical analysis of welfare implications. *Information Systems Research*, 17(1): 3–19.

Ghose, A., Mukhopadhyay, T., and Rajan, U. (2007). Impact of Internet referral services on supply chain. *Information Systems Research*, 18(3): 300–319.

Ghose, A. and Yang, S. (2007). An empirical analysis of search engine advertising: Sponsored search in electronic markets, Working Paper, New York University.

Goldfeld, S. and Quandt, R. (1973). A markov model for switching regressions. *Journal of Econometrics*, 1(3): 3–16.

Granados, N., Gupta, A., and Kauffman, R. (2005a). IT-enabled transparent electronic markets: The case of the air travel industry. *Information Systems and e-Business Management*, 5(1): 65–91.

Granados, N., Gupta, A., and Kauffman, R. (2005b). Market transparency strategy and multi-channel strategy: Modeling and empirical analysis. Working Paper, University of Minnesota.

Grobelnik, M., Mladenic, D., and Milic-Frayling, N. (2000). Text mining as integration of several related research areas. *Report on KDD'2000 Workshop on Text Mining*.

Gu, B., Konana, P., Rajagopalan, B., and Chen, H.M. (2007). Competition among virtual communities and user valuation: The case of investor communities. *Information Systems Research*, 18(1) 68–85.

Hann, I., Savin, S., and Terwiesch, C. (2005). Online haggling and price discrimination at a name-your-own price channel: Theory and application. *Management Science*, 51(3): 339–351.

Hann, I. and Terwiesch, C. (2003). Measuring the frictional costs of online transactions: The case of a name-your-own-price channel. *Management Science*, 49(11): 1563–1579.

Hatzivassiloglou, V. and McKeown, K.R. (1997). Predicting the semantic orientation of adjectives. *Proceedings of the 35th Annual Meeting of the Association for Computational Linguistics (ACL'97).*

Hearst, M. (1999). Untangling text data mining. *Proceedings of the 37th Annual Meeting of the Association for Computational Linguistics.*

Hennig-Thurau, T., Gwinner, K.P., Walsh, G., and Gremler, D. (2004). Electronic word-of-mouth via consumer-opinion platforms: What motivates consumers to articulate themselves on the Internet? *Journal of Interactive Marketing*, 18(1): 38–52.

Hill, S., Provost, F., and Volinsky, C. (2006). Network-based marketing: Identifying likely adopters via consumer networks. *Statistical Science*, 21(2): 256–276.

Hill, S., Provost, F., and Volinsky, C. (2007). Learning and inference in massive social networks. *Proceedings of the 5th International Workshop on Mining and Learning with Graphs.*

Hong, H. and Shum, M. (2006). Using price distributions to estimate search costs. *Rand Journal of Economics*, 37(2): 257–275.

Hu, M. and Liu, B. (2004). Mining and summarizing customer reviews. *Proceedings of the 2004 ACM SIGKDD International Conference on Knowledge Discovery and Data Mining (KDD'04).*

Huberman, B. A. and Wu, F. (2006). The economics of attention: maximizing user value in information-rich environments, *Proceedings of The First International Workshop on Data Mining and Audience Intelligence for Advertising (ADKDD'07).* San Jose, California, US.

Jank, W. and Kannan, P.K. (2006). Understanding geographical markets of online firms using spatial models of customer choice. *Marketing Science*, 24(4): 623–634.

Janssen, M. and Karamychev, V. (2002). Cycles and multiple equilibria in the market for durable lemons. *Economic Theory*, 20: 579–601.

Johnson, E. J., Moe, W., Fader, P., Steven, B., and Lohse, J. (2004). On the depth and dynamics of online search behavior. *Management Science*, 50: 299–308.

Kano, K. (2006). Menu costs, strategic interactions, and retail price movements. Working Paper, University of British Columbia.

Kim, S. and Hovy, E. (2004). Determining the sentiment of opinions. *Proceedings of COLING-04.*

King, G. and Zeng, L. (2001). Logistic regression in rare events data. *Political Analysis*, 9(2): 137–163.

Kumaran, G., Jones, R., and Madani, O. (2005). Biasing Web search results for topic familiarity. *Proceedings of the 14th ACM International Conference on Information and Knowledge Management.*

Levy, D., Bergen, M., Dutta, S., and Venebirgeo, D. (1997). The magnitude of price adjustment cost: Direct evidence from large U. S. supermarket chains. *Quarterly Journal of Economics*, 112(3): 791–825.

Levy, D., Ray, S., Chen, H., and Bergen, M. (2006). Asymmetric wholesale price adjustment: Theory and evidence. *Marketing Science*, 25(2): 131–154.

Liu, X., Croft, B., Oh, P., and Hart, D. (2004). Automatic recognition of reading levels from user queries. *Proceedings of the 27th Annual International ACM SIGIR Conference on Research and Development in Information Retrieval.*

Liu, B., Hu, M., and Cheng, J. (2005). Opinion observer: Analyzing and comparing opinions on the Web. *Proceedings of the 14th International Conference on the World Wide Web.*

Pang, B., Lee, L., and Vaithyanathan, S. (2002). Thumbs up? Sentiment classification using machine-learning techniques. *Proceedings of the Conference on Empirical Methods in Natural Language Processing.*

Pavlou, P. and Dimoka, A. (2006). The nature and role of feedback text comments in online marketplaces: Implications for trust building, price premiums, and seller differentiation. *Information Systems Research*, 17(4): 392–414.

Popescu, A. and Etzioni, O. (2005). Extracting product features and opinions from reviews. *Proceedings of HLT-EMNLP.*

Prentice, D.A., Miller, D.T., and Lightdale, J.R. (1994). Asymmetries in attachments to groups and to their members: Distinguishing between common-identity and common-bond groups. *Personality & Social Psychology Bulletin*, 20(5): 484–493.

Prince, J. (2006). The beginning of online/retail competition and its origins: An application to personal computers. *International Journal of Industrial Organization*, 25(1): 139–156.

Radner, R. (2003). Viscous demand. *Journal of Economic Theory*, 112(2): 189–231.

Reinstein, D. and Snyder, C. (2005). The influence of expert reviews on consumer demand for experience goods: A case study of movie critics. *Journal of Industrial Economics*, 53(1): 27–51.

Ren, Y., Kraut, R.E., and Kiesler, S. (2007). Applying common identity and bond theory to design of online communities. *Organization Studies*, 28(3): 377–408.

Resnick, P., Zeckhauser, R., Swanson, J., and Lockwood, K. (2006). The value of reputation on eBay: A controlled experiment. *Experimental Economics*, 9(2): 79–101.

Riloff, E. and Wiebe, J. (2003). Learning extraction patterns for subjective expressions. *Conference on Empirical Methods in Natural Language Processing (EMNLP-03), ACL SIGDAT.*

Rossi, P. and Allenby, G. (2003). Bayesian statistics and marketing. *Marketing Science*, 22, 304–329.

Scott-Morton, F., Zettlemeyer, F., and Risso, J. (2006). How the Internet lowers prices: Evidence from matched survey and auto transaction data. *Journal of Marketing Research*, 43(2): 168–181.

Shugan, S. (1980). The cost of thinking. *Journal of Consumer Research*, 7(2): 99–111.

Si, L. and Callan, J. (2006). A statistical model for scientific readability. *Proceedings of the Tenth International Conference on Information and Knowledge Management.*

Sinai, T. and Waldfogel, J. (2004). Geography and the Internet: Is the Internet a substitute or complement for cities? *Journal of Urban Economics*, 56: 1–24.

Slade, M. (1998). Optimal pricing with costly adjustment: Evidence from retail-grocery prices, *Review of Economic Studies*, 65: 87–107.

Turney, P.D. (2000). Types of cost in inductive concept learning. *Proceedings of the Workshop on Cost Sensitive Learning at the Seventeenth International Conference on Machine Learning (WCSL), ICML-2000.*

Turney, P.D. (2002). Thumbs up or thumbs down? Semantic orientation applied to unsupervised classification of reviews. *Proceedings of the 40th Annual Meeting of the Association for Computational Linguistics.*

Van Alstyne, M. and Brynjolfsson, E. (2005). Global village or cyber-balkans? Modeling and measuring the integration of electronic communities. *Management Science*, 51(6): 851–868.

Wiebe, J. (2000). Learning subjective adjectives from corpora. *Proceedings of the 17th National Conference on Artificial Intelligence (AAAI-2000).*

Zbaracki, M., Ritson, M., Levy, D., Dutta, S., and Bergen, M. (2004). The managerial and customer dimensions of the cost of price adjustment: Direct evidence from industrial markets. *Review of Economics and Statistics*, 86: 514–533.

4

IS PRIVACY PROTECTION FOR DATA IN AN E-COMMERCE WORLD AN OXYMORON?

STEPHEN E. FIENBERG
Department of Statistics, Machine Learning Department, and Cylab Carnegie Mellon University, Pittsburgh, Pennsylvania

4.1 INTRODUCTION: E-COMMERCE PRIVACY BREACHES ARE IN THE NEWS

On August 6, 2006, the media were abuzz when AOL posted on a webpage information on 20,000,000 searches from approximately 650,000 AOL users. The data included all searches from those users for a three-month period that year, as well as whether they clicked on a result, what that result was, and where it appeared on the result page. It was a 439 MB compressed download that expanded to just over 2 GB. In the released database, AOL replaced individuals' real IDs by UserID numbers (see Figure 4.1, which contains information on the released searches). Within 24 hours, thousands of people had downloaded the data file before AOL abruptly withdrew it, although publicly available copies appear to abound.[1] On August 8, 2006, the *New York Times* published an article which included an interview with an AOL subscriber whom the reporters claimed to have identified from the released files.[2] Although AOL publicly

[1]See, e.g., http://www.gregsadetsky.com/aol-data.
[2]Michael Barbaro and Tom Zeller, Jr., "The Face Behind AOL User 4417749," *New York Times*, August 9, 2006.

Statistical Methods in e-Commerce Research. Edited by W. Jank and G. Shmueli

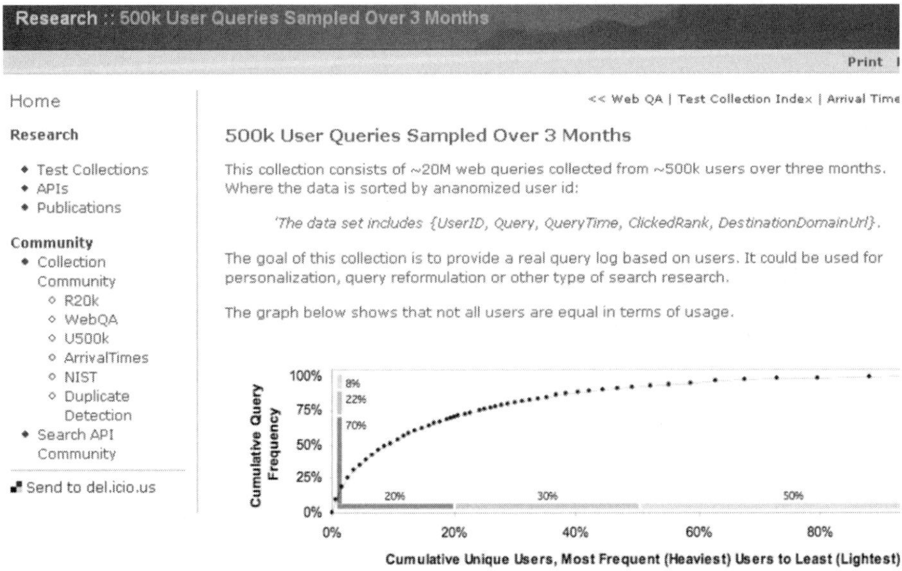

Figure 4.1 Copy of the AOL webpage describing the released search data. Source: http://blog.outer-court.com/files/aol-vs-privacy-large.png.

apologized,[3] resignations of several key staff members quickly followed, as did complaints filed with the FTC,[4] and the repercussions from this event are still being felt across the e-commerce industry.

The AOL search data illustrate why a history of "mere searches" in a search engine contains personal content with a high degree of private and associative information that many believe should be protected. There are continuing questions about why many Web-focused companies need to save detailed search queries in the first place, i.e., what business purpose is being served. And there are those who argue that if AOL was planning at any point to release its users' detailed search queries and search result clickstreams with detailed timestamps, it should have clearly and plainly informed its customers of this specific use in its privacy policy.

But perhaps even more disconcerting to many was the fact that the posted AOL search data were similar to those data sought by the U.S. Justice Department when

[3] "This was a screwup, and we're angry and upset about it. It was an innocent enough attempt to reach out to the academic community with new research tools, but it was obviously not appropriately vetted, and if it had been, it would have been stopped in an instant. . . . Although there was no personally-identifiable data linked to these accounts, we're absolutely not defending this. It was a mistake, and we apologize. We've launched an internal investigation into what happened, and we are taking steps to ensure that this type of thing never happens again." AOL spokesperson Andrew Weinstein, August 7, 2006. See http://sifaka.cs.uiuc.edu/xshen/aol/20060807_AOLSpokesmanApology.txt.

[4] In its complaint, the Electronic Frontier Foundation noted that it had identified 175 searches from 106 distinct users that appear to contain Social Security numbers, 8457 searches from 3739 distinct users that appear to contain phone numbers, and 10,835 searches from 4099 distinct users that appear to contain street addresses. http://www.eff.org/Privacy/AOL/aol_ftc_complaint_final.pdf.

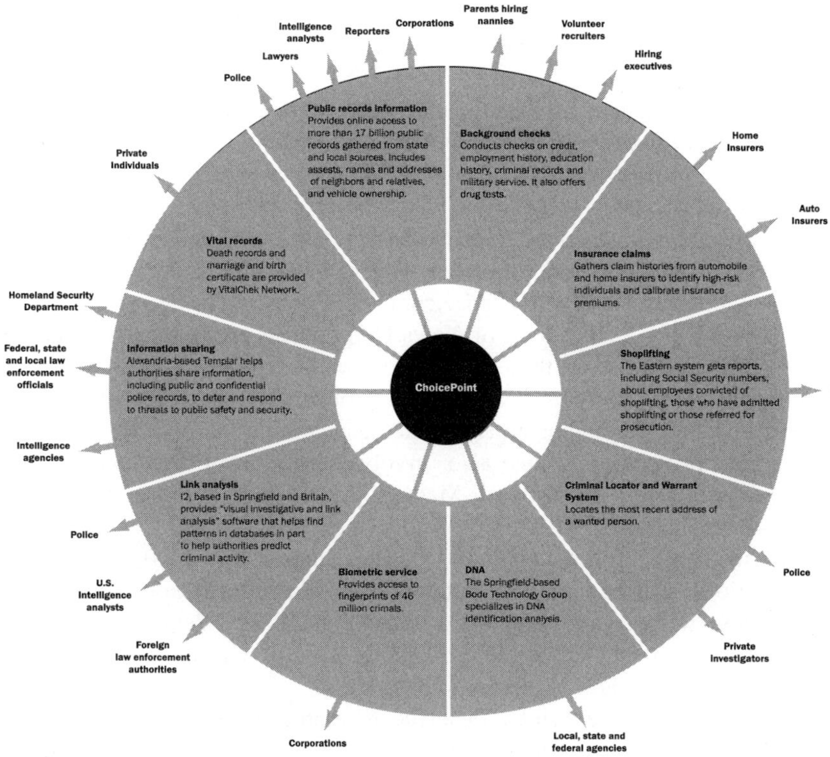

Figure 4.2 ChoicePoint data sources and clients. Source: Adapted from *Washington Post*, January 20, 2005.

it subpoenaed Internet companies, including AOL, a couple of years ago. AOL complied and handed over search terms that were not linked to individuals. Google fought the subpoena in court and won, but the broad concerns regarding privacy of data gathered in an e-commerce setting linger and are closely associated with concerns regarding the aggregation of commercial and other data in data warehouses and the subsequent use of those data by U.S. government departments.

Data warehousing companies such as Acxiom, ChoicePoint, and LexisNexis use their data to perform background checks on prospective applicants for employers, insurers, and credit providers. They also sell their data to state and federal governments. Figure 4.2 shows the array of data available from ChoicePoint and the types of clients who access them, as presented by the *Washington Post*.[5] If you go to the ChoicePoint website[6] and read the privacy policy, you are told "How we protect you," but if you want to check the accuracy of information on yourself that ChoicePoint sells to others, you need to provide your Social Security number! This

[5]http://www.washingtonpost.com/wp-srv/business/daily/graphics/choicepoint_012005.html.
[6]http://www.choicepoint.com.

means that if ChoicePoint didn't have your Social Security number before, it would now, and they make no promise about how it will (or will not) be used or shared in the future.

In part as a consequence of the data security breaches of the sort described above, some form of data breach legislation has been introduced in at least 35 states and signed into law in at least 22, according to data compiled by the National Conference of State Legislatures.[7]

Privacy concerns with e-commerce clearly run deeper and wider than the AOL breach, electronic transfers of funds, or the aggregation of data by Acxiom, ChoicePoint, and LexisNexis. MySpace.com, which was recently purchased by Fox Interactive Media, a subsidiary of News Corp., has a user base of around 180 million profiles, and many of these contain personal information of precisely the sort that the public is constantly advised to protect. MySpace also works with other online operations to aggregate and share data.[8]

Many other privacy breaches also involve the use of the Internet. For example, the University of Pittsburgh Medical Center (UPMC) recently admitted that a physician and medical school faculty member included the names and Social Security numbers of 80 patients in a presentation for a 2002 symposium. The information, which included detailed medical information about some of the patients, remained posted on a UPMC radiology department website for five years![9] And the list could go on and on, in virtually every area of e-commerce and those parts of our lives that abut it.

In the next section, we briefly describe a related set of government data-mining and data warehousing activities that came into the public eye following the terrorist attacks of September 11, 2001. Section 4.3 provides an overview of record linkage and its use for merging large data files from diverse sources, as well as its implications for the splitting of databases for privacy protection. Section 4.4 reviews some proposals that have surfaced for the search of multiple databases without compromising possible pledges of confidentiality to the individuals whose data are included and their link to the related literature on privacy-preserving data mining. In particular, we focus on the concept of *selective revelation* and its confidentiality implications. Section 4.5 relates these ideas to the recent statistical literature on disclosure limitation for confidential databases. Section 4.6 explains the problems with the privacy claims, and we discuss the special aspects of transactional data, summarized in the form of a graph or a network diagram. We conclude with some observations regarding privacy protection and e-commerce.

[7]Tom Zeller, Jr., "Link by Link; Waking Up to Recurring ID Nightmares," *New York Times*, January 9, 2006.

[8]Caroline McCarthy, "MySpace to Provide Sex Offender Data to State AGs," http://news.com.com/MySpace+to+provide+sex+offender+data+to+state+AGs/2100-1030_3-6185333.html.

[9]Mark Houser, "UPMC Admits Privacy Violation," *Tribune-Review*, April 13, 2007. http://www.pittsburghlive.com/x/pittsburghtrib/news/cityregion/s_502469.html.

4.2 HOMELAND SECURITY AND THE SEARCH FOR TERRORISTS

A report from the U.S. General Accounting Office (GAO) (2004) notes that at least 52 agencies are using or planning to use data mining, *factual data analysis*, or *predictive analytics* in some 199 different efforts. Of these, at least 29 projects involve analyzing intelligence and detecting terrorist activities, or detecting criminal activities or patterns. Notable among the nonresponders to the GAO inquiry were agencies like the Central Intelligence Agency (CIA) and the National Security Agency (NSA), and many more agencies and programs surely are engaged in such activities today.

Perhaps the most visible of these efforts was the *Total Information Awareness* (TIA) program initiated by the Defense Advanced Research Program (DARPA) in DARPA's Information Awareness Office (IAO), which was established in January 2002 in the aftermath of the September 11 terrorist attacks. The TIA research and development program was aimed at integrating information technologies into a prototype to provide tools to better detect, classify, and identify potential foreign terrorists. When it came under public scrutiny in 2003, TIA morphed into the *Terrorist Information Program* (still TIA) with essentially the same objectives, although it too was not implemented. TIA served as the model, however, for the *Multistate Anti-terrorism Information Exchange* system (MATRIX), which was used in seven states for a period of time during 2004 and 2005 and was intended to provide the capability to store, analyze, and exchange sensitive terrorism-related information in MATRIX data bases among agencies, within a state, among states, and between state and federal agencies.

According to a recent report from the Congressional Research Service (Relyea and Seifert (2005); footnotes omitted):

> The MATRIX project was initially developed in the days following the September 11, 2001, terrorist attacks by Seisint, a Florida-based information products company, in an effort to facilitate collaborative information sharing and factual data analysis. At the outset of the project, MATRIX included a component Seisint called the High Terrorist Factor (HTF), which was designed to identify individuals with high HTF scores, or so-called terrorism quotients, based on an analysis of demographic and behavioral data. Although the HTF scoring system appeared to attract the interest of officials, this feature was reportedly dropped from MATRIX because it relied on intelligence data not normally available to the law enforcement community and because of concerns about privacy abuses.
>
> ...The analytical core of the MATRIX pilot project is an application called Factual Analysis Criminal Threat Solution (FACTS), described as a "technological, investigative tool allowing query-based searches of available state and public records in the data reference repository." The FACTS application allows an authorized user to search "dynamically combined records from disparate datasets" based on partial information, and will "assemble" the results. The data reference repository used with FACTS represents the amalgamation of over 3.9 billion public records collected from thousands of sources. The data contained in FACTS include FAA pilot license and aircraft ownership records, property ownership records, information on vessels registered with the

> Coast Guard, state sexual offender lists, federal terrorist watch lists, corporation filings, Uniform Commercial Code filings, bankruptcy filings, state-issued professional license records, criminal history information, department of corrections information and photo images, driver's license information and photo images, motor vehicle registration information, and information from commercial sources that "are generally available to the public or legally permissible under federal law."
>
> . . .To help address the privacy concerns associated with a centralized data repository, some officials have suggested switching to a distributed approach whereby each state would maintain possession of its data and control access according to its individual laws.

The data reference repository is said to exclude data from the following sources:

- telemarketing call lists
- direct mail mailing lists
- airline reservations or travel records
- frequent flyer/hotel stay program membership information or activity
- magazine subscription records
- information about purchases made at retailers or over the Internet
- telephone calling logs or records
- credit or debit card numbers
- mortgage or car payment information
- bank account numbers or balance information
- records of birth certificates, marriage licenses, and divorce decrees
- utility bill payment information

MATRIX was officially abandoned as a multistate activity in April 2005, although individual states were allowed to continue with their parts of the program. This does not mean the demise of the TIA effort, however, as there are other federal initiatives built on a similar model:

- *Analysis, Dissemination, Visualization, Insight, and Semantic Enhancement* (ADVISE), is a research and development program within the Department of Homeland Security (DHS), part of its three-year-old "Threat and Vulnerability, Testing and Assessment" portfolio.[10]
- The *Information Awareness Prototype System* (IAPS), the core architecture that tied together numerous information extraction, analysis, and dissemination tools developed under TIA, including the privacy-protection technologies, was moved to the Advanced Research and Development Activity (ARDA), housed at NSA headquarters in Fort Meade, Maryland.[11]

[10]Mark Clayton, "US Plans Massive Data Sweep," *Christian Science Monitor*, February 9, 2006. http://www.csmonitor.com/2006/0209/p01s02-uspo.html.
[11]Shane Harris, "TIA Lives On," *National Journal*, Thursday, February 23, 2006.

In TIA, MATRIX, ADVISE, and IAPS, the data miner can issue queries to the multiple linked databases and receive responses that combine data on individuals across the databases, e.g., see Popp and Poindexter (2006). The goal is the identification of terrorists or criminals in a way that would not be possible from the individual databases. We distinguish between two aspects of this goal: (1) identification of known terrorists, which is a form of retro- or postdiction, and (2) identification of potential future terrorists, and profiling, which involves prediction. Prediction cannot be separated from uncertainty; postdiction might conceivably be. Most of the public outcry regarding TIA and MATRIX has focused on concerns regarding what has been described as *dataveillance* (Clarke 1988) and terrorist profiling, i.e., concerns both about the use of data for purposes other than those for which they were collected without the consent of the individual, and about the quality and accuracy of the mined data and the likelihood that they may help falsely identify individuals as terrorists.

The difficulty of prediction is not lost on government planners at NSA and DHS. As Chris Christopher, a science fiction fan and communications officer with DHS's Science and Technology division, described it recently: "If you think what you always thought, you'll get what you've always got!"[12]

Like TIA, ADVISE has run into difficulty over privacy issues. A March 2007 GAO Report (U.S. GAO 2007) touched off another public debate by noting that "until a privacy-impact assessment is conducted, little assurance exists that privacy risks have been rigorously considered and mitigating controls established." A current draft of the 2008 Homeland Security appropriations bill, would withhold funding for the ADVISE program until DHS submits a privacy-impact assessment for the program.[13]

In the next two sections, we explore some issues related to the creation and use of linked databases for the privacy of the individuals whose confidential information is contained in them.

4.3 MATCHING AND RECORD LINKAGE METHODS

More than 100 vendors offer record-matching systems, some of which sell for thousands of dollars, but most of the underlying methodology for such systems is proprietary and few details are publicly available. Matches can occur at random. For example, consider a pair of files, A and B, containing n records on the same individuals. The probability of correctly matching exactly r individuals by picking a random permutation for file B and linking to file A is

$$\frac{\sum_{\nu=0}^{n-r} \frac{(-1)^{n}-r}{\nu!}}{r!}. \tag{4.1}$$

[12]This was in the context of DHS asking science fiction writers for terrorist scenarios: "Science Fiction in the National Interest," *On the Media, National Public Radio*, July 1, 2007. http://www.onthemedia.org.
[13]Chris Strohm, "Lawmakers Move to Halt Funds for Data-Mining Plan," *National Journal's Technology Daily*, June 11, 2007. http://www.govexec.com/dailyfed/0607/061107tdpm1.htm.

Domingo-Ferrer and Torra (2003) derive this baseline and illustrate it numerically in an example with $n = 90$, where the expected number of correct matches is $O(10^{24})$. Working with actual data in the matching process can change this situation drastically.

Bilenko et al. (2003) provide an overview of the published literature on the topic, noting that most methods rely on the existence of unique identifiers or use some variation of the algorithm presented in Fellegi and Sunter (1969). Fellegi and Sunter's approach is built on several key components for identifying matching pairs of records across two files:

- Represent every pair of records using a vector of features (variables) that describe the similarity between individual record fields. Features can be Boolean (e.g., last-namematches), discrete (e.g., first-*n*-characters-of-name-agree), or continuous (e.g., string-edit-distance-between-first-names).
- Place feature vectors for record pairs into three classes: matches (M), nonmatches (U), and possible matches. These correspond to "equivalent," "nonequivalent," and "possibly equivalent" (e.g., requiring human review) record pairs, respectively.
- Perform record-pair classification by calculating the ratio $(P(\gamma \mid M))/(P(\gamma \mid U))$ for each candidate record pair, where γ is a feature vector for the pair and $P(\gamma \mid M)$ and $P(\gamma \mid U)$ are the probabilities of observing that feature vector for a matched and a nonmatched pair, respectively. Two thresholds based on desired error levels—T_μ and T_λ—optimally separate the ratio values for equivalent, possibly equivalent, and nonequivalent record pairs.
- When no training data in the form of duplicate and nonduplicate record pairs are available, matching can be unsupervised, where conditional probabilities for feature values are estimated using observed frequencies.
- Because most record pairs are clearly nonmatches, we need not consider them for matching. The way to manage this is to "block" the databases—for example, based on geography or some other variable in both databases—so that only records in comparable blocks are compared. Such a strategy significantly improves efficiency.

The first four components lay the groundwork for accuracy of record-pair matching using statistical techniques such as logistic regression, the EM algorithm, and Bayes networks (e.g., see Jaro 1995; Larson and Rubin 2001; Winkler 2002). Accuracy is known to be high when there is a one-to-one match between records in the two systems and deteriorates as the overlap between the files decreases, as well as with the extent of measurement error in the feature values. While the use of human review of possible matches has been an integral part of many statistical applications, it may well be infeasible for large-scale data warehousing. The fifth component provides for efficiently processing large databases but, to the extent that blocking is approximate and possibly inaccurate, the use of blocking decreases the accuracy of record pair matching. For a detailed description of both the basic technology of record linkage and its use in specific applications, see Herzog et al. (2007).

There are three potential lessons associated with this literature on matching and the methods it has produced:

1. If we are trying to protect against an intruder who would like to merge the data in a confidential database with an external database in his or her possession, then we need to assure ourselves and the intruder that the accuracy of matching is low and that individuals cannot be identified with high probability. We need to keep in mind that an intruder will have easy access to a host of identifiable public record systems. For example, as of July 1, 2007, SearchSystems.net[14] listed 38,672 free searchable public record data bases on its website!
2. One strategy for protecting a database against attack from an intruder is to split it into parts, perhaps overlapping, to decrease the likelihood of accurate matches. The parts should be immune from attack (with high probability) but of value for analytical purposes. For categorical data, this might correspond to reporting lower-dimensional margins from a high-dimensional contingency table (see Bishop et al. 1975; Dobra and Fienberg 2001, 2003; Fienberg and Slavkovic 2004). For continuous data, we might need to apply disclosure protection methods to the split components (e.g., see Duncan 2001 and Fienberg 2005a) for overviews. It is the uncertainty associated with efforts to concatenate the separate pieces that provides the confidentiality protection in both instances. The higher the uncertainty, the better the protection.
3. Unless Choicepoint and other data warehousers are adding data to their files using unique identifiers such as Social Security numbers (and even Social Security numbers are not really unique!), or using highly accurate addresses and/or geography, some reasonable fraction of the data in their files will be the result of inaccurate and faulty matches. Data quality for data warehouses is an issue we all need to worry about (see Winkler 2005).

4.4 ENCRYPTION, MULTIPARTY COMPUTATION, AND PRIVACY-PRESERVING DATA MINING

If you search the World Wide Web (WWW) for *e-commerce* and *data privacy protection*, you will find extensive discussion about firewalls, intrusion prevention, (IPS), intrusion detection systems (IDS), and secure socket layer (SSL) encryption technology. Indeed, these technological tools are important for secure data transmission, statistical production, and offline data storage (Domingo-Ferrer et al. 2000). But encryption cannot protect the privacy of individuals whose data are available in online databases.

[14]http://www.searchsystems.net.

Among the methods advocated to carry out such data-mining exercises are those that are described as privacy-preserving data mining (PPDM). PPDM typically refers to data-mining computations performed on the combined datasets of multiple parties without revealing each party's data to the other parties. The data consist of possibly overlapping sets of variables contained in the separate databases of the parties and overlapping sets of individuals. When the parties have data for the same variables but different individuals, the data are said to be *horizontally partitioned*; when the individuals are the same but the variables are different, the data are said to be *vertically partitioned.* Here we are concerned with the more complex case involving both overlapping variables *and* overlapping sets of individuals. PPDM research comes in two varieties. In the first, sometimes referred to as the construction of *privacy-preserving statistical databases*, the data are altered prior to delivery for data mining, for example, through the addition of random noise or some other form of perturbation. While these approaches have much in common with the methods in the literature on statistical disclosure limitation, they are of little use when it comes to the identification of terrorists. In the second variety, the problem is solved using what is known as *secure multiparty computation*, where no party knows anything except its own input and the results. The literature typically presumes that data are included without error and thus could be matched perfectly if only there were no privacy concerns. The methods also focus largely on situations where the results are of some computation, such as a dot product or the description of an association rule. See the related discussion in Fienberg and Slavkovic (2005).

A major problem with the PPDM literature involving multiparty computation is that the so-called proofs of security are designed not to protect the individuals in the database but rather the database owners, as in the case of two companies sharing information but not wanting to reveal information about their customers to one another beyond that contained in the shared computation. Once the results of the data mining consist of linked extracts of the data themselves, however, the real question is whether one of the parties can use the extra information to infer something about the individuals in the other party's data that would otherwise not be available.

Secure multiparty computation (SMC) is a technique for carrying out computations across multiple databases without revealing any information about data elements found only in one database. The technique consists of a protocol for exchanging messages. We assume the parties to be *semihonest*—that is, they correctly follow the protocol specification, yet they attempt to learn additional information by analyzing the messages that are passed. For example, Agrawal et al. (2003) illustrate the secure computation notion via an approach to the matching problem for parties A and B. They introduce a pair of encryption functions E (known only to A) and E' (known only to B) such that for all x, $E(E'(x)) = E'(E(x))$. A's database consists of a list $\mathbf{A}$ and B's consists of a list B. A sends B the message $E(\mathbf{A})$; B computes E' $(E(\mathbf{A}))$ and then sends to A the two messages $E'(E(\mathbf{A}))$ and $E'(\mathbf{B})$. A then applies E to $E'(\mathbf{B})$, yielding $E'(E(A))$ and $E'(E(B))$. A computes $E'(E(\mathbf{A})) \cap E'(E(B))$. Since A knows the order of items in $\mathbf{A}$, A also knows the order of items in E' $E(A))$ and can quickly determine $\mathbf{A} \cap \mathbf{B}$. The main problems with this approach are that (1) it is asymmetric, that is, B must trust A to send $\mathbf{A} \cap \mathbf{B}$ back, and (2) it presumes semihonest behavior.

Li et al. (2005) describe a variety of scenarios in which the Agrawal et al. protocol can easily be exploited by one party to obtain a great deal of information about the other's database, and they explain the drawbacks of some other secure computation methods, including the use of one-way hash-based schemes. As Dwork and Nissim (2004) note: "There is also a very large literature in secure multi-party computation. In secure multi-party computation, functionality is paramount, and privacy is only preserved to the extent that the function outcome itself does not reveal information about the individual inputs. In privacy-preserving statistical data bases, privacy is paramount." The problem with privacy-preserving data mining for terrorist detection is that it seeks the protection of the latter while revealing individual records using the functionality of the former. For more details on some of these and other issues in the context of using SMC protocols to do statistical calculations such as regression and logistic regression, see Karr et al. (2006) and Fienberg et al. (2007). The literature on SMC protocols is extensive and intricate, and includes secure approximation techniques as well (e.g., see Feigenbaum et al. 2006).

The U.S. Congress and various private foundations have taken up the issue of privacy protection from government datamining activities, especially in the post-9/11 world. For example, in its recent report, the U.S. Department of Defense Technology and Privacy Advisory Committee (TAPAC) (2004) stressed the existence of a broad array of government data-mining programs, as well as "disjointed," "inconsistent," and "outdated" laws and regulations protecting privacy. TAPAC recommended broad new actions to protect privacy, both within the Department of Defense and across agencies of the federal government.

The long-standing concern regarding surveillance of U.S. citizens and others by government agencies has been heightened during the war on terror (e.g., Kreimer 2004) and especially most recently with the controversy over unauthorized domestic spying.[15]

4.5 SELECTIVE REVELATION, THE RISK-UTILITY TRADE-OFF, AND DISCLOSURE LIMITATION ASSESSMENT

To get around the privacy problems associated with the development of the TIA and MATRIX systems Tygar (2003a, 2003b) and others have advocated the use of what has come to be called *selected revelation*, involving something like the risk-utility trade-off in statistical disclosure limitation. Sweeney (2005b) used the term to describe an approach to disclosure limitation that allows data to be shared for surveillance purposes "with a sliding scale of identifiability, where the level of anonymity matches scientific and evidentiary need." This corresponds to a monotonically increasing threshold for maximum tolerable risk in the R-U confidentiality map

[15]David Johnston and Neil A. Lewis, "Domestic Surveillance: The White House: Defending Spy Program, Administration Cites Law," *New York Times*, December 23, 2005.

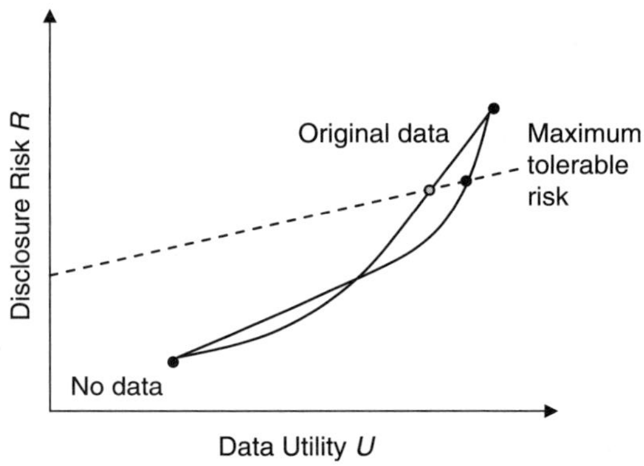

Figure 4.3 R-U confidentiality maps for two different disclosure limitation methods with varying parameter settings. Adapted from Duncan and Stokes (2004).

framework previously described in Duncan et al. (2001), as depicted in Figure 4.3. See also Duncan and Stokes (2004) and Duncan et al. (2004). There are some related ideas emanating from the computer science literature, but most authors attempt to demand a stringent level of privacy, carefully defined, and restrict access through the addition of noise and limitation on the numbers of queries allowed (e.g., see Chawla et al. 2005).

Figure 4.4 depicts the basic selective revelation scheme as described in a committee report on TIA privacy methodology of the Information Science and Technology Study Group on Security and Privacy (2002).

The TIA privacy report suggests that

> Selective revelation works by putting a security barrier between the private data and the analyst, and controlling what information can flow across that barrier to the analyst. The analyst injects a query that uses the private data to determine a result, which is a high-level sanitized description of the query result. That result must not leak any private information to the analyst. Selective revelation must accommodate multiple data sources, all of which lie behind the (conceptual) security barrier. Private information is not made available directly to the analyst, but only through the security barrier.

One effort to implement this scheme was dubbed *privacy appliances* by Lunt (2003), and it was intended to be a stand-alone device that would sit between the analyst and the private data source so that private data stay in authorized hands. These privacy controls would also be independently operated to keep them isolated from the government. According to Lunt (2003), the device would provide

- *inference control* to prevent unauthorized individuals from completing queries that would allow identification of ordinary citizens
- *access control* to return sensitive identifying data only to authorized users
- *Immutable audit trail* for accountability

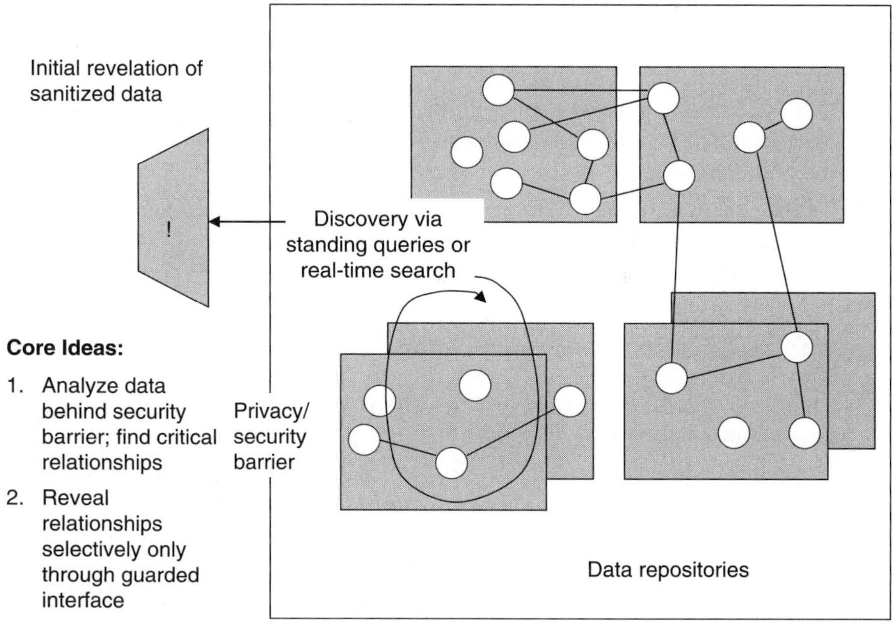

Figure 4.4 Idealized selective revelation architecture. Adapted from Information on Science and Technology Study Group on Security and Privacy (2002).

Implicit in the TIA report and in the Lunt approach was the notion that linkages across databases behind the security barrier would utilize identifiable records and thus some form of multiparty computation method involving encryption techniques.

The real questions of interest in *inference control* are: (1) What disclosure limitation methods should be used? (2) To which databases should they be applied? and (3) How can the inference control approaches be combined with the multiparty computation methods? Here is what we have in the way of answers:

1. Both Sweeney (2005b) and Lunt et al. (2005) refer to Sweeney's version of micro-aggregation, known as *k-anonymity*, but with few details on how it could be used in this context. This methodology combines observations in groups of size k and reports either the sum or the average of the group for each unit. The groups may be identified by clustering or some other statistical approach. Left unspecified is what kinds of users might work with such aggregated data. Further, neither k-anonymity nor any other confidentiality tool does anything to cope with the implications of the release of exactly-linked files requested by "authorized users."
2. Much of the statistical and operations research literature on confidentiality fails to address the risk-utility trade-off, largely by focusing primarily only on privacy or on technical implementations without understanding how users wish to analyze a database (e.g., see Gopal et al. 2002).

3. A clear lesson from the statistical disclosure limitation literature is that privacy protection in the form of *safe releases* from separate databases does not guarantee privacy protection for a merged database. A figure in Lunt et al. (2005) demonstrates recognition of this fact by showing privacy appliances applied for the individual databases and then again independently for the combined data.
4. To date, there have been a limited number of crosswalks between the statistical disclosure limitation literatures on multiparty computation and risk-utility trade-off choices for disclosure limitation. Yang et al. (2005) provide a starting point for discussions on k-anonymity. There are clearly a number of alternatives to k-anonymity, and ones which yield *anonymized* databases of far greater statistical utility!
5. The hype associated with the TIA approach to protection has abated, but largely because TIA no longer exists as an official program. But similar programs continue to appear in different places in the federal government, and no one associated with any of them has publicly addressed the provacy concerns raised here regarding the TIA approach.

When the U.S. Congress stopped the funding for DARPA's TIA program in 2003, Lunt's research and development effort at the PARC Research Center was an attendant casualty. Thus, to date, there are no publicly available prototypes of the privacy appliance, nor are there likely to be in the near future. The claims of privacy protection and selective revelation continue with MATRIX and other data warehouse systems, but without an attendant research program, and the federal government continues to plan for the use of data-mining techniques in other federal initiatives such as the Computer Assisted Passenger Profiling System II (CAPPS II). Similar issues arise in the use of government, medical, and private transactional data in bio-terrorism surveillance (e.g., see Fienberg and Shmueli 2005; Sweeney 2005a).

4.6 ANALYZING NETWORK DATA BASED ON TRANSACTIONS

Transactions can often be described in terms of networks—for example, in the form of graphs $G(V,E)$ with vertices (nodes), V, corresponding to units or individuals and edges, E, corresponding to the transactions. If unit i sends something to unit j, then there is a directed edge connecting them. Much of the DARPA initiative linked to TIA was focused on link mining of transaction databases (e.g., see the summary in Senator 2005 as well as Popp and Poindexter 2006). For discussion of related machine learning methodologies, see the papers in the special issue of *SIGKDD Explorations* on link mining described by Getoor and Diel (2005) and the workshop papers in Airoldi et al. (2007). Especially challenging from a statistical perspective is the development of dynamic models for network evolution.

This problem has been in the news recently because of the controversy over the NSA's access (or attempted access) to the telephone log records of millions of Americans.[16] Clearly, such log records can be represented in network form. The claim has been made that such data will be used to discover linkages among individuals in a terrorist cell. This can typically be done in one of two ways:

1. By looking at telephone links to and from known or suspected terrorists. Issue: how far out do we look for links, and how do we distinguish between terrorist links and links that represent everyday life—ordering take-out meals from a pizza parlor or calling neighbors or work colleagues at home?
2. By using a *template* or signature for a terrorist cell and sifting the entire database of linkages, looking for cliques that resemble the signature. Issues: One needs to discover one or more terrorist network templates and distinguish them from other subnetworks and cliques with similar pattern.

Much has been made of efforts to retrospectively discover linkages among the twenty 9-11 terrorists, using publicly available information from newspapers and the WWW, as well as from various transaction databases. Krebs (2005) gives a detailed discussion of the use of publicly available data to construct the terrorist network reproduced here in Figure 4.5, but he also notes that he only needed to look retrospectively at links between known individuals in order to construct it.

Krebs provides little information on how many others there are to whom these twenty are linked in multiple transaction databases and what one might do with those links if we didn't have definitive information on which individuals composed a terrorist cell. He discusses how one might move toward this kind of prospective analysis, but much of what he suggests is speculation at best. See also the related discussions in the other papers that appear in the 2005 special issue of *Connections* on this topic, available online.

Despite the extensive literature on link mining and social network analysis, there seems to be limited progress at best in understanding how we can use transaction databases to identify terrorist cells or networks. The size of the telephone log records databases makes searching for signatures especially problematic. Schneier has claimed that "Finding terrorism plots is not a problem that lends itself to data mining. It's a needle-in-a-haystack problem, and throwing more hay on the pile doesn't make that problem any easier."[17] Viewed from the perspective of the risk-utility tradeoff, the use of large-scale transaction databases appears to have high privacy risk and low utility, at least given the current state of data mining and privacy-protection methodology. For a discussion of the latter, see Backstrom et al. (2007).

But many other kinds of network data pose far greater privacy problems and affect a broader array of individuals worldwide. A good example is the social network companies like Facebook and MySpace, where millions of individuals provide others

[16] "Bush Administration's Warrantless Wiretapping Program," *Washington Post*, Tuesday, May 15, 2007. http://www.washingtonpost.com/wp-dyn/content/article/2007/05/15/AR2007051500999_pf.html.

[17] Bruce Schneier, "We're Giving up Privacy and Getting Little in Return," *Minneapolis Star Tribune*, May 31, 2006. http://www.startribune.com/562/v-print/story/463348.html.

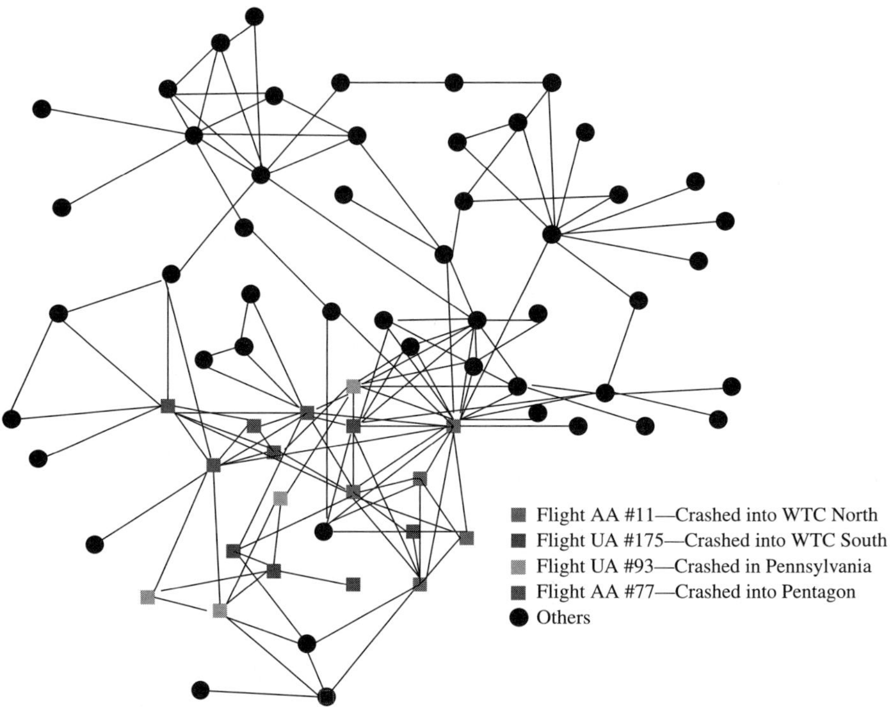

Figure 4.5 Network map of the 9-11 terrorist networks based on public information. The hijackers are color coded by the flight they were on. The dark grey nodes are others who were reported to have had direct, or indirect, interactions with the hijackers. Source: Adapted from Krebs [36].

with enormous amounts of potentially harmful information (Gross and Acquisti 2005; Acquisti and Gross 2006; Romanosky et al. 2006). Two other chapters in this volume also deal with network data and differing methods of statistical analysis but do not include discussions of privacy concerns (see Dass and Reddy, this volume, and Warren et al., this volume).

4.7 CONCLUSIONS

Data privacy protection is a major issue for e-commerce, cf. Muralidhar et al. (2001). While solutions like SSL encryption may help companies protect confidential data transmission, the privacy pitfalls of marketing data as part of e-commerce are many. In this chapter, we have focused on large-scale data warehousing in part because the repeated announcements of security breaches in systems operated by the major vendors such as Acxiom, ChoicePoint, and LexusNexus have filled our morning newspapers during the past several years. The public and civil rights groups have argued that this is just the tip of the privacy-violation iceberg, and they have called for government intervention and legal restrictions on both public and private organizations with respect to data warehousing and data mining. The lessons from such privacy breaches extend easily to virtually all electronically accessible databases. Companies need to take data security seriously and implement best practices, and they need to rethink their policies on data access by others. As the recent thefts of 26.5 million records from the Department of Veterans Affairs and of 1500 records at a National Nuclear Security Administration center in Albuquerque, New Mexico, make clear, government agencies are just as vulnerable (see U.S. GAO 2006b).

The giant data warehouses described in this chapter have been assembled through the aggregation of information from many separate databases and transactional data systems. They depend heavily on matching and record linkage methods that are intrinsically statistical in nature and whose accuracy deteriorates rapidly in the presence of serious measurement error. Data-mining tools can't make up for bad data and poor matches, and someone beyond wronged consumers will soon begin to pay attention.

Should you worry about these data warehouses? With very high probability they contain data on you and your household, but you will never know what the data are or how accurate the information is. And soon the data may be matched with a government-sponsored terrorist search systems such as the one being set up by the Transportation Security Administration (TSA) to match passenger lists with a consolidated watch list of suspected terrorists. On September 19, 2005, the "Secure Flight" Working Group to the Transportation Security Administration (TSA) submitted a report questioning TSA's secrecy regarding what data it plans to use and how (Secure Flight Working Group 2005):

> The TSA is under a Congressional mandate to match domestic airline passenger lists against the consolidated terrorist watch list. TSA has failed to specify with consistency whether watch list matching is the only goal of Secure Flight at this stage....
>
> Will Secure Flight be linked to other TSA applications?...
>
> How will commercial data sources be used? One of the most controversial elements of Secure Flight has been the possible uses of commercial data. TSA has never clearly defined two threshold issues: what it means by "commercial data"; and how it might use commercial data sources in the implementation of Secure Flight. TSA has never clearly distinguished among various possible uses of commercial data, which all have different implications.

The story continues, however, since a few months later, it was revealed that TSA had purchased a database from ChoicePoint to be matched against the watch list.[18] Williams (2007) even suggested farming out such programs to ChoicePoint, ignoring most of the privacy issues described in this chapter!

[18] "TSA Chief Suspends Traveler Registry Plans," *Associated Press*, February 9, 2006.

We emphasize that the discussion in this chapter is not intended to imply that the government should not be sharing databases across agencies, either for statistical purposes or to support homeland security initiatives, (see U.S. GAO 2006a). Rather, our argument is that such data sharing needs to be done with care and due attention to privacy and confidentiality concerns.

Finally, we need new computational and statistical technologies to protect linked multiple databases from privacy invasion in the face of commercial and government queries. Terms like *selective revelation* without technical backup are not enough. This might be provided by the serious integration of research ideas emanating from the statistical disclosure and cryptography communities. The technologies that result from such collaborative research must be part of the public domain, because only then can we evaluate their adequacy.

As a reviewer of an earlier version of this chapter pointed out, we have covered only the tip of the e-commerce data disclosure iceberg, and the lack of privacy protection in the commercial sector is probably much more rampant than the issues described here. Loyalty cards and the prevalence of Radio Frequency Identification (RFID) chips are just two examples of data collection devices that are used regularly to profile and target customers across different databases. They raise new issues and and new opportunities for the loss of privacy that also require the attention of the e-commerce industry, government, and those whose data remain at risk. Until then, privacy protection for the data in the world of e-commerce will truly remain an oxymoron.

ACKNOWLEDGMENTS

The research reported here was supported in part by NSF Grant IIS-0131884 to the National Institute of Statistical Sciences and by Army Contract DAAD19-02-1-3-0389 to CyLab at Carnegie Mellon University. This chapter is an updated version of materials from an article in *Statistical Science*, and a related book chapter (Fienberg 2008). I have benefited from conversations with Chris Clifton, Cynthia Dwork, Alan Karr, and Latanya Sweeney about the material described here, but they bear no responsibility for the way I have represented their input.

REFERENCES

Acquisti, A. and Gross, R. (2006). Imagined communities: Awareness, information sharing, and privacy on the Facebook. *Workshop on Privacy-Enhancing Technologies (PET).*

Agrawal, R., Evfimievski, A., and Srikant, R. (2003). Information sharing across private databases. *Proceedings of the 2003 ACM SIGMOD International Conference on Management of Data.*

Airoldi, E.M., Blei, D., Fienberg, S.E., Goldenberg, A., Xing, E., and Zheng, A. (2007). *Statistical Network Analysis: Models, Issues and New Directions, ICML 2006 Workshop on Statistical Network Analysis, Pittsburgh, PA, USA, June 2006, Revised Selected*

Papers. E., Airoldi, Lecture Notes in Computer Science, vol. 4503. Heidelberg: Springer-Verlag.

Backstrom, L., Dwork, C., and Kleinberg, J. (2007). Wherefore art thou R3579X? Anonymized social networks, hidden patterns, and structural steganography. *Proceedings of the 16th International Conference on World Wide Web.*

Bilenko, M., Mooney, R., Cohen, W.W., Ravikumar, P., and Fienberg, S.E. (2003). Adaptive name-matching in information integration. *IEEE Intelligent Systems*, 18: 16–23.

Bishop, Y.M.M., Fienberg, S.E., and Holland, P.W. (1975). *Discrete Multivariate Analysis: Theory and Practice*. Cambridge, MA: MIT Press. Reprinted 2007, Springer-Verlag, New York.

Chawla, S., Dwork, C., McSherry, F., Smith, A., and Wee, H. (2005). Toward privacy in public databases. *Proceedings of the 2nd Theory of Cryptography Conference (TCC'05).*

Clarke, R. (1988). Information technology and dataveillance. *Communications of the ACM*, 31: 498–512.

Dobra, A. and Fienberg, S.E. (2001). Bounds for cell entries in contingency tables induced by fixed marginal totals. *Statistical Journal of the United Nations ECE*, 18: 363–371.

Dobra, A. and Fienberg, S.E. (2003). Bounding entries in multi-way contingency tables given a set of marginal totals. *Foundations of Statistical Inference: Proceedings of the Shoresh Conference 2000.*

Domingo-Ferrer, J.M., Mateo-Sanz, J.M., and Sánchez del Castillo, R.X. (2000). Cryptographic techniques in statistical data protection. *Proceedings of the Joint UN/ECE-Eurostat Work Session on Statistical Data Confidentiality.*

Domingo-Ferrer, J.M. and Torra, V. (2003). Statistical data protection in statistical microdata protection via advanced record linkage. *Statistics and Computing*, 13: 343–354.

Duncan, G.T. (2001). Confidentiality and statistical disclosure limitation. *International Encyclopedia of the Social and Behavioral Sciences*, 2521–2525. Amsterdam: Elsevier.

Duncan, G.T., Fienberg, S.E., Krishnan, R., Padman, R., and Roehrig, S.F. (2001). Disclosure limitation methods and information loss for tabular data. In *Confidentiality, Disclosure and Data Access: Theory and Practical Applications for Statistical Agencies* (P. Doyle, J. Lane, J. Theeuwes and L. Zayatz, eds.). Amsterdam: Elsevier.

Duncan, G.T., Keller-McNulty, S.A., and Stokes, S.L. (2004). Database security and confidentiality: Examining disclosure risk vs. data utility through the R-U confidentiality map. Technical Report Number 142, National Institute of Statistical Sciences.

Duncan, G.T. and Stokes, S.L. (2004). Disclosure risk vs. data utility: The R-U confidentiality map as applied to topcoding. *Chance*, 17: 16–20.

Dwork, C. and Nissim, K. (2004). Privacy-preserving data mining in vertically partitioned databases. *Proceedings of CRYPTO 2004, 24th International Conference on Cryptology.*

Feigenbaum, J., Ishai, Y., Malkin, T., Nissim, K., Strauss, M.J., and Wright, R.N. (2006). Secure multiparty computation of approximations. *ACM Transactions on Algorithms*, 2(3): 435–472.

Fellegi, I.P. and Sunter, A.B. (1969). A theory for record linkage. *Journal of the American Statistical Association*, 64: 1183–1210.

Fienberg, S.E. (2005a). Confidentiality and disclosure limitation. *Encyclopedia of Social Measurement*, 463–469. Amsterdam: Elsevier.

Fienberg, S.E. (2005b). Homeland insecurity: Datamining, terrorism detection, and confidentiality. *Bulletin of the International Statistical Institute, 55th Session, Sydney.*

Fienberg, S.E. (2006). Privacy and confidentiality in an e-commerce world: Data warehousing, matching, and disclosure limitation. *Statistical Science*, 21: 143–154.

Fienberg, S.E. (2008). Homeland insecurity: Data mining, privacy, disclosurel limitation, and the hunt for terrorists. In *Terrorism Informatics: Knowledge Management and Data Mining for Homeland Security* (H. Chen, Edna Reid, J. Sinai, A. Silke, and B. Ganor, eds.). New York: Springer-Verlag.

Fienberg, S.E., Karr, A.F., Nardi, Y., and Slavkovic, A. (2007). Secure logistic regression with distributed databases. *Bulletin of the International Statistical Institute.*

Fienberg, S.E. and Shmueli, G. (2005). Statistical issues and challenges associated with rapid detection of bio-terrorist attacks. *Statistics in Medicine*, 24: 513–529.

Fienberg, S.E. and Slavkovic, A.B. (2004). Making the release of confidential data from multiway tables count. *Chance*, 17: 5–10.

Fienberg, S.E. and Slavkovic, A.B. (2005). Preserving the confidentiality of categorical statistical data bases when releasing information for association rules. *Data Mining and Knowledge Discovery*, 11: 155–180.

Getoor, L. and Diehl, C.P. (2005). Introduction: Special issue on link mining. *SIGKDD Explorations*, 7(2): 76–83.

Gopal, R., Garfinkel, R., and Goes, P. (2002). Confidentiality via camouflage. *Operations Research*, 50: 501–516.

Gross, R. and Acquisti, A. (2005). Information revelation and privacy in online social networks. *Workshop on Privacy in the Electronic Society (WPES).*

Herzog, T.N., Scheuren, F.J., and Winkler, W.E. (2007). *Data Quality and Record Linkage Techniques*. New York: Springer-Verlag.

Information Science and Technology Study Group on Security and Privacy (chair: J.D. Tygar). (2002). *Security with Privacy.* December 13, Briefing.

Jaro, M.A. (1995). Probabilistic linkage of large public health data files. *Statistics in Medicine*, 14: 491–498.

Karr, A.F., Lin, X., Sanil, A.P., and Reiter, J.P. (2006). Secure statistical analysis of distributed databases. In *Statistical Methods in Counterterrorism* (D. Olwell and A.G. Wilson, eds.). New York: Springer-Verlag.

Krebs, V.E. (2005). Mapping networks of terrorist cells. *Connections*, 24(3): 43–52. Available at http://www.insna.org/Connections-Web/Volume24-3/Valdis.Krebs.web.pdf.

Kreimer, S.F. (2004). Watching the watchers: Surveillance, transparency, and political freedom in the war on terror. *Journal of Constitutional Law*, 7: 133–181.

Larsen, M.D. and Rubin, D.B. (2001). Alternative automated record linkage using mixture models. *Journal of the American Statistical Association*, 79: 32–41.

Li, Y., Tygar, J.D., and Hellerstein, J.M. (2005). Private matching. In *Computer Security in the 21st Century* (D. Lee, S. Shieh, and J.D. Tygar, eds.). New York: Springer-Verlag.

Lunt, T. (2003). Protecting privacy in terrorist tracking applications. Presentation to the Department of Defense Technology and Privacy Advisory Committee, September 29, 2003. Available at http://www.sainc.com/tapac/library/Sept29/LuntPresentation.pdf.

Lunt, T., Staddon, J., Balfanz, D., Durfee, G., Uribe, T., et al. (2005). Protecting privacy in terrorist tracking applications. PowerPoint presentation. Available at http://research.microsoft.com/projects/SWSecInstitute/five-minute/Balfanz5.ppt.

Muralidhar, K.R., Parsa, K., and Sarathy, R. (2001). An improved security requirement for data perturbation with implications for e-commerce. *Decision Sciences*, 32: 683–698.

Popp, R. and Poindexter, J. (2006). Countering terrorism through information and privacy protection technologies. *IEEE Security and Privacy*, 4(6): 18–27.

Relyea, H.C. and Seifert, J.W. (2005). *Information Sharing for Homeland Security: A Brief Overview*. Congressional Research Service, Library of Congress (updated January 10).

Romanosky, S., Acquisti, A., Hong, J., Cranor, L., and Friedman, B. (2006). Privacy patterns for online interactions. *Proceedings of the Pattern Languages of Programs Conference (PLOP)*.

Secure Flight Working Group (2005). *Report of the Secure Flight Working Group*. Presented to the Transportation Security Administration, September 19.

Senator, T.E. (2005). Link mining applications: Progress and challenges. *SIGKDD Explorations*, 7(2): 76–83.

Sweeney, L. (2005a). Privacy-preserving bio-terrorism surveillance. *AAAI Spring Symposium, AI Technologies for Homeland Security*.

Sweeney, L. (2005b). Privacy-preserving surveillance using selective revelation. LIDAP Working Paper, Carnegie Mellon University.

Tygar, J.D. (2003a). Privacy architectures. Presentation at Microsoft Research, June 18. Available at http://research.microsoft.com/projects/SWSecInstitute/slides/Tygar.pdf.

Tygar, J.D. (2003b). Privacy in sensor webs and distributed information systems. In *Software Security* (M. Okada, B. Pierce, A. Scedrov, H. Tokuda, and A. Yonezawa, eds.). New York: Springer-Verlag.

U.S. Department of Defense Technology and Privacy Advisory Committee (TAPAC) (2004). *Safeguarding Privacy in the Fight Against Terrorism*.

U.S. General Accounting Office (2004). *Data Mining: Federal Efforts Cover a Wide Range of uses*. GAO-04-548, Report to the Ranking Minority Member, Subcommittee on Financial Management, the Budget, and International Security, Committee on Governmental Affairs, U.S. Senate. U.S. Government Printing Office, Washington, DC.

U.S. Government Accountability Office (2006a). *Information Sharing: The Federal Government Needs to Establish Policies and Processes for Sharing Terrorism-Related and Sensitive but Unclassified Information*. GAO-06-385. Washington, DC: U.S. Government Printing Office.

U.S. Government Accountability Office (2006b). *Privacy: Preventing and Responding to Improper Disclosures of Personal Information*. GAO-06-833T. Washington, DC: U.S. Government Printing Office.

U.S. Government Accountability Office (2007). *Datamining: Early Attention to Privacy in Developing a Key DHS Program Could Reduce Risks*. GAO-07-293. Washington, DC: U.S. Government Printing Office.

Williams, M. (2007). Confusing Osama bin Laden with Johnny Rotten. *Technology Review*. MIT, http://www.technologyreview.com/Infotech/18484/?a=f.

Winkler, W.E. (2002). Record linkage and Bayesian networks. *Proceedings of the Section on Survey Research Methods*. American Statistical Association, CD-ROM.

Winkler, W.E. (2005). Data quality in data warehouses. In Encyclopedia of Data Warehousing and Data Mining. (J. Wang, ed.). Hershey, PA: Idea Group Publishing.

Yang, Z., Zhong, S., and Wright, R.N. (2005). Privacy-enhancing k-anonymization of customer data. *24th ACM SIGMOD International Conference on Management of Data/Principles of Database Systems (PODS 2005)*.

5

NETWORK ANALYSIS OF WIKIPEDIA

ROBERT H. WARREN

School of Computer Science, University of Waterloo, Waterloo, Ontario, Canada

EDOARDO M. AIROLDI

Computer Science Department and Lewis-Sigler Institute for Integrative Genomics, Princeton University, Princeton, New Jersey

DAVID L. BANKS

Department of Statistics, Duke University, Durham, North Carolina

5.1 INTRODUCTION

Wikipedia is an online encyclopedia created by the volunteer efforts of Internet users all over the world. Although it is not a commercial enterprise, it has relevance to e-commerce activities. In particular, companies with business models in which value is created by users who link sites and share content should be interested in the dynamics of network growth seen in the Wikipedia data.

Some companies of this kind generate revenue through advertising or provide such valuable functionality that their enterprise is bought by others. Others do not offer their network directly to the world, but use it internally to manage and annotate memos, files, and accounts. Snapfish, for photo sharing, is an example of a for-profit company that generates revenue chiefly through advertising. Facebook is an example of a social networking tool that has become so popular that Google thinks it will draw people to them, although their specific business plan is unclear. And John Negroponte, the U.S. intelligence czar, says that his analysts use an in-house system called Intellipedia to manage their internal information sharing (Reuters 2006). There are many other examples, and no doubt more are coming.

Statistical Methods in e-Commerce Research. Edited by W. Jank and G. Shmueli

5.1.1 Motivation

This chapter uses data on the initiation and editing of Wikipedia entries to understand growth, revision, and linkage in a complex multiuser network.

Specifically, organizations considering the use of wikis will want to weight the benefits of a wiki with the preparation required. Beyond hardware and software requirements, there is the question of the amount of initial bootstrapping content that the organization must provide in order to make the wiki usable and attractive to the target audience.

This initial content varies in its quantity and sophistication, and we wished to know the relationship between the initial content provided by the organization and the content that would be added by the user.

Additionally, most wikis allow for the creation of category data that classify and link the different content pages in a taxonomy. Since the difference between category and content is under continual debate in the expert fields, we thought it interesting to look at the category taxonomy constructed by the end users. Would they use the skeleton provided by the organization, provide their own categorical data, or ignore the categories altogether?

Finally, there exists a debate on whether wiki methods are a worthwhile new means of representing information, separate from the online journals (blogs) and commented lists of links (web logs) that are prevalent on the Internet. If wikis are expected to be online references, then they should be more than just lists of links, and should provide new and unique user-generated content.

Looking forward, we believe that e-commerce business enterprises that attempt to emulate Wikipedia's strategy for creating value will want to benchmark their own networking projects against the growth dynamics of Wikipedia; it is likely that the rapid growth of Wikipedia reflects a fortunate confluence of circumstances that deserves study and replication. Two key questions are:

- What is the balance between the amount of effort required to create new entries compared to editing and correcting entries, and how does this balance change over time?
- Are there economies of scale in managing a collaborative project? Does the growth of Wikipedia suggest a cartoon model with typical phases of growth?

Answering these questions will provide insight into the growth of a major new Internet phenomenon and perhaps guidance to those who attempt to mimic its success.

As a caveat, one of the challenges in this kind of research is that the best sources are online, and Wikipedia itself maintains some of the most useful information and analyses. But these sites can be revised at any time; in particular, posted information may be updated or removed. The material used in this chapter is current as of May 8, 2007. A related problem, much less important but aesthetically irksome, is the fact that the use of URLs in citation causes word processing systems to balk, producing line overruns or introducing potentially misleading dashes. The world needs a convention for hyphenating URLs; we propose and use the ⊕ sign, since that cannot appear as a character in any URL and thus avoids ambiguity.

5.2 BACKGROUND ON WIKIPEDIA

There are many Wikipedias, divided according to language and largely independent of one another. This chapter focuses on the first and largest, which is the English Wikipedia. The English Wikipedia was conceived in 1999, the creation of Jimmy Wales and Larry Sanger as a project at Wales' company Bomis. Bomis built Nupedia, an online encyclopedia with free content, but the articles were produced by selected experts and refereed for content. The recruitment of experts and the refereeing of the articles led to substantial delays; thus, Wales and Sanger decided to drop that project entirely and create a software system that allowed volunteers to create and post articles, which could then be revised and improved by other volunteers. The term *wiki* is derived from the Hawaiian word for "quick" and presumably relates to the faster production time.

On January 15, 2001, Wikipedia went public. Its innovative approach had several immediate consequences. First, Wikipedia quickly spread beyond traditional encyclopedia topics to become a guide to popular culture—it contains articles on television shows, movies, and minor players on the world stage. Second, the accuracy of Wikipedia entries is less than that of hard-copy encyclopedias (though not by much; see Giles 2005). Third, there are opportunities for mischief and vandalism. The comedian Stephen Colbert was banned from Wikipedia posting after he entered false information and encouraged his audience to do likewise (Pava 2006). Also, the German version of Wikipedia was hacked so as to distribute copies of the Blaster worm (Leyden 2006). Fourth, Wikipedia has become immensely popular; as of January 2007 it has grown to include more than 1.5 million articles in English and about 5 million articles in about 250 other languages (http://en.wikipedia.org/wiki/Wikipedia).

These kinds of issues and pressures meant that Wikipedia had to carefully track content creation, editing, and cross-linking. The key technology supporting wikis was developed by Ward Cunningham in 1995 (Leuf and Cunningham 2001). The main feature of this technology is that each page has an "edit this topic" link that sends users to a control site that allows them to make changes to the topic. As part of that process, people can register with Wikipedia, creating a user profile (or, in about 25% of the cases, choose to remain anonymous). The version control system for Wikipedia tracks all changes to each topic and which user (or IP address for anonymous users) made those changes. This version control data is one of the main data sources for the analyses in this chapter, along with the link structure for each topic and information from the user profiles.

The legal structure supporting the distribution of Wikipedia text is the GNU Free Documentation License. This permits anyone to use, modify, and distribute source code without limitation. That license has provision for *invariant sections* that cannot be changed except by the creator, even if they are inaccurate or plagiarized.

Those portions of Wikipedia that are invariant pose content-management problems. But most content is not so governed, and contributors have no ownership; indeed, one of the facilitators of growth is the flexibility with which volunteers can correct and revise each other's work.

The thousands of Wikipedia volunteer editors collaborate to build a consensus on change. They review new entries, identify conflicts, and negotiate agreement on changes in content. The custom in the community is to avoid majority voting, although straw polls are used to get a sense of how the collective editorship stands. When disputes arise, as commonly happens, and no consensus emerges, the matter can be referred to a mediation committee; if that fails, Jimmy Wales has the authority to make the final decision. Individuals who gain prestige within the Wikipedia community through their discussion, mediation, and other contributions can obtain higher levels of privilege, such as the power to delete or freeze pages or attain administrator status.

The development of this system of consensus building means that social networks among the editors play a large role in guiding the growth and content of the Wikipedia.

Section 5.2 has described the simple statistical features of Wikipedia growth and has tempts to interpret those features in terms of a corporate growth model. Section 5.3 focuses on the network features of Wikipedia and how those have changed over time Section 5.4 looks at the role of the social network within the Wikipedia community and examines how that has affected growth. Section 5.5 draws general conclusions that may apply to similar efforts.

5.3 THE GROWTH OF THE ENGLISH WIKIPEDIA

Wikipedia is a fast-growing database. The success of the collaborative effort at the core of Wikipedia rests, in part, on its popularity as a source of information. Although such content does not go through the thorough vetting process that information in printed encyclopedias has to pass, the information available through Wikipedia has four crucial characteristics: it is quick to find, thanks to the many search engines that index its pages; it is good enough to give a reader the big picture about an event, a person, or a difficult mathematical concept; it provides lots of useful pointers; and it evolves (by updating and self-correction) over time.

There are many metrics for growth, such as database size, the number of users, and the number of hits. Wikipedia itself is the primary source of data on its growth. Looking first at the number of megabytes over time, as available at http://stats.wikimedia.org/EN, Figure 5.1 shows a classic exponential growth pattern.

The fitted model is $M = 11.39 \exp(T) - .00000383$, where M is the number of megabytes and T is time. The adjusted R-squared is .974, and the root mean squared error is 271.28; this indicates a very good fit. The natural interpretation is that the rate of growth of the Wikipedia is proportional to its size.

Figure 5.2 shows a similar pattern in the number of articles. The fitted model is $N(t) = N(0) \exp(t/\tau)$, where the estimated value is $\tau = 499.7$ and $N(0)$ is the starting

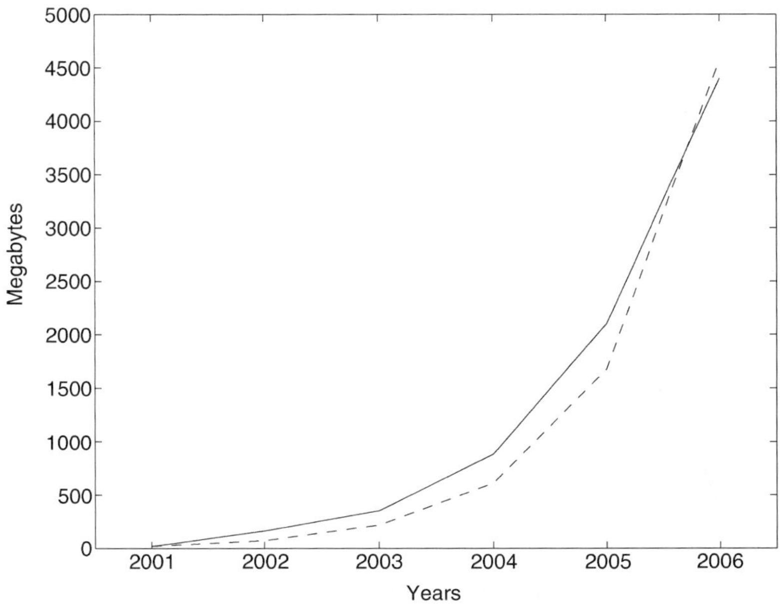

Figure 5.1 A plot of the number of megabytes in the English Wikipedia over time. The solid line represents the observed values, and the dashed line is a fitted exponential function.

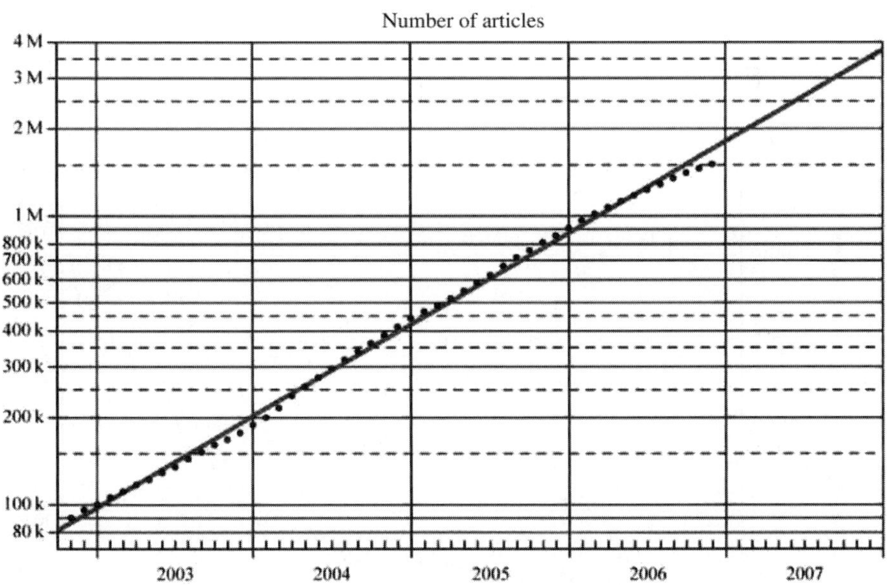

Figure 5.2 A plot of the number of articles in the English Wikipedia over time. The number of articles is on the log scale, and the line shows the best exponential fit. The image is from http://en.wikipedia.org/wiki/Wikipedia:Modelling_Wikipedia's_growth.

value of 80 kilobytes in October 2002. Note the small drop-off that occurs at the end of 2006, as the number of articles falls below the unsustainable growth of the exponential model.

From a commercial perspective, one of the most important metrics is *reach*, an estimate of the number of people who use (or are specifically aware of) a product. Alexa Internet, Inc., is a company that captures statistics on site usage (through their toolbar product), and their data provide a picture of the growth of Wikipedia according to several different criteria (see http://www.alexa.com for more details). Figure 5.3 presents an estimate of the reach of Wikipedia from January 2003 to January 2006. The estimate is based on the the number of unique Alexa toolbar users who visit a site on a given day, with some smoothing over a rolling three-month period. (Thus, the occasional downturn in the Figure 5.3 does not represent people who forgot that Wikipedia exists, but rather a transient drop in the number of hits.) As of January 2006, Alexa estimated that 1 person in 50 knew about Wikipedia.

A second measure of success is traffic rank. Alexa's rankings are estimated from the proportion of Alexa toolbar users accessing top-level domains, with a rolling three-month weighting scheme to smooth out short-term effects. Figure 5.4 presents the traffic rank of Wikipedia, as estimated by Alexa Internet, Inc., with a log scale for the rank axis. Note that on the log scale the growth trend is almost linear over this period and that Wikipedia has a rank of about 30.

To provide context for the graphs in Figures 5.3 and 5.4, Table 5.1 lists the top-ranked domains, in terms of traffic, for the week of March 2, 2007. For these sites, the table also reports the percentages of population reached (a measure of awareness)

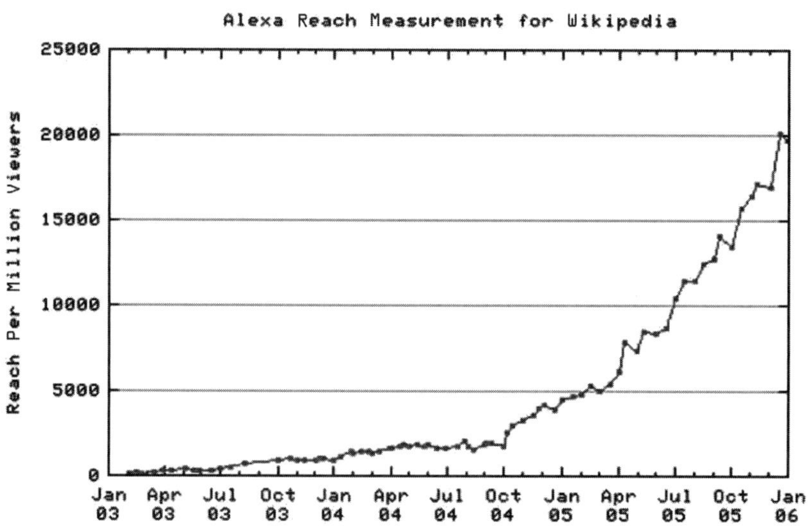

Figure 5.3 Reach per million measurement for wikipedia.org. (Source: Wikipedia/Alexa.)

Figure 5.4 Traffic ranking for the English Wikipedia site. A rank equal to 1 would mean that Wikipedia is the most popular site on the Internet—according to the sampled traffic data. (Source: Wikipedia/Alexa.)

and the percentage of pages viewed (a measure of active engagement), as well as the same statistics rescaled with Wikipedia's corresponding values—so that Wikipedia's R/W and V/W would equal 1. Obviously, for an encyclopedia such as Wikipedia, the percentage of pages viewed will be relatively small on any given day. And note that Wikipedia's rank is 10, which accords well with the linear trend seen in Figure 5.4 (that trend is clearly unsustainable, and the March 2007 data show some flattening, but it is clear that steady growth in rank has been a stable feature of the Wikipedia phenomenon).

TABLE 5.1 Snapshot of the Top Traffic Sites, as Estimated for the Week of March 2, 2007

Rank	Site	Reach	Views	R/W	V/W
1	yahoo.com	26.8	6.0	4.25	13.3
2	msn.com	30.1	3.6	4.78	8.0
3	google.com	25.1	2.3	3.98	5.1
4	youtube.com	8.9	1.6	1.41	3.6
5	myspace.com	4.4	2.4	0.70	5.3
6	live.com	14.6	0.63	2.32	1.3
7	baidu.com	6.1	1.2	0.97	2.7
8	qq.com	5.3	0.73	0.84	1.6
9	orkut.com	2.6	1.4	0.41	3.1
10	wikipedia.com	6.3	0.45	1	1
11	yahoo.co.jp	2.8	0.93	0.44	2.1

Note: The columns "Reach" and "Views" (pages viewed) are percentages of the estimated totals. Columns R/W and V/W are relative to Wikipedia's corresponding statistics.
Source: Wikipedia/Alexa.

With a nod to self-conscious postmodern reflexivity, Wikipedia collects some of its own traffic and usage data. These can be found at http://en.wikipedia.org/⊕ wiki/WP:AS.

Reach and rank represent only two of many metrics for growth. Table 5.2 presents a selection of additional statistics that suggests how the database has evolved since its creation in 2001. In particular, the rapid increase in the number of non-English Wikipedias that contain more than 1000 articles (the last line in the table) is strong evidence of the popularity, portability, and perceived utility of the collaborative business model. As of March 2007 there were 242 non-English Wikipedia sites that contained at least 10 articles (http://en.wikipedia.org/wiki/Wik ⊕ ipedia:Multilingual_statistics).

As Table 5.2 suggests, a key aspect of the dynamics of the Wikipedia growth concerns its structure. Images were introduced after a slight delay in 2002. The introduction of categories, however, was even slower and did not occur until May 2003. The late introduction of categories may be due to two tightly coupled sets of issues: On the one hand, it is good to wait and see what kind of content users contribute before using management resources to define a tree of labels (otherwise, many labels may remain unused). On the other hand, the size of the encyclopedia in its early stages may not have required a formal tree of labels. But the number of categories has increased quickly over the last two years, from about 23,000 to about 176,000. Furthermore, the internal structure of the categories itself has evolved greatly over the years; compare, for example, the high-level structure in November 2005 described by Holloway et al. (2006) to the current one shown at http://stats.wikimedia.org/EN/CategoryOverview_EN_Concise.htm.

The different metrics for growth all tell a similar story. Until 2002, Wikipedia evolution was driven by a relative handful of insiders and enthusiasts. The mechanism for growth was erratic, and the relative variation in any performance measure,

TABLE 5.2 Quoted Statistics Measured at the End of October in Each Indicated Year

Statistic	2001	2002	2003	2004	2005	2006
No. of articles	11k	90k	168k	379k	791k	1.4M
No. of internal links	87k	1.1M	2.7M	7.0M	16.7M	32.1M
No. of external links	2.7k	20k	76k	300k	996k	2.6M
No. of words	2.4M	26.2M	52.0M	121M	289M	609M
No. of images		3.9k	24k	122k	388k	876k
No. of contributors	238	1077	4282	17542	56142	151934
Contributors (>5 edits)	110	324	1122	4853	14923	43001
Contributors (>100 edits)	10	100	198	779	1964	4330
Mean edits per article	1.8	4.8	9.2	15.7	24.1	38.0
No. of categories				23k	76k	176k
Categorized articles				61%	80%	86%
No. of Wikipedias	1	7	17	48	76	113

Source: http://stats.wikimedia.org/EN/TablesWikipediaEN.htm and http://stats.wikimedia.org/EN/⊕ TablesArticlesTotal.htm.

compared to the average level, was fairly high. Sometime after 2002, it appears to have reached a critical mass that drove something similar to self-sustaining exponential growth with respect to almost any metric one wants to consider. Then, toward the end of 2006, the growth fell to a subexponential rate, possibly reflecting saturation of the pool of volunteer contributors, or completion of topic areas that enthusiasts wanted to pursue, or diminishing novelty, or the inevitable loss of perceived prestige (coolness) when the number of contributors is very large.

The first phase of growth is only hinted at in the figures, since for many metrics the time scale does not exist before 2002. But from a management standpoint, it seems inevitable, and the next section will discuss the role of management in more detail. The exponential growth phase is well supported in all of the figures shown. The third phase is very recent, and the long-term trend cannot be discerned from the available information. It is certainly possible that growth will tick back up, especially if Wikipedia leadership introduces new functionalities that attract fresh contributions. But it seems more likely that the rapid early spurt is in the past, and the new management model should be one of consolidation and steady, but not explosive, growth.

5.3.1 Micro-Growth

Besides the long-term growth phases, there is also interesting variation that occurs on short time scales. There are clear holiday effects in the submission of contributions, and a regular drop in September that people speculate is associated with the distractions attending the start of the school year (cf. http://en.wikipedia.org/wiki/⊕ Wikipedia:Modelling_Wikipedia's_growth).

Also, there are claims that the revision times for Wikipedia constitute a self-similar process (Almeida et al. 2007). Such processes exist in Internet traffic, but the mechanism for such behavior in Wikipedia postings, though provocative, is unclear.

5.4 CREATING AND CORRECTING CONTENT

From the standpoint of e-commerce, managers want to understand how to replicate the growth mechanisms of Wikipedia. There is no explicit recipe for exporting its success, but some general principles are evident. This section also considers strategies for ensuring future growth through the creation of new kinds of value, as has happened at various times during the evolution of the Wikipedia.

Below the take-off point, we believe that management had to invest significant resources in creating infrastructure, content, and enthusiasm. This section looks at some of those topics in detail. The other main feature of Wikipedia is the flatness of the organization and its relatively permissive power-sharing. This has fostered a sense of community among the contributors that causes Wikipedia to stand apart from traditional corporate environments and has helped to engage and empower a broad base of user-contributors.

5.4.1 The Number of Contributors

Over time, Wikipedia's increasing popularity drew more and more users to contribute, creating a positive feedback loop on the quantity and quality of its content. Table 5.2 hinted at some of this. On the one hand, the number of active contributors (users who contributed more than five edits) as a fraction of the total number of contributors (users who contributed at least one edit) has stabilized over time. This, together with the observation that the pool of contributors is growing exponentially fast, suggests that modeling the reach of Wikipedia as a multiplicative process, with saturation, is reasonable. On the other hand, Table 5.2 shows that the pool of contributors who are *very* active is much smaller. The number of very active contributors (users who contribute more than 100 edits), as a fraction of the total number of contributors, has been slowly decreasing since 2003. This suggests that there was an early stage of forced growth driven by management investment in and cultivation of highly active contributors, but that Wikipedia has now reached a point at which the initial pool of these supererogatory contributers is exhausted.

To explore this further, Figure 5.5 shows that a very small number of people make a great many contributions but that many people make a small number of contributions. The very smooth curve strongly suggests that a simple behavioral law governs this relationship. If so, then business plans for creating collaborative content can anticipate specific distributions for the degree of volunteer involvement (the parameters of the curve may depend upon the project, but if there is a generalizable behavioral law, the properties of the distribution should be stable). It is likely that the super-contributors are wordsmithing to enforce

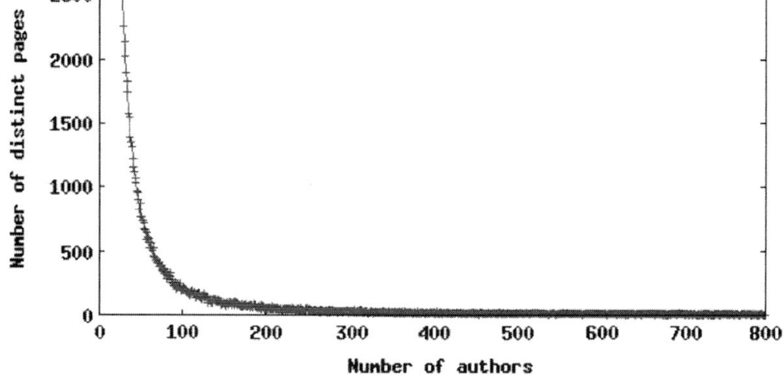

Figure 5.5 Histogram of the number of articles contributed to by individuals.

style conventions, and in a commercial enterprise they would require special compensation or recognition.

5.4.2 The Overhead for Content Maintenance

As of January 1, 2007, Wikipedia had 2,463,839 pages related to administration and 3,806,878 pages related to content. The latter includes 855,427 pages of content discussion, in which contributors point out gaps or raise questions about accuracy or style. Clearly, there is a significant overhead associated with the generation and maintenance of Wikipedia content.

For example, there exist nine different types of pages that have the title "Censorship," including a content page, a content discussion page, a category page, a category discussion page, a template page, a template discussion page, a (deleted) user's discussion page, a demonstration page, and the discussion about the demonstration page. All of these pages deal with different facets of the content and its maintenance. This is an extreme example; most topics have fewer pages. But it illustrates the complexity of the underlying structure.

Importantly, there is much less revision activity on the administrative pages than on the content pages. For the content pages, there were 58,636,873 revisions; for the content discussion, there were 6,049,233 separate postings/revisions. In contrast, the administrative pages had only 15,715,896 revisions. In both cases, most revisions are relatively minor, but it certainly appears that the administrative pages are more stable and persistent than the content pages. The discussion pages are more complex; some topics are controversial and generate a great deal of discussion, but most get little attention. But the short message is that 61% of the pages carry content, but 80% of the revisions are about content. From a business standpoint it seems that the administrative pages carry significant overhead, but note that Wikipedia has successfully offloaded much of that burden to a decentralized volunteer community.

To assess the balance between content creation and content maintenance, we examined the relationship between the number of discussion points on a topic and the number of content revisions. We found that a unit of discussion produces slightly less than a unit of content, both at the aggregate level for all of Wikipedia and for the topic category of mathematics. Figure 5.6 plots the amount of content and the amount of discussion for a random sample of 50,000 topics. The 45-degree line corresponds to the case in which one discussion entry corresponds to one content revision.

The vertical lines in Figure 5.6 mostly correspond to holding pages in high-level categories where new articles are entered prior to indexing and linking. For these pages the content changes rapidly, but there is relatively little discussion. Note that this kind of enrollment process requires regular management and attention from domain experts.

For comparison, we also created a similar plot for the first 200 Wikipedia articles indexed under mathematics (Figure 5.7). We chose this topic because it was numerous and because mathematics is not intrinsically controversial in the way that articles on politics, history, or *Star Trek* might be. Surprisingly, the same general pattern

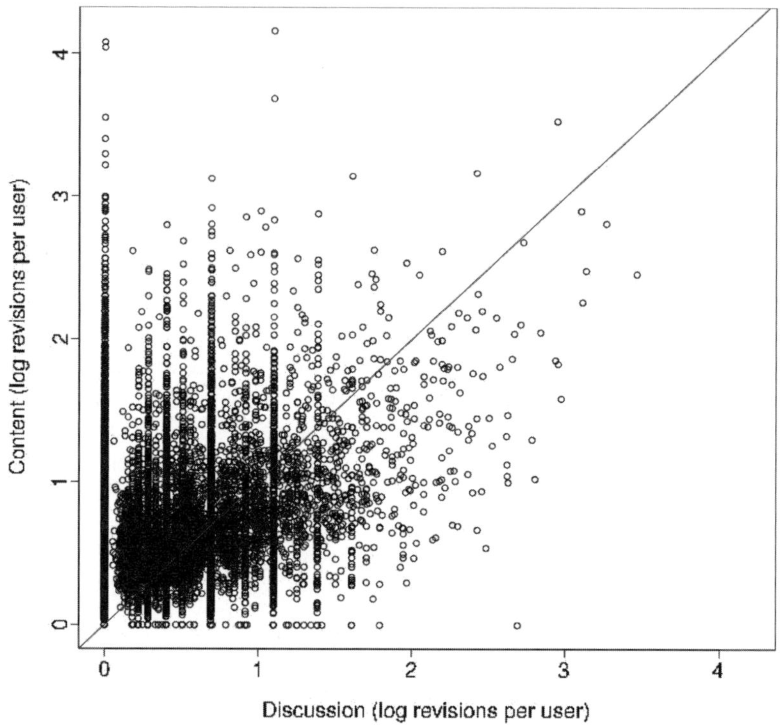

Figure 5.6 The amount of revision of discussion pages compared with the number of revisions of content pages for 50,000 topics. (The 45-degree line is shown for convenience in comparison.)

noted for the sample of 50,000 articles from all of Wikipedia holds for the 200 articles in mathematics.

These findings are compatible with two quite different views of the database. Either the high quality of first drafts leaves little room for disagreement, or the amount of content supervision is very low, or both.

5.4.3 Content Protection

In any business plan with content creation by an open community, there is an explicit management responsibility to protect content quality. In Wikipedia, the most visible aspect of this problem is protection of content from vandalism. Section 5.2 listed some of the more famous instances of deliberate content destruction. Wikipedia has two main defenses: they can restore the original content or they can freeze the entry so that unauthorized contributors cannot change the contents.

To get a sense of the scope of the problem, Figure 5.8 plots the number of times that inappropriate content has been deleted from the Wikipedia system (plotted by $\times$), as well the number of times that contents were protected from future changes

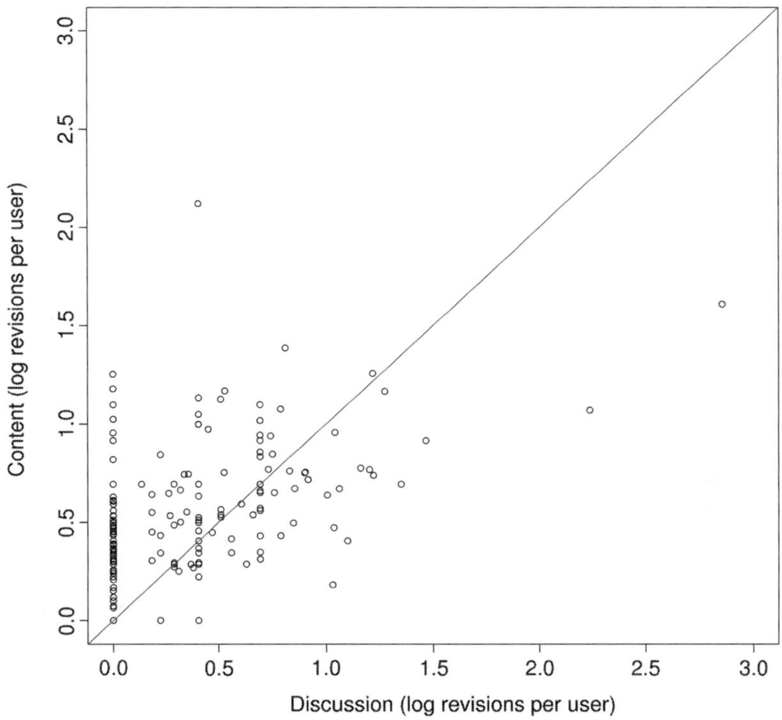

Figure 5.7 The amount of revision of discussion pages for 200 mathematics articles compared with the number of revisions of the corresponding content pages. (The 45-degree line is shown for convenience.) As before, a unit of discussion produces slightly less than a unit of content.

(plotted by +). Both are good indicators of the amount of vandalism present in the system. The increase over time in these variables is linear rather than the exponential trend seen in almost every other plot of Wikipedia activity. This probably reflects the fact that this deletion and protection are administrative tasks that require executive attention, which is a limiting resource. And it is worth noting that many minor instances of vandalism are probably never noticed.

For emerging businesses, it would make sense to develop a text-mining system that scrutinizes entries or edits for possible violations of the social compact. Certainly there are keywords that help flag problems, and revisions by anonymous contributors could also raise cautions. More subtle signals are also possible; the success of Bayesian methods for spam filtering (Madigan 2005) suggests that a great deal of progress is possible.

A related topic concerns effective management strategies for content control. We do not know what criteria Wikipedia administrators use in deciding whether to freeze a particular entry, but it is likely that whatever (possibly informal) guidelines exist depend upon the history of attacks upon the text. Figure 5.9 shows a frequency

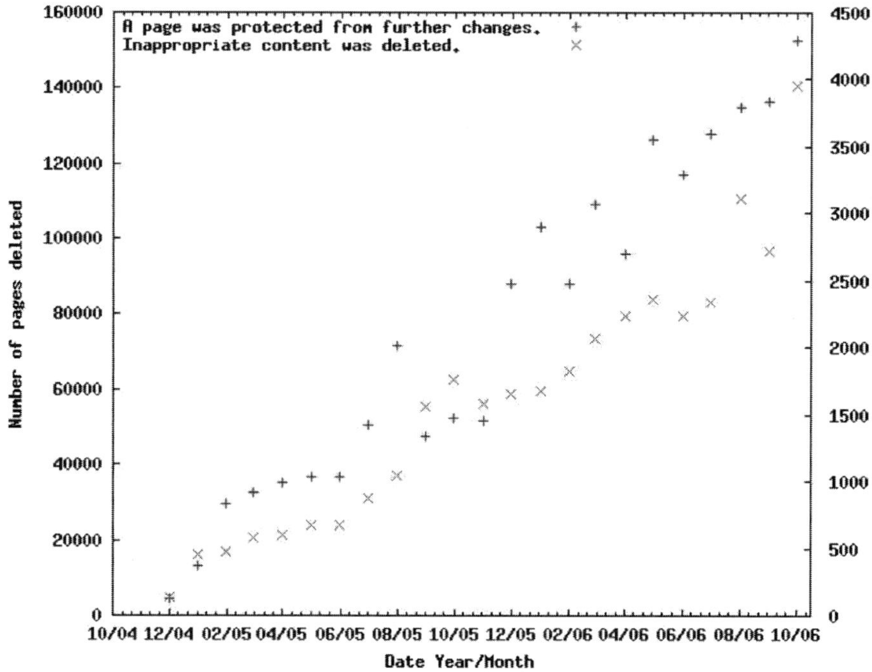

Figure 5.8 Amount of content deleted and protected over time. The + indicates the number of protected pages, and the + indicates the number of times content was deleted.

plot of the number of times the content of a page was protected and the number of times the page was revised. It appears that most protections occur when there has been no revision; these are probably administrative pages or obviously controversial topics. A small number of protections occur after substantial revisions; these may represent honest intellectual disagreements for which a community consensus slowly crystallizes. In between is the domain in which management policies might have impact; for example, a policy of protecting any entry that is vandalized once would be a reasonable approach.

5.4.4 Revision Management

Revision is a balancing act when handling user-contributed content. If there is too much, then the product is unstable. People want to know that what they saw a month ago is probably still there. But if there is too little revision, then some of the key benefits of open collaboration are lost. Wikipedia has tended to indulge revision and this has worked well, but how well this policy generalizes to other applications is unclear.

In order to study the role of revision, we considered a subset of 900,000 randomly chosen articles. For each page we counted the number of revisions to the discussion page, the number of unique users participating in the discussion, the number of

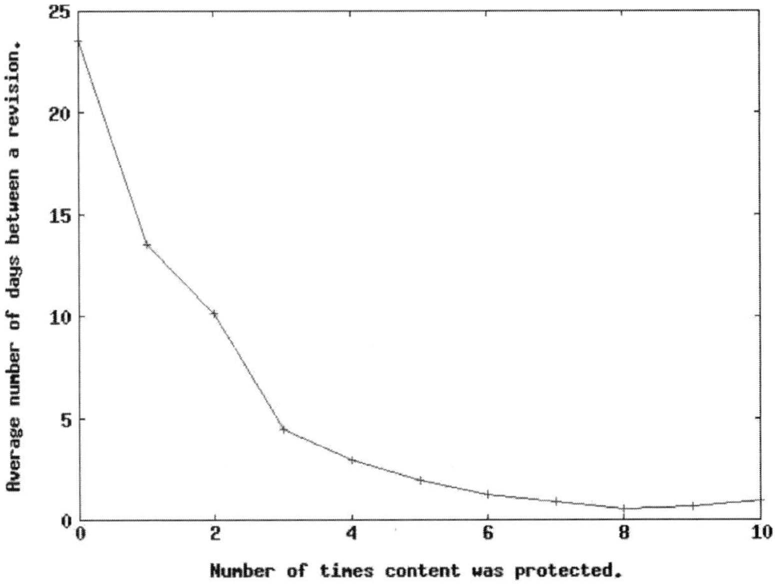

Figure 5.9 Relationship between the rate of revisions and the number of times that content was protected.

revisions to the content page, and the number of unique users providing content. We found that, on average, a modification to to a page occurs every 23 days, with seven different individuals providing inputs over 13 revisions. But the tail behavior is extreme.

In Figure 5.10, each point corresponds to an article. The x-axis measures the number of revisions to the content, and the y-axis measures the number of unique contributors to the discussion and/or the revision. As it shows, it is not uncommon for some articles to be touched by hundreds of hands. At the same time, the number of revisions tends to be more than the number of participating users. This points up the wisdom of the laissez-faire approach taken by Wikipedia; people make many small edits, and in general the quality improves.

Obviously, for most articles, the rate at which changes are made should diminish over time as errors are removed, key facts are validated, consensus is reached, and the attention of the administrative community shifts to other topics. Figure 5.11 shows how the number of actively edited articles and stale articles changes over time (we define an active chapter as one that has been revised within the last 6 months). We also plot the number of active contributors over time.

The behavior of the number of stale entries and the number of active contributors is reasonable. But we have no explanation for the sudden drop in the number of actively edited entries around January 1, 2006. We suspect this indicates some kind of change in the management of the revision process, and we hope that Wikipedia insiders can clarify this.

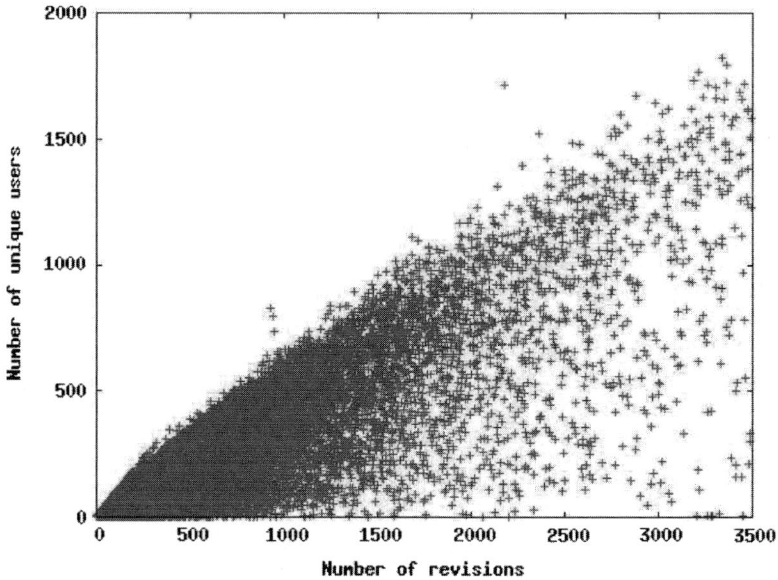

Figure 5.10 The number of contributors as a function of the number of revisions. Note that some entries have thousands of revisions and contributors; it is likely that many of these are pages with lists.

5.4.5 Linkages

A key functionality that Wikipedia provides is internal links between articles and links to external webpages. Obviously, the latter pose a management problem in terms of potential instability, but the convenience has more than compensated for the occasional dead link.

To get a sense of the scale, as of January 1, 2007, there were 2,528,868 articles with outside links, 87,593,800 nonduplicative links between articles within Wikipedia, and 14,079,567 category links that were explicitly defined. The number of category links that were also represented as content-to-content links was 813,176. Thus, there has been significant effort by contributors to differentiate between the semantic linkages of the content and the taxonomy used to classify it. Obviously, this has implications for a business model that includes multiple kinds of links.

In terms of designing such systems, there is the question of whether the taxonomy should be provided mostly by administrators or mostly by the content providers themselves in a self-organizing manner. For Wikipedia, traversing the category linkages from the top-level concepts shows that only about 10,000 categories are related to administrative matters, whereas the remaining categories were generated by the end users themselves. Thus, it would seem that within the Wikipedia experiment, the taxonomy being used is not only created on the fly by the community, but it is also constructed by a community that clearly differentiates between content and classification.

For Wikipedia, many articles now include user-contributed links to other websites on the Internet. Specifically, 535,750 content pages link to one or more URLs outside

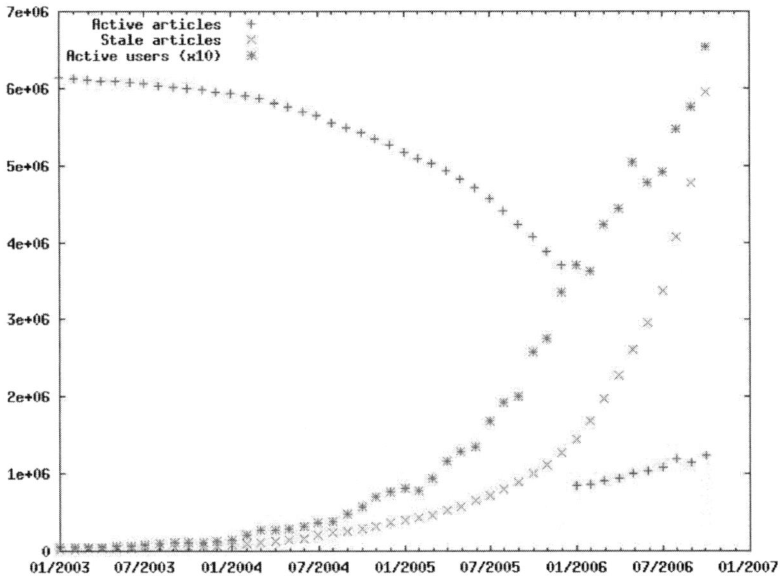

Figure 5.11 Tracing the number of entries being actively edited (+), the number of entries no longer being edited (×), and the number of active contributors (*) over time.

the Wikipedia namespace. This means that about 81% of Wikipedia content pages have no links to the outside Web. However, it could still be the case that most of the content pages are actually pointers to other pages, with little actual written prose to analyze. In this case, 1,538,983 pages (or slightly more than half of the content pages) reference at least one other Wikipedia page. Hence the data within Wikipedia seem to be highly self-referenced while still containing substantial user-generated content.

For additional perspective on the linkage patterns within Wikipedia, Figure 5.12 plots the relationship between the length in characters of a Wikipedia page and the number of web links to pages outside of Wikipedia. The figure suggests a superposition of several different clusters of pages. The pages that have small numbers of characters but many internal links are probably administrative lists; pages that are less extreme, but still have many links per character, could be indexes. The structures suggested in the plot require further study but indicate a wide range of internal-link behavior.

Symmetrically, Figure 5.13 plots the relationship between the number of external links to the Internet on a page and its length in characters. Note that this figure shows even more structure, with clearly defined rays emanating from the origin. The most vertical ray probably corresponds to sites that are essentially lists of URLs, but the other rays invite further study. It certainly appears that there are distinct clusters of pages with respect to external links.

Note that Figures 5.12 and 5.13 both contain artifacts that imply a linear relationship between the number of links and the size of the page. The reason for this is that

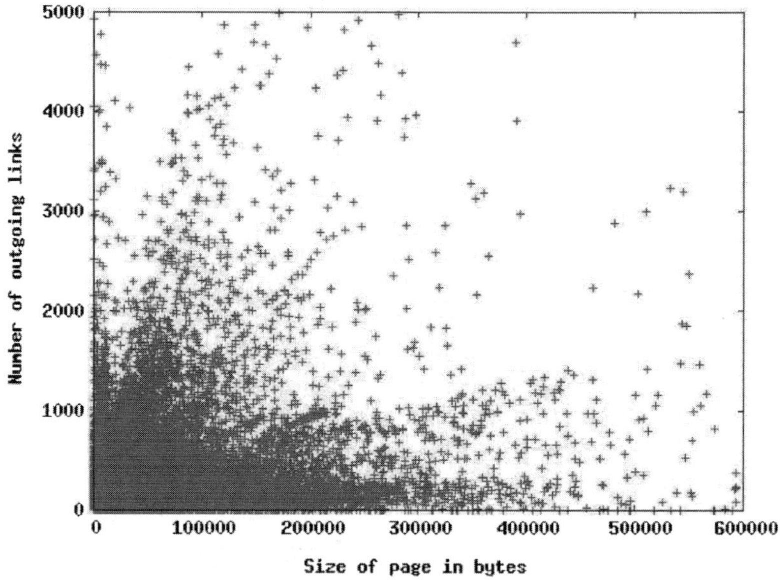

Figure 5.12 Plot of the relationship between the number of outgoing links to other pages and page length.

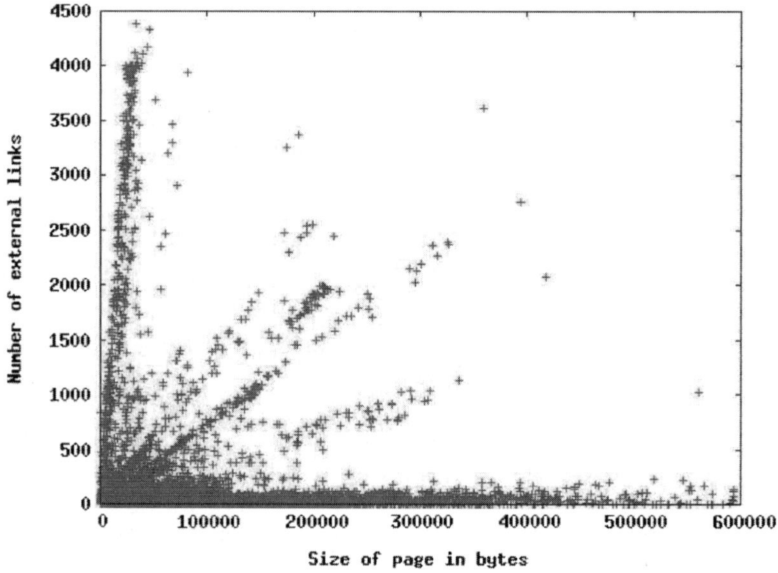

Figure 5.13 Plot of the relationship between the number of external links to the Internet and page length.

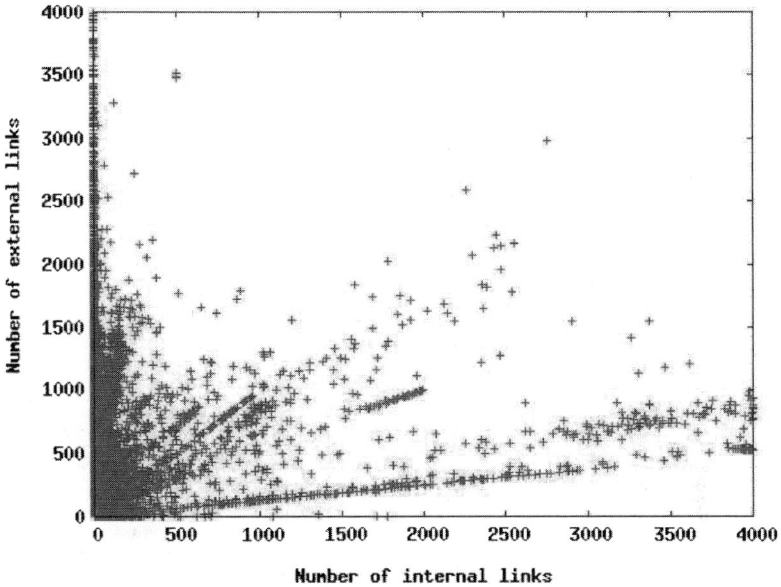

Figure 5.14 Plot of the number of links external to Wikipedia against the number of internal links for a sample of Wikipedia articles.

the links themselves occupy space within the page, and therefore the number of links directly influences the size of the page. This is more evident in Figure 5.13 because the average length of a URL is 58 characters, versus an internal Wikipedia link that is usually 18 characters long. Nonetheless, these artifacts only partially account for the observed structure.

To complete the narrative loop, Figure 5.14 plots the number of external links against the number of internal links. There are many pages that have large numbers of external URL links but few internal links. As a generalization, pages with very large numbers (over 1000) of external links tend to be automatically generated tools pages meant to be used for maintenance. Pages with significant numbers of links (100s) tend to be lists of places, people, or objects. Again, there is clear structure in the graph but only partial explanation.

To close, consider the external perspective. Figure 5.15 shows the relationship between the number of external pages that point to a specific Wikipedia article and the number of links from that article to pages outside the Wikipedia namespace. The mass at the left is simple to understand; many Wikipedia pages have lists of external links but do not have pointers from the outside. The long tail to the right is surprising.

5.4.6 New Functionality

Like any successful enterprise, the Wikipedia has not only grown—it has changed. Key innovations were the enabling of categories, external links, and the extensive

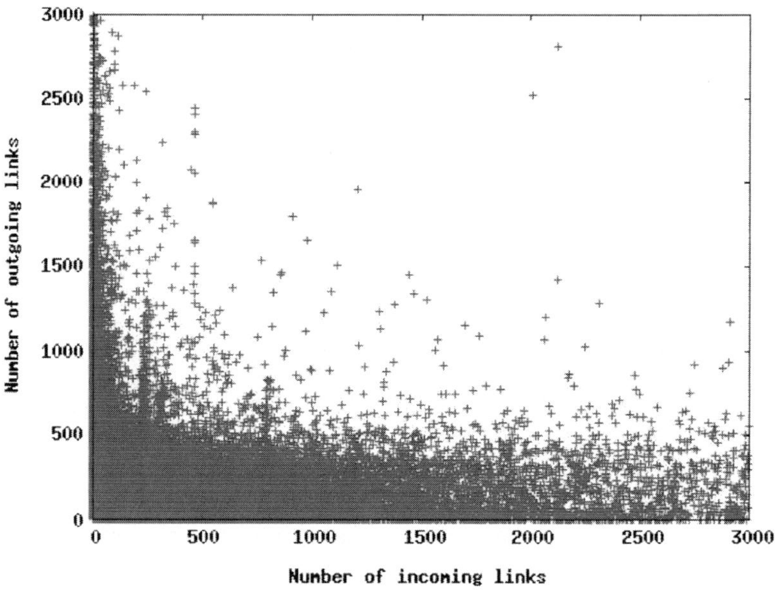

Figure 5.15 Comparison plot of the relationship between the number of incoming and outgoing links to other pages within Wikipedia.

use of images, all of which emerged after the initial format had been developed. It seems likely that audio and video capability will someday be added, as well as transparent links to external software packages.

But other kinds of functionality may have even larger implications. Alexander Wissner-Gross, as a Ph.D. physics student at Harvard, developed software to help Wikipedia users find related information on a general topic. The algorithm uses text mining, as well as information on the popularity of particular paths through the Wikipedia network of links. In some sense, this is rather like the popular recommender system used by Amazon to point out books a customer might enjoy, based on their purchase and browsing history.

Also, Luca de Alfaro at the University of California at Santa Cruz has developed software that estimates, phrase by phrase, the trustworthiness of text in Wikipedia articles (Powell 2007). The procedure is based on tracking the number of times that a particular content contributor's work has been removed or revised. Color coding flags text by people with high rates of reversal as potentially less reliable than other portions of the same article. This capability is another example of the fresh and unforeseen potential of the Wikipedia data archives.

More broadly, the organic growth of links between Wikipedia topics demands network analysis. There are probably deep questions about information structures that could be addressed. For example, what are the empty spots in the Wikipedia system, and how would one notice them? Do different fields have similar internal connection structures, or do some fields show very different kinds of linkage?

How might one segment the Wikipedia network into meaningful cliques, and would these correspond to a recognizable ontology?

With regard to clique segmentation, there are a number of traditional approaches developed in the social network community, but these are mostly ad hoc. If one wanted to estimate the size of the clique corresponding to, say, mathematics, it would useful to adapt methods developed for estimating the size of the World Wide Web (Bradlow and Schmittlein 2000; Dobra and Fienberg 2003) that are based on capture-recapture models and Markov chain explorations.

5.5 DISCUSSION

Although the Wikipedia is not a for-profit enterprise, it is a unique example of a novel approach to constructing value. As such, its evolution and management structures hold important lessons for e-commerce.

In the first part of this chapter, we focused on the growth history of Wikipedia. The mathematical picture of exponential growth in the middle phase is well established, according to many different metrics of growth. In the late phase, there is emerging evidence that growth has become subexponential, and the causes for this (aside from mathematical inevitability) are unclear. Bold development of new functionality could easily reestablish exponential growth for a while. Our data have least to say about the first phase, during which the Wikipedia founders established the infrastructure and recruited an initial team of enthusiastic content creators. But it seems clear that critical ingredients were a social network within the encyclopedia community, building on the Nupedia connections, and a flat, decentralized management system that invited self-paced contribution and recognized volunteerism.

The second part of the chapter focused on the technical mechanisms of content creation. This addressed growth in the number of contributors, the administrative costs of content maintenance (as inferred from administrative pages), the balance between open editing and content protection (as indicated by trends in the number of protected pages and editing histories), revision management, the different kinds of links needed to support the Wikipedia functionalities, and prospects for new kinds of service in the future.

As a research area, Wikipedia science is exciting. There is an enormous amount of data, and whenever one looks closely, there are research problems. It is a rich example of the evolution of a self-organizing system, and its processes inform many aspects of organizational theory.

ACKNOWLEDGMENT

Robert Warren is partially supported by NSERC and Ontario Graduate Student Scholarships. Edoardo Airoldi is supported by NIH Grant R01 GM071966 and NSF Grant IIS-0513552, both to Olga Troyanskaya at Princeton University. David Banks was partially supported by NSF Grant DMS-0437183.

REFERENCES

Almeida, R., Mozafari, B., and Cho, J. (2007). On the evolution of Wikipedia. International Conference on Weblogs and Social Media. Available at http://www.icwsm.org/⊕ papers/paper2.html.

Bradlow, E. and Schmittlein, D. (2000). The little engines that could: Modeling the performance of World Wide Web search engines. *Marketing Science*, 19: 43–62.

Dobra, A. and Fienberg, S. (2003). How large is the World Wide Web? In *Web Dynamics* (M. Levene and A. Poulovassilis, eds.). New York: Springer-Verlag.

Giles, J. (2005). Internet Encyclopedias Go Head to Head. *Nature*, 438: 900–901.

Holloway, T., Božičević, M., and Börner, K. (2006). Analyzing and visualizing the semantic coverage of Wikipedia and its authors. Available at http://arxiv.org/pdf.cs.IR/0512085⊕ .

Leuf, B. and Cunningham, W. (2001). *The Wiki Way: Quick Collaboration on the Web*. New York: Addison-Wesley.

Leyden, J. (2006). Wikipedia blaster "Fix" points to malware. Available at 11/3, http://⊕ www.theregister.co.uk/2006/11/03/wikipedia_blaster_attack.

Madigan, D. (2005). Statistics and the war on spam. In *Statistics, a Guide to the Unknown* (R. Peck, G. Casella, G. Cobb, R. Hoerl, and D. Nolan, eds.). Belmont, CA: Duxbury-Brooks/Cole.

Pava, A. (2006). Colbert banned from Wikipedia. Civic Actions, August 2. Available at http://⊕ www.civicactions.com/node/405.

Powell, H. (2007). New program color-codes text in Wikipedia entries to indicate trustworthiness. University of California, Santa Cruz, press release. Available at http://⊕ www.ucsc.edu/news_events/press_releases/text.asp?pid–1471.

Reuters (2006). Intelligence czar unveils spy version of Wikipedia. Available at http://⊕ news.zdnet.com/2100-1009_22-6131309.html.

SECTION II

E-COMMERCE APPLICATIONS

6

AN ANALYSIS OF PRICE DYNAMICS, BIDDER NETWORKS, AND MARKET STRUCTURE IN ONLINE ART AUCTIONS

Mayukh Dass

Area of Marketing, Rawls College of Business, Texas Tech University, Lubbock, TX

Srinivas K. Reddy

Department of Marketing and Distribution, Terry College of Business, University of Georgia, Athens, Georgia

6.1 INTRODUCTION

One of the greatest contributions of e-commerce and online auctions is their ability to provide detailed bidding records in the form of a bid history. These data sources not only furnish dynamic information regarding the status of auctions, but also capture the essence of auction activities that was not accessible from live auctions. In this chapter, we explore three broad online auction issues from these bidding data: the effects of various factors on price dynamics, effects of inter-bidder interactions on bidder dynamics, and underlying relationships of auctioned objects. From the auction house manager's perspective, insights from issues like how price formation takes place, how competitive bidding among bidders evolves, and which items are perceived as similar and dissimilar by the bidders are vital in setting up future auctions. Here we present results of the above investigations using data from online auctions of fine arts.

Statistical Methods in e-Commerce Research. Edited by W. Jank and G. Shmueli

Fine art auctions present a unique context to our study. Unlike most functional products such as computers and electronic devices, whose auction data are used in prior studies, art objects are more hedonic. Their consumption is driven more by affective experience such as the aesthetic pleasure that one derives from them than by their utilitarian or functional benefits (Reddy and Dass 2006). Furthermore, the prices of artworks sold in these auction houses range from a few thousand to a few million dollars, resulting in higher stakes on the auction outcome. Therefore, crucial information on price dynamics, bidder dynamics, and market structure is greatly valued by both auctioneers and bidders.

Prior research on fine arts and art markets has mostly focused on the price realization of artworks in auctions and on their potential as an investment (Baumol 1986; Ashenfelter 1989; Pesando 1993; Mei and Moses 2002; Ashenfelter and Graddy 2003).[1] However, very little is known about the issues frequently faced by art auction house managers. In this chapter, we present some new approaches and investigate the relevant issues.

To help us understand how bid history can facilitate our studies, we consider a bid history snapshot (Figure 6.1) taken from an online auction of fine arts. Like any typical bid history, it consists of three important elements: bidder information, bid amount, and bid time (the time at which the bid is placed). Such information constitutes the fundamental basis for online auction research, which can be represented in a three-dimensional data cube (Figure 6.2). This data cube illustration streamlines the information available from the bid history. The x-axis of the plot represents the lots (auctioned items), the y-axis represents the bid time, and the z-axis represents the

♦25
G.Ravinder Reddy

Untitled

Polyester resin fiberglass painted and gilded
Height : 33.5 in (85 cm)
Width : 20 in (51 cm)
Depth : 32 in (81.5 cm)

$30,000 - 35,000
Rs 1,290,000 - 1,505,000

Next Valid Bid : $ 71,500 (Rs 3,074,500)

Rank	Nick Name	Amount($)	Amount(Rs)	Bid Type	Date & Time (US EST)
Current Highest Bid 1	A1	66,500	2,859,500		Mar 16 2006 7:19:25 AM
2	Anonymous 81	61,500	2,644,500		Mar 16 2006 7:07:54 AM
3	A1	56,500	2,429,500		Mar 15 2006 8:59:30 PM
4	Anonymous 93	51,500	2,214,500		Mar 15 2006 8:16:18 PM
5	Rumplestiltskin	49,000	2,107,000		Mar 15 2006 2:07:25 PM
6	Kyozaan	46,500	1,999,500		Mar 15 2006 2:03:48 PM
7	Rumplestiltskin	44,000	1,892,000	Proxy	Mar 15 2006 2:03:33 PM
8	Kyozaan	43,000	1,849,000		Mar 15 2006 2:03:33 PM
9	Rumplestiltskin	40,500	1,741,500	Proxy	Mar 15 2006 2:03:14 PM
10	Kyozaan	38,000	1,634,000		Mar 15 2006 2:03:14 PM
11	Rumplestiltskin	35,500	1,526,500	Proxy	Mar 15 2006 6:29:38 AM
12	Kyozaan	33,000	1,419,000		Mar 15 2006 5:59:15 AM
13	Anonymous 57	30,500	1,311,500		Mar 15 2006 5:10:29 AM
14	Rumplestiltskin	28,000	1,204,000		Mar 15 2006 12:52:43 AM
15	Anon 12	25,500	1,096,500		Mar 15 2006 12:50:45 AM
	Start price	24,000	1,032,000		Mar 14 2006 10:30:00 PM

Figure 6.1 Bid history snapshot of an online art auction.

[1]For example, in their analysis of 15 studies, Ashenfelter and Graddy (2003) report an estimated average real return of art as investment of 2.6%, with a range of 0.55% to 5%.

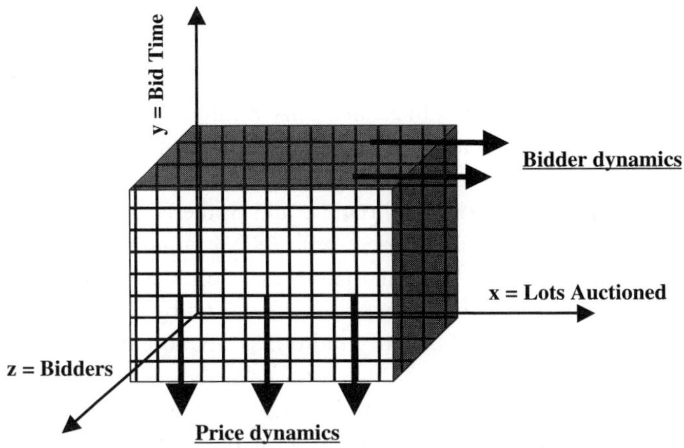

Figure 6.2 Dimensions of online auction information.

bidders who participated in the auction. Such an illustration is vital to determine which characteristics of the bid history to use for which analysis. For example, if we focus only on the auctioned lots (x-axis) and the bid time (y-axis), we study price dynamics research (Bapna et al. 2008; Reddy and Dass 2006). If we focus on individual bidders (z-axis) across all the auctioned lots (x-axis), we study bidder dynamics research (Dass et al. 2007a).

We also present some innovative and state-of-the-art approaches in some of the analysis. First, we use functional data analysis to investigate price dynamics. Second, we examine bidder dynamics by considering a network framework among the bidders. In particular, we create a network among bidders and then use social network analysis to fulfill our goal. Finally, we develop an artist network between the artists whose artwork is auctioned and then use the purchase intent of the bidders to determine the underlying art market structure.

This chapter is organized as follows. First, we discuss the context of our study, i.e., a contemporary Indian art auction, and describe the online auction data used in our studies. We used two different datasets, one for the price and bidder dynamics and another for market structure analysis. We discuss each of them in Section 6.3. Second, in Section 6.4, we explain our work on price dynamics of online art auctions and discuss the findings. In Section 6.5, we present our methods and the results of the bidder dynamics analysis. In Section 6.6, we explain our approach to studying the art market structure and discuss the results. Finally, in Section 6.7, we discuss the managerial implications and future research opportunities.

6.2 CONTEMPORARY INDIAN ART

The fine arts market is one of the fastest-growing markets in the world. In 2005, the turnover for fine arts sales in auctions exceeded over \$4 billion (up from \$3.6 billion in 2004, a growth rate of 10%), and the total world art market is estimated to be over

$30 billion (www.artprice.com). Two esteemed auction houses, Christie's and Sotheby's, have dominated the art market since the eighteenth century, but due to the recent popularity of the Internet, they are facing fierce competition from some newly established auction houses that have only online operations. One example is the emerging market for contemporary Indian art. With over $100 million in auction sales[2] in 2006, contemporary Indian art is now one of the leading emerging art markets in the world. Although Christie's and Sotheby's started auctioning this Art in 1995, in 2000 the market exploded, with 68.7% annual revenue growth (coincidentally, this is when SaffronArt.com, the source of our data, started its operations). In 2006, online auction sales of contemporary Indian art by SaffronArt.com ($20.80 million) were comparable to those of traditional auction houses like Sotheby's ($35.29 million) and Christie's ($33.08 million). Further, SaffronArt.com sold 390 art items, whereas Sotheby's and Christie's sold 484 and 329 items respectively in that year.[3] The top 10 Indian artists sold 31% of the lots and contributed to 57% of the total value realized at the auctions since 1995. Two of these artists are now ranked among the top 100 artists in the world based on their auction sales in 2005. A new set of emerging artists (the new trendsetters, typically born after 1955) have contributed 2% in value and 3% in lots and are becoming increasingly popular, commanding ever-higher prices.[4]

6.3 ONLINE AUCTION DATA

SaffronArt.com data are ideal for our study, as this auction comparable with house in the mainstream auction houses, provides rich auction data, and gives a good representation of the contemporary Indian art market. The auctions are held over a multiday period (typically three days). Bid histories of the auctioned items are available from the auction house's website during the auction. Figure 6.1 is a snapshot of a bid history from the auction website. This auction uses an ascending-bid format with a fixed ending time and date set by the auction house. Further, to discourage snipping behavior by the bidders, the auction adds three more minutes to the time clock if the last bid is placed in the last three minutes of the auction[5] (Roth and Ockenfels 2002). The auction also uses a *proxy-bid* system similar to the one used by eBay, where the bids are automatically updated on behalf of the bidders. Proxy bidding is a common feature in most online auction houses, where bidders set the maximum amount they are willing to pay for the auctioned item and let the auction house place proxy bids on their behalf until that price is reached. Bidders using this facility have a predetermined value for the item and use it to stay within

[2]Total sales of contemporary Indian art in both online and offline auctions in 2006 were $136 million.

[3]In 2005, online auction sales of contemporary Indian art by SaffronArt.com were $18.06 million, more than those of Sotheby's ($10.49 million) and Christie's ($14.89 million). SaffronArt.com also sold more art items (390) than Sotheby's (276) and Christie's (248) in 2005.

[4]For more information on the contemporary Indian art market, visit http://www.modernindianart.net.

[5]Roth and Ockenfels (2002) examined the difference in the last-minute bidding strategies of bidders in these types of auctions and in eBay auctions, where the auction ends exactly at a particular time.

that limit (Bapna et al. 2004). We also collected other item-specific information for our analysis. A complete list of these items will be discussed later.

For our analysis, we use two different datasets. The first dataset is taken from an online auction where 199 lots (each lot typically is a unique piece of art—namely, a painting, a drawing, or a sculpture) were auctioned in December 2005. In this particular auction, works of 70 artists were auctioned, with an average of three lots per artist. The average realized price per lot was $62,065 and ranged from a low of $3,135 to a high of $1,486,100. Overall, 256 bidders participated in this online auction and placed 3080 bids. The number of bids per lot averaged 15.47 and ranged from 2 to 48. On average, 6 bidders participated in each lot, ranging from 2 to 14 bidders across the auction. The mean number of bids per bidder was 4.93, with a range from 1 to 65. Some of the key descriptive information about the auction is presented in Table 6.1. We used this dataset to investigate the price dynamics and bidder dynamics issues.

For our art market structure study, we used a dataset from another auction held in March 2005. Only 44 lots from nine artists were auctioned, and 66 bidders participated. As one of our primary goals in this study is to illustrate the process of determining the underlying market structure, we decided to use this smaller dataset

TABLE 6.1 Summary Description of Dataset 1

	Mean (SD)	Median	Min.	Max.
No. of unique bidders/lot	6.35 (2.47)	6	2	14
No. of unique lots bid/bidder	4.93 (7.95)	3	1	65
No. of bids/lot	15.47 (7.46)	15	2	48
Opening bid in $	$19,343 ($36,663)	$6400	$650	$300,000
Preauction low estimates of the lots	$24,128 (45,747)	$8000	$795	$375,000
Preauction high estimates of the lots	$31,065 (60,351)	$10,230	$1025	$475,000
Realized value of the lots in USD($)	$62,065 (133,198)	$22,000	$3135	$1,486,100
Realized sq. inch price of the lots in USD($)/sq. inch	$108.77 (225.49)	$45.12	$1.40	$1865.42

TABLE 6.2 Summary Description of Dataset 2

	Mean (SD)	Median	Min.	Max.
No. of unique bidders/lot	4.25 (1.50)	4	2	8
No. of unique lots bid/bidder	2.83 (3.14)	2	1	17
No. of bids/lot	8.66 (3.58)	9	3	18
Opening bid in $	$23,308 (20,346)	$18,000	$3040	$91,230
Preauction low estimates of the lots	$27,563 (23,939)	$23,260	$3490	$104,660
Preauction high estimates of the lots	$33,536 (28,993)	$27,910	$4500	$127,910
Realized value of the lots in USD($)	$44,397 (39,703)	$35,750	$4125	$176,000
Realized sq. inch price of the lots in USD($)/sq. inch	$55.25 (48.00)	$38.24	$13.22	$225.64

(the works of only 9 artists were auctioned compared to 70 in the first dataset) based solely on its size. Sixty-six bidders posted 381 bids with 8.66 bids/lot. The average value of the lots auctioned was \$44,397 (min = \$4125, max = \$176,000). The average number of bidders per lot was 4.25, with an average time of auction entry of 0.0861 minute. Some of the key descriptive information about the auction is presented in Table 6.2.

6.4 PRICE DYNAMICS

Investigation of price formation dynamics provides vital information regarding the factors contributing to the evolution of the bid dynamics. Such information is vital for auction house managers, as it provides a basis to determine how prices change during the auction. The factors whose effect on price dynamics we explore are artist characteristics (established or emerging artist; prior sales history), art characteristics (size; painting medium—canvas or paper), auction design characteristics (opening bids, preauction estimates; position of the lot in the auction), and competitive characteristics (number of bidders; number of bids). In this section, we will first discuss our modeling process and then present the results.

Price formation analysis of online auctions of hedonic heterogeneous products is both complex and challenging. The uniqueness and scarcity of the art objects and the genre differences among the artists result in high variability among the lots, making the development of a generalized model for estimating price dynamics a challenging task. Prior studies by Jank, Shmueli, and their associates (Shmueli et al. 2004; Bapna et al. 2008; Shmueli and Jank 2005a, 2005b) have developed sophisticated statistical methods to visualize and analyze the dynamics of online auctions. Recently, Wang and her colleagues (2006) explored ways to apply such dynamics to forecast the final prices of the auctioned items on an ongoing basis. These studies form the methodological background for our approach to analyze price dynamics in online auctions of fine arts.

From the modeling perspective, online auction data present challenges that make applying traditional econometric/regression methods difficult. The data consist of records of bid sequences placed at evenly spaced intervals, thus precluding the use of traditional time series methods in our analysis. Moreover, as in eBay auctions, a bidding frenzy is observed toward the beginning and near the end of the auction. Functional data analysis (FDA) (Ramsay and Silverman 2005), which at its core is the analysis of curves rather than points, is well suited to analyze this type of data. Using this technique, we analyze the price dynamics (velocity and acceleration) in online auctions after recovering the underlying price curves using a nonparametric curve-fitting technique such as splines (Simonoff 1996). Whereas the traditional regression methods are useful in modeling the final price points in the auctions, FDA provides the required tool to model the price dynamics and determine their relationship with the strategic variables during the entire auction. In this process, we first smooth the bid data for each lot and recover the underlying price curves. Then we model the heterogeneity of these price paths using the above-mentioned characteristics to provide insights into the relationship of these covariates in the price dynamics during the auction.

6.4.1 Method

As the first step toward our goal, we must smooth the given data. This involves some preprocessing steps. First, we standardize the auction time by scaling it within 0 to 1; thus, $0 \leq t_{ij} \leq 1$, where t_{ij} represents the *j*th bid in lot *i*. Then we accommodate the irregular spacing of the bid arrivals by linearly interpolating the raw data and sampling them at a common set of time points t_i, $0 \leq t_i \leq 100$, $i = 1, 2 \ldots$ total number of lots. For each of the bid times, we compute the corresponding log-transformed bid values to reduce the skewness of its distribution.

As part of the second stage in our analysis, we use penalized smoothing splines (Simonoff 1996; Ramsay and Silverman 2005) to recover the underlying price curves. These splines effectively capture the local variation in the dataset and readily provide different derivatives of the smoothed price curves. This functionality allows us to analyze higher-order functions of the auction price, namely, price velocity (first-order derivative) and price acceleration (second-order derivative). To recover the underlying price curve, we consider a polynomial spline of degree *p*.

$$f(t) = \beta_0 + \beta_1^t + \beta_2 t^2 + \beta_3 t^3 \cdots + \beta_p t^p + \sum_{l=1}^{L} \beta_{pl}[(t - \tau)_+]^p \tag{6.1}$$

where $\tau_1, \tau_2 \ldots \tau_L$ is a set of *L* knots and $u_+ = uI_{[u \geq 0]}$. The choice of *L* and *p* determines the departure of the fitted function from a straight line, with higher values resulting in a rougher *f*. This may result in a better fit but a poorer recovery of the underlying trend, as it has a tendency to overfit the given data. To avoid this problem, the following a roughness penalty function (*PEN*) is imposed to measure the degree of departure from the straight line:

$$PEN_m = \int [D^m f(t)]^2 dt \tag{6.2}$$

where $D^m f$, $m = 1, 2, 3\ldots$ is the *m*th derivative of the function *f*. The goal is to find a function $f^{(j)}$ (the *j*th bid in lot *i*) that minimizes the penalized residual sum of squares (*PENSS*)

$$PENSS_{\lambda,m}^{(j)} = \sum^{n} [y_i^j - f^{(j)}(t_i)]^2 + \lambda \times PEN_m^{(j)} \tag{6.3}$$

where the smoothing parameter λ provides the trade-off between the fit $[(y_i^{(j)} - f^{(j)}(t_i))^2]$ and variability of the function (roughness) as measured by PEN_m. We used the b-spline module developed by Ramsay (2003) for minimizing $PENSS^{(j)}{}_{\lambda,m}$.

Finally, to analyze the effects of different covariates on price dynamics, we apply functional regression with the price functions as our response variable and the determinants such as artist characteristics (established or emerging artist; prior sales history), art characteristics (size; painting medium—canvas or paper), auction design characteristics (opening bids, preauction estimates; position of the lot in the

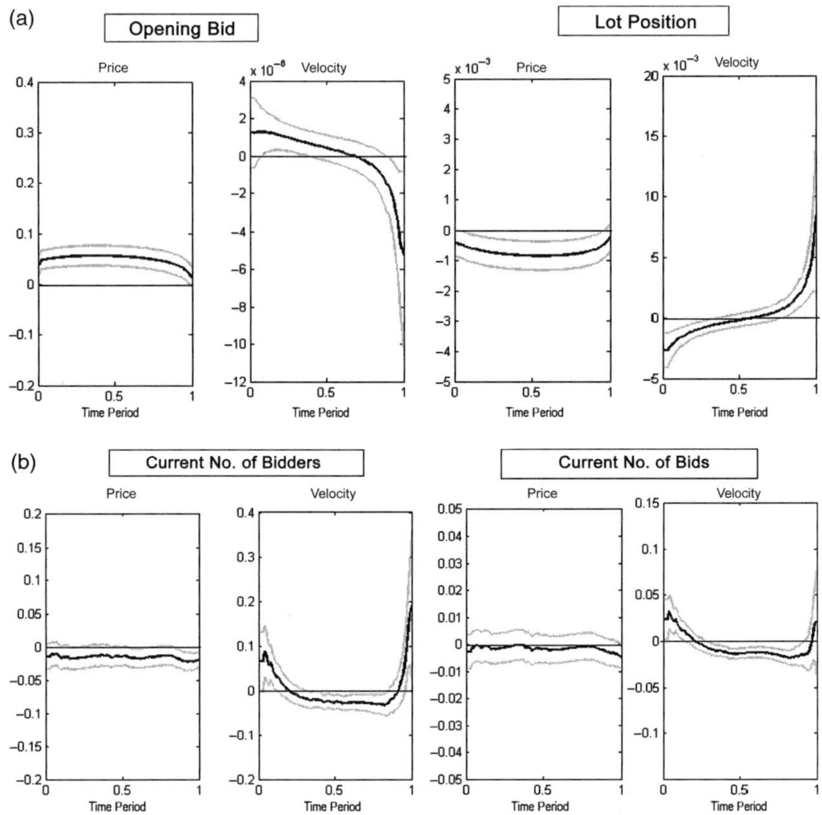

Figure 6.3 (a) Auction design characteristics; (b) competition characteristics; (c) artist characteristics; (d) art characteristics.

auction), and competitive characteristics (number of bidders; number of bids) as our explanatory variables. Functional regression is ideal for our case, as unlike a typical regression setting, it allows the functional form of variables in the analysis. For example, the response variables in our case are the price curve $f_j(t)$ and the price-velocity curve $f_j'(t)$[6] that capture the price formation process during the auction. Functional regression models allow us to understand the influence of covariates on price dynamics over time. As Ramsay and Silverman (2005) point out, this is achieved by estimating $\beta(t_i)$ for a finite number of points in time t (in our case, $t = 100$) and constructing a continuous parameter curve by simply interpolating between the estimated values $\hat{\beta}(t) \ldots \hat{\beta}(t_n)$. To capture the effects of the explanatory variables on each of the price dynamic variables, we run a regression for each time period $(1-100)$ for data from all the lots ($n = 199$). The parameter estimates

[6]For an auction house manager, practical use of price dynamics is limited to price velocity. Therefore, we did not perform any analysis on price acceleration.

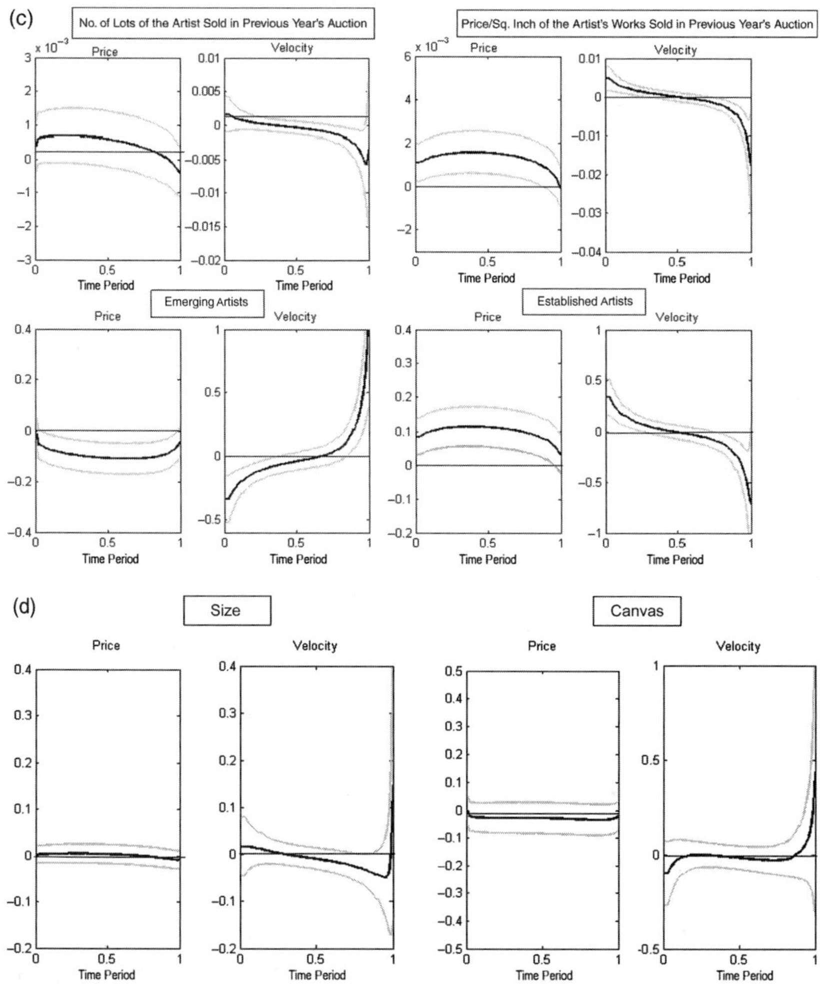

Figure 6.3 *Continued.*

associated with each explanatory variable are plotted along with confidence bands to indicate the impact and its significance over the entire auction. Figures 6.3a–d illustrate the results of our analysis.

6.4.2 Results

We find auction design characteristics such as opening bid for the lots, to have a positive effect on price formation throughout the auction, with its effect decreasing toward the end of the auction. Its effect on price velocity is also positive at the beginning of the auction but becomes negative by the end. This indicates that lots with higher opening bids show less price velocity at the end of the auction or, alternatively, that lots with lower opening bids show greater price velocity at the end of the auction. We also find that lots auctioned later in the auction exhibit lower price levels during the auction. Such items have less price velocity during the early stages of the auction, but it increases toward the end of the auction. The results are illustrated in Figure 6.3a.

Interestingly, the current number of bids and the number of bidders during the auction are found to have no significant effect on price formation but considerable effect on price velocity. It is high at the beginning and near the end of the auction, with an increasing number of bids and bidders, but low during the middle of the auction. The results are illustrated in Figure 6.3b.

Results of the historical auction activities of artists (the price realized and the number of lots sold in the previous year) show that their effect on price velocity diminishes as the auction progress (Figure 6.3c). Price level is also found to be positively affected by the artist's historical price records, with the effect being strongest during the middle of the auction. Comparing the price dynamics of emerging and established artists, we find that the price levels of lots painted by emerging artists are lower throughout the auction, whereas those of lots painted by established artists are high throughout the auction. Furthermore, price velocity is low for emerging artists early in the auction but high near the end. The opposite is true for established artists. No significant effects of art characteristics (size; painting medium—canvas or paper) are found in our study. The results are illustrated in Figure 6.3d.

A complete summary of the results is shown in Table 6.3.

TABLE 6.3 Summary of Findings

	Price Level	Price Velocity
	Auction Design Characteristics	
Opening Bid	Opening bid has positive effect on price level throughout the auction, with the effect decreasing toward the end of the auction.	The impact of the opening bid on price velocity is positive at the beginning of the auction. This effect is negative by the end of the auction, indicating that lots with higher opening bids show less price velocity at the end of the auction. Alternatively, lots with lower opening bids show greater price velocity at the end of the auction.
Lot Position	Lots auctioned later in the auction have lower price levels during the auction.	Change in price is slow for lots having a higher lot position during the early stages of the auction. Price velocity is rapid at the end of the auction.

(Continued)

TABLE 6.3 *Continued*

	Price Level	Price Velocity
	Competition Characteristics	
Number of Bidders	Not significant.	Price velocity is greater at the beginning and end of the auction, with increasing numbers of bidders. The impact of the number of bidders on price velocity during the middle of the auction is lower.
Number of Bids	Not significant.	Price velocity is greater at the beginning and end of the auction, with increasing numbers of bids. The impact of the number of bids on price velocity during the middle of the auction is lower.
	Artist Characteristics	
Established Artists	Price level of lots painted by established artists is higher throughout the auction.	For established artists, price velocity is reduced as the auction progresses.
Emerging Artists	Price level of lots painted by emerging artists is lower throughout the auction.	For emerging artists, price velocity is low during early in the auction and greater at the end.
Historical Value of Artist (price/sq. inch)	Price level is positively affected by the artist's historical price record, with the effect being strongest during the middle of the auction.	Historical auction activity of artists on price velocity diminishes as the auction progresses.
No. of Lots of the Artist Sold in the Previous Year's Auction	Not significant.	Historical auction activity of artists on price velocity diminishes as the auction progresses.
	Art Characteristics	
Size of the Painting	Not significant.	Not significant.
Canvas	Not significant.	Not significant.

6.5 BIDDER DYNAMICS

Analyzing competitive bidding is vital to understand bidder dynamics in online auctions. Fortunately, the available bid history provides sufficiently detailed information on these activities to facilitate our study. Competitive bidding, i.e., repeated outbidding between two specific bidders, is common in most auctions (Gupta 2002). At a fundamental level, such consecutive bids on common items among bidders can be conceptualized into a dyadic relationship between them. This exposes not only the bidding patterns of the bidders but also private information such as their purchase intention, depth of pocket, and private value of the lots they are bidding on. In this section, we focus on this phenomenon and determine how such dyadic bidder relations are formed during the auction. We also identify the bidder subgroups in the auctions based on such interdependence. To facilitate our goal, we introduce a new approach of forming a bidder network based on these dyadic bidder activities and analyze it with social network analysis (SNA). A bidder network, like any other network, is defined as a set of bidders with connections or relationships between them. Therefore, in our case, the nodes of the network represent the bidders, and the strength of the link between them corresponds to the intensity of competitive bidding between them. We use SNA (Wasserman and Faust 1994) to perform our investigation.

6.5.1 Method

The concepts of *social network* (Wasserman and Faust 1994) and *network analysis* have found a wide variety of applications in sociology, marketing, and statistics. This powerful tool has been used in the investigation of interorganizational communications (Hutt et al. 1988; Gloor et al. 2004), buying centers (Bagozzi 1978; Johnston and Bonoma 1981), channels (Dwyer et al. 1987; Stern and Scheer 1991), brand-switching behavior in the auto industry (Iacobucci et al. 1996), the World Wide Web (Katona and Sarvary 2005), and relationships among family members (Corfman and Lehmann 1987; Qualls 1987). More recently, researchers have focused on the dynamic aspect of network evolution, with Barabasi and his colleagues (2002) leading the research stream. The popularity of this concept has even crept into modern culture. A game called "Six Degrees of Kevin Bacon," where the challenge is to connect any given actor or actress directly or indirectly to Kevin Bacon, a renowned actor, has become very popular.[7] This game considers a link between two social entities (actor or actress) if they have worked together in the same movie. Another social network application is the work of a motivated photographer, Andy Gotts, who developed a photographic collection of movie actors and actresses called "Degrees," where each entry is a result of another entry's referral (Gotts 2005). SNA has also recently been used as a powerful tool in national security applications. For example, Krebs (2001) used such analysis to map terrorists' participation in the 9-11 attack and determined the central player in the event. Further, SNA is widely used by the U.S. security agencies to scan telephone databases for possible threats to national security (Dryer 2006).

SNA focuses on relationships among social entities and on the patterns and implications of these relationships (Wasserman and Faust 1994). In particular, it investigates how the interactions among social entities constitute a framework to

[7]For more information, visit http://oracleofbacon.org./

Bidder →

Previous bidder ↓	Bidder 1	Bidder 2	Bidder 3	.... Bidder 256
Bidder 1	0	x_{12}	x_{13}	
Bidder 2	x_{12}	0	x_{23}	
Bidder 3	x_{13}	x_{23}	0	
.				
.				
Bidder 256				

256 x 256 Adjacency Table

Figure 6.4 Sociomatrix of the bidder network.

understand various roles played by them in the network. We capture competitive bidding as dyadic interactions between a pair of bidders. We define the bidder network as a set of g bidders whose relationship strength is based on the number of times bidder i and bidder j bid sequentially on a lot where $i, j \varepsilon N$. We define $N = \{1, 2, \ldots g\}$ as the set of g bidders and X_m as a bidding relation of type m. X_m is a set of ordered pairs recording the extent of a relationship of type m between pairs of bidders. In this case, we define the extent of a relationship as the number of times p that bidder i and bidder j bid sequentially on the lots in the auction. X_m is represented as a $g \times g$ matrix (Figure 6.4), where $(X_m)_{ij} = p$; $p = \{0, 1, 2, \ldots P\}$, where P is the maximum number of consecutive bids placed in the auction. We create this nondirectional, symmetric matrix X_m for $g = G$ bidders in the online auction which forms the basis for the network analysis.

Let us illustrate the process of the network formation with an example (Figure 6.5). Consider the two bid histories shown in Figures 6.5a and 6.5b. In the first bid history (Figure 6.5a), Anonymous 3 and Kyozaan are found to bid sequentially three times. Therefore, we consider two nodes in the network, one representing Anonymous 3 and

(a)

Rank	Nickname	Amount($)	Amount(Rs)	Date & Time (US EST)
Current Highest Bid 1	Poker	71,500	3,146,000	Dec 6 2005 3:03:48 AM
2	Kyozaan	66,500	2,926,000	Dec 6 2005 1:27:43 AM
3	Poker	61,500	2,706,000	Dec 6 2005 1:26:17 AM
4	Kyozaan	56,500	2,486,000	Dec 6 2005 1:26:10 AM
5	Anonymous 3	51,500	2,266,000	Dec 6 2005 1:25:58 AM
6	Kyozaan	49,000	2,156,000	Dec 6 2005 1:25:58 AM
7	Anonymous 3	46,500	2,046,000	Dec 5 2005 10:30:00 PM
	Start price	44,000	1,936,000	Dec 5 2005 10:30:00 PM

(b)

Rank	Nickname	Amount($)	Amount(Rs)	Date & Time (US EST)
Current Highest Bid 1	Kyozaan	121,500	5,346,000	Dec 6 2005 9:53:24 AM
2	Socrates	111,500	4,906,000	Dec 6 2005 9:52:25 AM
3	Kyozaan	101,500	4,466,000	Dec 6 2005 9:52:25 AM
4	Socrates	94,000	4,136,000	Dec 6 2005 4:42:36 AM
5	Poker	86,500	3,806,000	Dec 6 2005 4:16:25 AM
6	Anonymous 38	79,000	3,476,000	Dec 6 2006 4:14:37 AM
7	Kyozaan	74,000	3,256,000	Dec 6 2005 1:16:56 AM
8	Anonymous 38	69,000	3,036,000	Dec 5 2005 10:40:07 PM

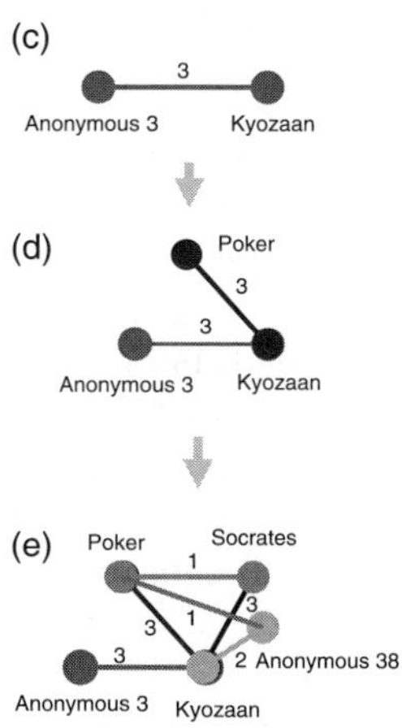

Figure 6.5 Formation of a bidder network.

the other Kyozaan, and link them with an arc having the value of 3 (Figure 6.5c). In the same bid history, we find that another bidder, Poker, has also bid against Kyozaan three times. Therefore, we include another node in our network to represent this bidder and link it to Kyozaan with an arc of value 3 (Figure 6.5d). Now, let's consider the second bid history (Figure 6.5b). Here we find that Socrates and Anonymous 38 have bid against Poker and Kyozaan. We add two more nodes in our network to represent these bidders and link them to related bidders with appropriate values (Figure 6.5e). In this manner, we consider the complete bid history of the 199 items sold in the auction, consider each of the sequential bids between the bidders, and develop the network. If two bidders appear sequentially in more than one item, we add all their appearances to compute the strength between them. Further, to explore the evolution of the bidder network and the presence of bidder subgroups, we measure the network centrality indices Degrees and Bonacich's Power of the individual bidders and the overall bidder network. Both of these measures not only specify the extent of connectivity among bidders, but also indicate the role they play during the auction.

Degree refers to the number of links an actor has with other actors in a network (Freeman 1979). In our case, it is the number of interdependence relationships each bidder has with other bidders. The greater the number of bidder links, the more centrally located the bidder will be in the network. A central bidder, with the advantage of his or her location in the network, is more capable of playing an influential role in the auction than others. Bidders having minimum degree reside on the periphery of the network and have few interdependence relationships. The degrees are normalized and computed as (Freeman 1979)

$$C'_D(n_i) = d(n_i)/(g - 1) \tag{6.4}$$

where C'_D = degree of a bidder i, $d(n_i)$ is the total number of bidders linked to bidder n_i, and g is the total number of members in the network. We also compute average degree indices of the overall network at various time periods in the auction as

$$C_D = \sum_{i=1}^{g} [C_D(n^*) - C_D(n_i)]/[(g - 1)(g - 2)] \tag{6.5}$$

where $C_D(n_i)$ is the degree of bidder i, $C_D(n^*)$ is the largest observed degree in the network, and g is the total number of bidders in the network. This provides the normalized degree measure and compares networks formed at different auction times.

Bonacich's Power measures the total influence power of particular bidders over other bidders based on the possibility of influencing other bidders during auctions. Consider a hypothetical scenario where bidder A is connected to bidder B. Further, bidder A is connected to three more bidders (X, Y, and Z) and bidder B is connected to only one more bidder, say H. In this case, bidder A's influence over bidder B will be greater than that of bidder B over bidder A. This is because connectivity of bidder B is shared only between bidders A and H, but that of bidder A is shared between bidders B, X, Y, and Z. This is the fundamental notion of Bonacich's Power

(Bonacich 1987). Bonacich further argues that although a bidder's connection to other bidders makes that bidder central, this does not necessarily make him or her powerful. Therefore, a bidder connected to other minimally connected bidders is powerful, whereas a bidder connected to a well-connected bidder is not.[8]

Bidders with similar product choices and sometimes with similar bidder behavior typically engage in competitive bidding and thus tend to form dyadic relations. Given the heterogeneous nature of the products we are studying, maximally complete (well-connected) bidder subgroups will indicate the number of bidder clusters defined by bidders' bidding characteristics and the linkages in the bidder network. We used Bron and Kerbosch's (1973) algorithm to compute such subgroups in our bidder networks and a popular social network analysis program called Ucinet (Borgatti and Freeman 2002) to perform our analysis. We reconfirmed our analysis using the "sna" library in R (Butts 2006). We also divided the auction time into 10 equal periods and then analyzed the bidder interdependence network at each of them to investigate its evolution.

6.5.2 Results

6.5.2.1 Overall Bidder Network. The evolution of the bidder network in time periods 1, 5, and 10 shows a very interesting changing competitive bidding pattern (Figure 6.6).[9] Our initial observation indicates that as the auction progresses, more bidders arrive and create a densely connected network. By the first time period in the auction, there are 309 bidder relations in the network. This number increases to 1463 at the end of the auction. Although graphically our bidder network looks

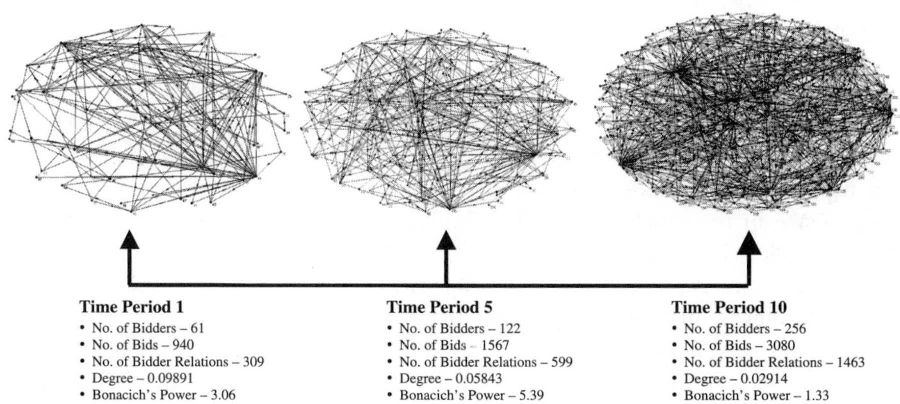

Figure 6.6 Evolution of a bidder network over the duration of the auction.

[8]See Bonacich (1987) for details on how Bonacich's Power of individual actors in a network is estimated.

[9]This is a Fruchterman Reingold's three-dimensional plot obtained from the SNA software called Pajek.

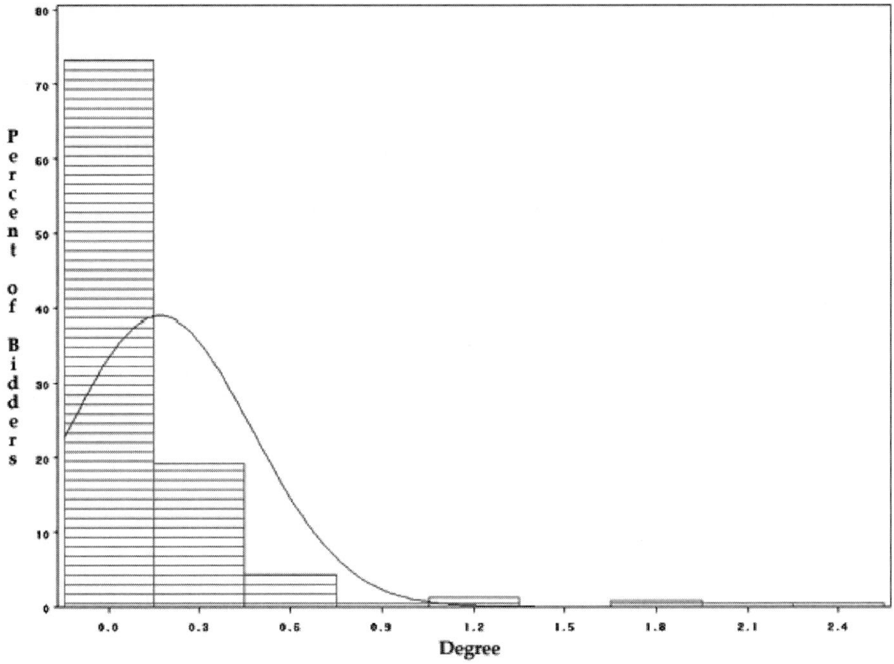

Figure 6.7 Degree of distribution of the interdependence bidder network.

well connected, we find that in reality, it is sparse $[2L/g(g-1) = 0.04482]$[10] and the average degree centrality of the bidders is only 2.914, which is small compared to the number of possible degrees: $g-1 = 255$. This indicates that our observed bidder network is neither regular (where all bidders are equally connected) nor random (where most of the bidders' degrees are concentrated around the mean degree). We further find that degree distribution (Figure 6.7) is left skewed, indicating that most bidders have fewer bidder linkages than a handful of active bidders. Decline in network centrality (degree) with progression in the auction indicates that the network becomes fragmented over time. This network fragmentation with the increase in the number of bidders implies that the central role played by the average bidders is reduced.

One of the central theses of bidder interdependence is that it results in bidder familiarity during auctions. Such familiarity is formed due to the transparency and repeated meetings among the bidders. If the bidder network characteristics support faster information flow from one peripheral end node to another, bidders may rapidly become familiar with other bidders. In network theory, networks exhibiting *small-world* properties are ideal for faster information dissemination. A network is said to have small-world properties when it is highly clustered, like a regular graph

[10]The sparse network test is illustrated by Braha and Bar-Yam (2004) with L as the number of bidder relationships and g as the number of nodes/bidders.

($C_{\text{real}} \gg C_{\text{random}}$), but possesses a small path length, like a random graph ($l_{\text{real}} \approx l_{\text{random}}$) (Watts and Strogatz 1998; Watts 1999). Our bidder network has a high clustering coefficient ($C_{\text{bidder}} = 0.881$) compared to a random network with the same number of bidders (256) and bidder relationships (1463) ($C_{\text{random}} = 0.022$), but a similar path length like a random network ($l_{\text{bidder}} = 3.097 \approx l_{\text{random}} = 3.391$), thus showing that it is possible that bidder information was disseminated rapidly in the network.

Average Bonacich's Power of the network also increased as the auction progressed (from 3.06 in the first time period to 11.33 at the end). This interesting finding suggests that as the auction progresses, power is not uniformly distributed and some key bidders are more powerful than others. Such bidder characteristics indicate that network characteristics of bidders are useful in defining their bidding behavior and thus may be a way to classify them.

6.5.2.2 Bidder Subgroup Analysis. Cohesive subgroups in a network are groups of actors who are more strongly connected to members belonging to the same subgroup than to members belonging to other subgroups (Wasserman and Faust 1994). The strength of a subgroup is determined by the level of interconnectivity among its members. In our case, bidder subgroups represent bidders who bid on similar lots frequently and compete against each other. Investigation of such bidder groups is vital in determining various types of bidders and, more importantly, various types of bidding activities. We used the algorithm developed by Bron and Kerbosch (1973) to compute the subgroups in the bidder interdependence network.

We find seven bidder subgroups in our bidder network, each represented by a unique set of bidders. The strongest subgroup contains 30 bidders, mostly the active bidders of the network. These bidders participated in the auction of the largest number of lots (190 lots), illustrating *participatory* behavior (Bapna et al. 2004) (active bidding throughout the auction, without a significantly low winning percentage). To compare the characteristics of the strongest subgroup with those of a weaker subgroup, we analyzed the bidders in these groups and their bidding activity (Figure 6.8). Unlike bidders associated with the strongest group, bidders in the weaker groups tend to be *opportunists*[11] who join the auction late, bid on a small number of lots, and bid less frequently. Such bidder behavior is quite similar to the *sniping* behavior of bidders in eBay auctions, although such activity is discouraged by the flexible closing time of these auctions. Moreover, these bidders have a higher winning percentage than those in the strongest group.

In summary, our bidder dynamics model illustrates some important issues regarding the bidder behavior in online auctions:

1. With more bidders joining the auction, the bidder network becomes more fragmented, thus reducing the average degree of centrality (bidder connectivity) in the network.

[11]Opportunism is also another bidding style found by Bapna and his colleagues (2004).

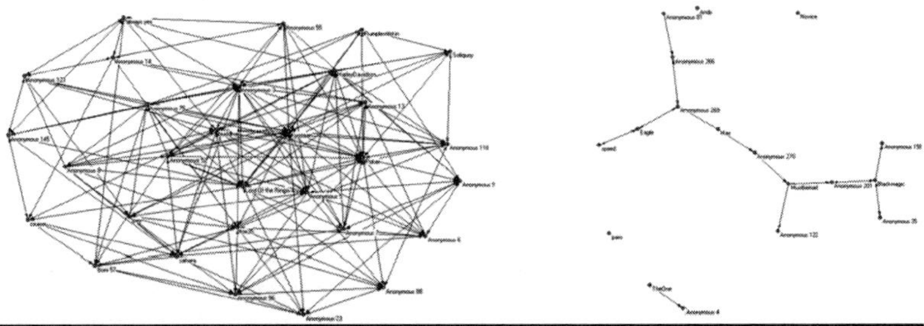

	Strongest Subgroup	Weaker Subgroup
No. of Bidders	30	18
Average No. of Lots Bid	14.11	5.00
No. of Unique Lots Bid	190	55
Total No. of Wins	60	8
Average Efficiency (% Win)	9.71%	13.69%
Average No. of Bids	51.90	10.42
Average Entry (First Bid) Time	0.323	0.568
Average Exit (Last Bid) Time	0.549	0.692
Average Total No. of Proxy Bids	28.63	4.00
Average (Median) Final Price of Lots Bid	$62,507 ($22,077)	$86,410 ($18,525)
Average (Median) Final Price of Lots Won	$57,010 ($18,454)	$228,916 ($37,900)
Highest Value of Lot Won	$566,000	$1,351,000

Figure 6.8 Comparison of the strongest bidder subgroup and a weaker bidder subgroup.

2. The average power in the network increases, with few bidders having more power than others.
3. There are seven bidder subgroups in the auction based on their bidding behavior.
4. Bidders in the strongest subgroup exhibit participatory bidding behavior, whereas those in the weaker subgroup show opportunist bidding behavior.

6.6 MARKET STRUCTURE

Finally, in this section, we illustrate an important use of bid history information to derive insights into the market structure. At a fundamental level, bid history represents the preference information of the bidders who participated in the auction. Specifically, using the available data, we can now identify the lots a bidder is interested in or not. Considering such preference information about all the participating bidders, in this section we investigate the market structure of modern Indian art using the second dataset discussed in Section 6.3.

As with any heterogeneous market, the structure of the art market is difficult to analyze. Contemporary Indian art especially is represented by a variety of artists[12] with a diverse set of techniques and styles. Further, these artists have had different types of training, and most of them have fashioned their own forms of creation.

[12]From the time of its introduction as an emerging art market (1995), works of more than 750 artists have been sold in various auctions.

Therefore, any traditional method of determining the underlying market structure is not appropriate in this market. Our approach of observing bidder preferences to determine the market structure is both innovative and insightful. We assume that the number of common bidders on two lots indicates the degree of similarity between the artworks. In other words, we are segmenting the market based on the popularity of the artworks. For example, consider the bid history illustrated in Figure 6.1. It looks like A1 is interested in purchasing lot 25. If she also bids on lots 3 and 6, then in her preference space, lots 3, 6, and 25 are close to each other but far from other lots listed in the auction. Now if other bidders also show an inclination to acquire lots 3, 6, and 25, then these lots are considered closer to each other. If we consider the preferences of all the participating bidders, we will be able to create a perceptual map of lot similarity and dissimilarity. In our analysis, the number of dimensions and the location of the art objects on these dimensions on the perceptual map are derived solely from the preference/choice data of bidders; thus, ours is a kind of internal analysis of the market structure as defined by Elrod and DeSabro and their colleagues (DeSabro and Rao 1986; Elrod 1991; DeSabro et al. 1993).

6.6.1 Method

Using the second dataset from an online auction held in March 2005, we construct a sociomatrix of the artists (instead of the bidders, as in the bidder network). Therefore, in other words, we create an artist network where nodes are two artists and the strength of the link indicates the pairwise demand for these two artists. There are few ways to determine the artist network for the market structure analysis. One of them is to create links between artists whenever a bidder has shown interest in purchasing their works. Considering the purchase intent of all the bidders in the auction, we will be able to form an artist network. This approach is feasible, but it fails to capture the ranges of similarity and dissimilarity among artists in the auction. Another reasonable approach is to assume a dyadic relationship between artists when a bidder posts a bid on their artworks sequentially. Although this approach captures the level of similarity/dissimilarity among the lots, we can further refine the measure by considering the bid time difference of the bidder on these lots. For example, if a bidder bids on the works of two artists within a short period, it is more likely that these two artists are closer in the bidder's preference space than far apart.

We construct an asymmetric 9×9 matrix[13] (Figure 6.9), with each element computed to facilitate the above concept of lot similarity. Taking a conservative measure, we calculate each element $y_{i,j}$ in the sociomatrix as the total number of common bidders y bidding on lot i and lot j. We perform nonmetric multidimensional scaling (Kruskal 1964) to determine the underlying perceptual map. We further verify the market structure obtained from the previous stage by clustering the lots with the hierarchical clustering technique (Johnson 1967). Once again, we use the popular SNA program Ucinet (Borgatti and Freeman 2002) to perform our analysis.

[13]Forty-four works of nine artists were auctioned, and 66 bidders participated in the auction.

Artists →

Artists	Artist 1	Artist 2	Artist 3	... Artist 9
Artist 1	**0**	x_{12}	x_{13}	
Artist 2	x_{21}	**0**	x_{23}	
Artist 3	x_{31}	x_{32}	**0**	
⋮				
Artist 9				

Figure 6.9 Sociomatrix for the artist network.

6.6.2 Results

The resulting perceptual map obtained from the nonmetric multidimensional scaling of the artist sociomatrix (Figure 6.10) shows some interesting patterns of consumer preference for modern Indian art. The map shows that bidders who are interested in purchasing works of J. Swaminathan are not interested in purchasing works of other artists. Works of J. Swaminathan are very different from those of other artists. They are typically low-priced (average price = \$19,575) compared to those of other artists (average price = \$46,818) and of low value (\$27.05/square inch compared to \$52.49/square inch) for other artists. We also find that the artist pairs Ram Kumar–M.F. Husain and F.N. Souza–A. Padamsee are plotted close to each other. One reason for such close links is the similar content of these artists. For example, works of Ram Kumar and M.F. Husain are both abstract. Works of

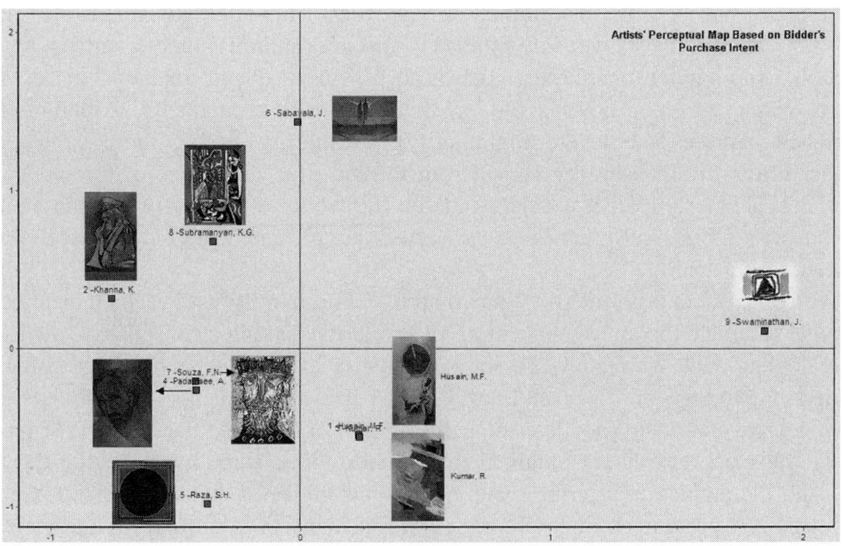

Figure 6.10 Perceptual map of contemporary Indian art.

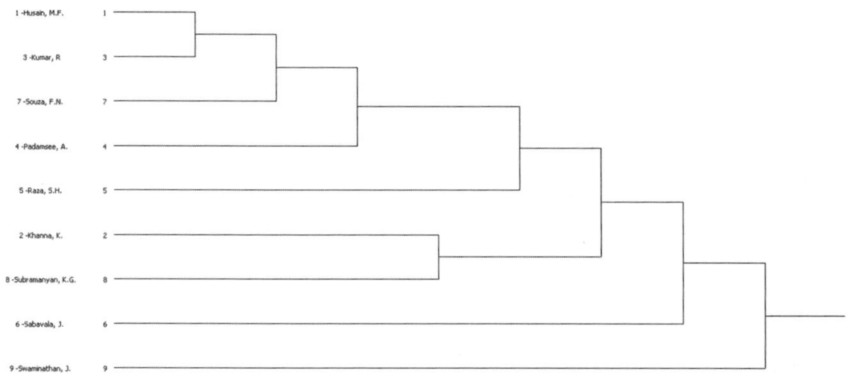

Figure 6.11 Hierarchical cluster of the artists.

F.N. Souza and A. Padamsee are mostly head shots. We also find S.H. Raza and J. Sabavala to be plotted opposite to each other. Raza's works are mostly painted with bright primary colors, whereas Sabavala's paintings contain light mixed colors. Considering all the works of these artists, we suggest that the dimension of the *y*-axis is "abstract," where it varies from "figures" (top) to "abstract" (bottom). The *x*-axis tends to be the realized price of the artworks, with low-priced paintings located on the right and high-priced paintings on the left.

To verify the concluding perceptual map from the nonmetric multidimensional scaling, we cluster the given artist network with a hierarchical clustering technique. The resulting dendrogram (Figure 6.11) not only validated our earlier results, but also provided insights into lot similarity.

6.7 CONCLUSION AND FUTURE DIRECTIONS

This chapter has investigated three vital issues in online auctions. Using the available bid history, we study price dynamics, bidder dynamics, and market structure. We also applied three innovative approaches—FDA, SNA, and multidimensional scaling—to auction data to achieve our goals. Our approach of representing bidding data in the form of a network provides a new paradigm for looking at auction data. From the auction managers' perspective, our analysis addresses some of the concerns they face when organizing a new auction event. These issues can be broadly categorized into three questions: what to sell, how to sell, and to whom to sell.

Before designing new auction events, auction managers need to come up with the item lineup, i.e., what items to sell and how to organize them. Most of the time they have a mixed set of art inventory, which they must use to make their selection. In our price dynamics analyses, we find that a preauction estimate has a positive effect on the price formation process. Since these estimates provide information about item quality, they are highly regarded as value signals by the bidders. The positive effect of this variable suggests that managers may consider auctioning only high-end items. Unfortunately, they do not have enough control over the available inventory. Most of the time, they also have art items from emerging/less popular artists in their auction lineup. Therefore, the issue of item organization becomes vital in such a situation. We found that the price dynamics of works by established and emerging artists are opposite to each other. This suggests that managers should present works of emerging artists right after those of established artists. This may result in a spillover effect of the high price dynamics of the established artist to the emerging artist, thus increasing the overall auction revenue. Our market structure analyses also provide some practical suggestions for managers. This approach illustrates which artists are similar and which are different based on the bidders' intent to purchase. For example, we find that the artist groups Ram Kumar–M.F. Husain and F.N. Souza–A. Padamsee are similar, thus forming substitutes for each other. The perceptual map (Figure 6.9) will also be beneficial if the managers desire to create theme-based auctions in the future.[14]

Finally, our study helps managers decide whom to invite to the auction. Before the auction starts, auction houses send a printed catalog and an invitation to all the bidders on their client list. Still, auction managers desire to concentrate on a smaller group of bidders who ultimately play a crucial role in the auction process. Using the results from the analysis of the bidder subgroups, managers can identify the most active bidders, their tastes, and their bidding strategies. This information may also be used to classify different bidders in the auctions.

From the research perspectives, our use of FDA to analyze price dynamics may be extended to develop models to predict final prices in online auctions. Although Wang and her colleagues (2006) have used similar techniques to predict the results of eBay auctions, extending them to include auctions of hedonic items like works of art will be useful. Further, using our approach of determining the dynamics of competitive bidding with a bidder network, future studies can now investigate their effect on price dynamics. One of the important contributions of our chapter is this new approach of examining bidder dynamics. Although we stop short of exploring the effects of these dynamics on price dynamics, future studies should investigate this subject in detail.

We used Bonacich's Power to determine the aggregate influence level of the bidders. Further investigation may be performed to determine the exact amount of influence of a bidder over another bidder in the auction. Using social network models developed by Hoff (2005) for our bidder framework, we will be able to get such information about bidders, similar to the recent works of Dass and his colleagues (2007b). The concept of *value affiliation*[15] among bidders (Milgrom and Weber 1982) in auctions of hedonic products like works of has existed art for

[14]Although theme-based auctions are not very common in traditional auction houses, recently newly established auction houses such as Osians and SaffronArt have been organizing auctions with specific themes. For example, SaffronArt's March 2005 auction focused only on nine established artists producing contemporary Indian art. Osians' November 2006 auction was themed as "Historical Series."

[15]Meaning that a high value of a bidder's estimate makes high values for other bidders' estimates more likely.

two decades, but there has been no empirical investigation of how such affiliation takes place or how the bidders process the value information during the auction. With our network approach, we can now examine the evolution of a bidder's valuation. Finally, our use of a traditional multivariate technique to determine the underlying market structure is unique, and more advanced models such as latent structure modeling (Hoff et al. 2002) can be used to obtain more detailed market information.

The studies presented in this chapter provide an alternate way of analyzing bidding data on online auctions. We hope that these studies will motivate other researchers to further our advance understanding of online auctions.

ACKNOWLEDGMENTS

Both authors have contributed equally to this research. We acknowledge the support of the Coca-Cola Center for Marketing Studies, Terry College of Business, University of Georgia.

REFERENCES

Ashenfelter, O. (1989). How auctions work for wine and art. *Journal of Economic Perspectives*, 3(3): 23–36.

Ashenfelter, O. and Graddy, K. (2003). Auctions and the price of art. *Journal of Economic Literature*, 41, 763–786.

Bagozzi, R. (1978). Exchange and decision processes in the buying center. In *Organizational Buyer Behavior* (T. Bonama and G. Zaltman, eds.). Chicago: American Marketing Association.

Bapna, R., Jank, W., and Shmueli, G. (2008). Price formation and its dynamics in online auctions, *Decision Support Systems*, 44(3): 641–656.

Bapna, R., Goes, P., Gupta, A., and Jin, Y. (2004). User heterogeneity and its impact on electronic auction market design: An empirical exploration. *MIS Quarterly*, 28(1): 21–43.

Barabasi, A.L., Jeong, H., Neda, Z., Ravasz, E., Schubert, A., and Vicsek, T. (2002). Evolution of the social network of scientific collaborations. *Physica*, A311: 590–614.

Baumol, W.J. (1986). Unnatural value: Or art investment as floating crap game. *American Economic Review*, 76(2): 10–14.

Bonacich, P. (1987). Power and centrality: A family of measures. *American Journal of Sociology*, 92(5): 1170–1182.

Borgatti, S.P. and Freeman, L.C. (2002). Ucinet 6 for Windows, 6.0 ed. Cambridge, MA: Harvard University, Analytic Technologies.

Braha, D. and Bar-Yam, Y. (2004). Information flow structure in large-scale product development organizational networks. *Journal of Information Technology*, 19(4): 244–253.

Bron, C. and Kerbosch, J. (1973). Finding all cliques of an undirected graph. *Communications of the ACM*, 16: 575–577.

Butts, C.T. (2006). Tools for social network analysis. Available at http://erzuli.ss.uci.edu/R.stuff.

Corfman, K.P. and Lehmann, D.R. (1987). Models of cooperative group decision-making and relative influence—an experimental investigation of family purchase decisions. *Journal of Consumer Research*, 14(1): 1–13.

Dass, M., Reddy, S.K., and Du, R. (2007a). Dyadic bidder interactions and key bidders in online auctions. University of Georgia, Athens, Working Paper.

Dass, M., Seymour, L., and Reddy, S.K. (2007b). Investigating overbidders in online art auctions using bilinear mixed model. University of Georgia, Athens, Working Paper.

DeSabro, W., Manrai, A.K., and Manrai, L.A. (1993). Non-spatial tree models for the assessment of competitive market structure: An integrated review of the marketing and psychometric literature. In *Marketing*, Vol. 5. (J. Elaishberg and G.L. Lilien, eds.). New York: North-Holland.

DeSabro, W. and Rao, V.R. (1986). A constrained unfolding methodology for product positioning. *Marketing Science*, 5(1): 1–19.

Dryer, A. (2006). How the NSA does "social network analysis": It's like the Kevin Bacon game. *USA Today*, May 15.

Dwyer, F.R., Schurr, P.H., and Oh, S. (1987). Developing buyer-seller relationships. *Journal of Marketing*, 51(2): 11–27.

Elrod, T. (1991). Internal analysis of market structure: Recent developments and future prospects. *Marketing Letters*, 2(3): 253–266.

Freeman, L.C. (1979). Centrality in social networks: I. Conceptual clarification. *Social Networks*, 1: 215–239.

Gloor, P., Laubacher, R., Zhao, Y., and Dynes, S. (2004). Temporal visualization and analysis of social networks. *North American Association for Computational Social and Organizational Science Conference*.

Gotts, A. (2005). *Degrees*. London: Dewi Lewis Media Ltd.

Gupta, S. (2002). Competition and collusion in a government procurement auction market. *Atlantic Economic Journal*, 30(1): 13–25.

Hoff, P.D. (2005). Bilinear mixed-effects models for dyadic data. *Journal of the American Statistical Association*, 100(469): 286–295.

Hoff, P.D., Raftery, A.E., and Handcock, M.S. (2002). Latent space approaches to social network analysis. *Journal of the American Statistical Association*, 97(460): 1090–1098.

Hutt, M.D., Reingen, P.H., and Ronchetto, J.R. (1988). Tracing emergent processes in marketing strategy formation. *Journal of Marketing*, 52(1): 4–19.

Iacobucci, D., Henderson, G., Marcati, A., and Chang, J. (1996). Network analysis of brand switching behavior. *International Journal of Research in Marketing*, 13(5): 415–429.

Johnson, S.C. (1967). Hierarchical cluster schemes. *Psychometrika*, 32(3): 241–254.

Johnston, W.J. and Bonoma, T.V. (1981). The buying center—structure and interaction patterns. *Journal of Marketing*, 45(3): 143–156.

Katona, Z. and Sarvary, M. (2005). *Network Formation and the Structure of the World Wide Web*. Fontainebleau, France: INSEAD.

Krebs, V. (2001). Mapping networks of terrorist cells. *Connections*, 24(3): 43–52.

Kruskal, J.B. (1964). Multidimensional scaling by optimizing goodness-of-fit to a non-metric hypothesis. *Psychometrika*, 29(1): 1–27.

Mei, J. and Moses, M. (2002). Art as an investment and the underperformance of masterpieces. *American Economic Review*, 92(5): 1656–1668.

Milgrom, P.R. and Weber, R.W. (1982). A theory of auctions and competitive bidding. *Econometrica*, 50(5): 1089–1122.

Pesando, J.E. (1993). Art as an investment: The market for modern prints. *American Economic Review*, 83(5): 1075–1089.

Qualls, W.J. (1987). Household decision behavior—the Impact of husbands' and wives' sex-role orientation. *Journal of Consumer Research*, 14(2): 264–279.

Ramsay, J.O. (2003). Matlab, R, and S-PLUS functions for functional data analysis. Available at ftp://ego.psych.mcgill.ca/pub/ramsay/FDAfuns.

Ramsay, J.O. and Silverman, B.W. (2005). *Functional Data Analysis*. New York: Springer-Verlag.

Reddy, S.K. and Dass, M. (2006). Modeling online art auction dynamics using functional data analysis. *Statistical Science*, 21(2): 179–193.

Roth, A.E. and Ockenfels, A. (2002). Last-minute bidding and the rules for ending second-price auctions: Evidence from eBay and Amazon auctions on the Internet. *American Economic Review*, 92(4): 1093–1103.

Shmueli, G. and Jank, W. (2005a). Modeling the dynamics of online auctions: A modern statistical approach. In *Economics, Information Systems and Ecommerce Research II: Advanced Empirical Methods* (R. Kauffman and P. Tallon, eds.). Armonk, NY: M.E. Sharpe.

Shmueli, G. and Jank, W. (2005b). Visualizing online auctions. *Journal of Computational and Graphical Statistics*, 14(2): 299–319.

Shmueli, G., Russo, R.P., and Jank, W. (2004). Modeling bid arrivals in online auctions. University of Maryland, Working Paper.

Simonoff, J.S. (1996). *Smoothing Methods in Statistics*. New York: Springer-Verlag.

Stern, L.W. and Scheer, L.K. (1991). Power and influence in marketing channel research: Observations on the state of the art. In *Advances in Distribution Channel Research* (G.L. Frazier, ed.). Greenwich, CT: JAI Press.

Wang, S., Jank, W., and Shmueli, G. (2006). Explaining and forecasting online auction prices and their dynamics using functional data analysis. *Journal of Business and Economic Statistics*, forthcoming.

Wasserman, S. and Faust, K. (1994). *Social Network Analysis: Methods and Applications*. New York: Cambridge University Press.

Watts, D.J. (1999). Networks, dynamics, and the small-world. phenomenon. *American Journal of Sociology*, 105(2): 493–527.

Watts, D.J. and Strogatz, S.H. (1998). Collective dynamics of "Small-World" networks. *Nature*, 393: 440–442.

7

MODELING WEB USABILITY DIAGNOSTICS ON THE BASIS OF USAGE STATISTICS

Avi Harel
Ergolight Ltd., Haifa, Israel

Ron S. Kenett*
KPA Ltd., Raanana, Israel, and Department of Applied Mathematics and Statistics, University of Torino, Torino, Italy

Fabrizio Ruggeri
CNR IMATI, Milano, Italy

7.1 BACKGROUND ON E-COMMERCE USABILITY

This chapter presents a method for usability diagnosis of webpages based on time analysis of clickstream data. The resulting diagnostic reports enable website managers to learn about possible usability barriers. Different website design deficiencies are associated with different patterns of exceptional navigation. This chapter presents a method based on the integration of stochastic Bayesian and Markov models with models for estimating and analyzing visitors' mental activities during their interaction with a website. Based on this approach, a seven-layer model for data analysis is proposed and an example of a log analyzer that implements this model is presented. The chapter describes state-of-the-art techniques and tools implementing these methods and maps areas for future research. We begin with some definitions and

*Corresponding author; email: ron@kpa.co.il.

Statistical Methods in e-Commerce Research. Edited by W. Jank and G. Shmueli

an introduction to key concepts in e-commerce usability. The web analytics models are presented later, followed by a case study.

7.1.1 Usability

Usability is a term used to denote the ease with which people can employ a particular tool, or other human-made objects, in order to achieve a particular goal. The International Organization for Standardization document ISO 9241-11 (1998), Guidance on Usability defines usability as

> the extent to which a product can be used by specified users to achieve specified goals with effectiveness, efficiency and satisfaction in a specified context of use.

In commercial websites, usability is about the choices of site visitors. Bucklin et al. (2002) provide a two-by-two categorization of applications delineating search versus purchase, on the one hand, and within-site versus across-site choices, on the other. This chapter can be associated with the category of research on within-site choices, namely, site navigation. In e-commerce, we are concerned about visitors purchasing the website's deliverables. A 1999 study of Web users asked respondents to list the five most important reasons to shop on the Web. Even though low prices definitely attract customers, pricing was only the third most important issue for respondents. Most of the answers were related to making it easy, pleasant, and efficient to buy. The top reason was "Easy to place an order" for 83% of the respondents (Nielsen et al. 1999). The increasing importance of e-commerce is apparent in a study conducted at the Georgia Institute of Technology (GVU 1997). Over half of the 10,000 respondents to questions in eight separate surveys reported having purchased items online. The most frequently cited reason for using the Web for personal shopping was convenience (65%), followed by availability of vendor information (60%), no pressure from salespeople (55%), and saving time (53%). The purchase process follows a stage of users' market research (Moe 2006a). Desired actions during market research browsing may be viewing a key page on the site or downloading a whitepaper. Desired actions at the purchase stage include submitting a sales lead and making a purchase. Usability in e-commerce includes both stages: market research browsing and final order submission.

7.1.2 Usability and Marketing: Beyond Conversion Rates

Usability contributes to both short- and long-range profitability of websites. Short-term profitability is about site visits that end with the purchase of a product or service. The percentage of short-term successes is known in the e-commerce literature as the *conversion rate*. Long-term profitability is about site visits in which the site visitor is satisfied with the navigation results. The distinction between short-term and long-term profitability is very important. Short-term profits can be achieved through banners and popup promotions, intended to attract (distract) the visitors during their original surfing intentions, according to specific marketing goals. Most research is about short-term attributes. Setting a banner to achieve to a particular e-commerce goal typically hampers usability (Rhodes 2001). Many task-oriented visitors might find banners distracting and might abandon the site too early. Subsequently, users who deviated from their original intention might be unwilling to visit the same site again. Moe (2006a) summarized the state of the art about the effect of interruption. Speier and Valacich (1999) found negative effects on visitors and proposed an explanation based on the theory of limited mental capacity in situations of high workload. On the other hand, Zijlstra et al. (1999) found positive effects, and explained the results by a model of overcompensation and by extending mental capacity. However, we note that in spite of the maturity of e-commerce, there is no substantial research proving that the benefits of marketing campaigns involving banners and popup promotions exceed the long-term loss of income due to the negative effects of these methods.

7.1.3 Barriers to Website Usability

Users may abandon a website after being dissatisfied with its content or behavior. Major design problems found in an IBM study of the usability of e-commerce sites (Tilson et al. 1998) included the following:

- The sites did not indicate effectively where and how to add an item to the shopping list (the list of items the user plans to buy).
- The sites did not provide effective feedback when items were and were not saved in the shopping list.
- The sites did not indicate effectively if and when users needed to register or log in to order.
- The sites did not facilitate easy navigation from the shopping list to other parts of the site.

7.1.4 Usability Assurance

The term *usability assurance* refers to methods for improving ease of use during the development process. As an example of the benefits of usability assurance, consider the two designs presented in Kohavi (2006). Figure 7.1a shows the website before usability assurance and Figure 7.1b shows the website afterward.

Design B involves nine usability changes in design A that produced a higher conversation rates. The changes are as follows:

1. The space between the top "Proceed to Checkout" button line and the next line was closed.
2. The top "Continue Shopping" button was removed.
3. The "Update" button underneath the quantity box was removed.
4. The "Total" box was moved down a line. Text and amount appear in different boxes.
5. Above the "Total" box is a "Discount" box, with the amount in a box next to it.
6. Above the "Shipping Method" line is "Enter Coupon Code" with a box to enter it.

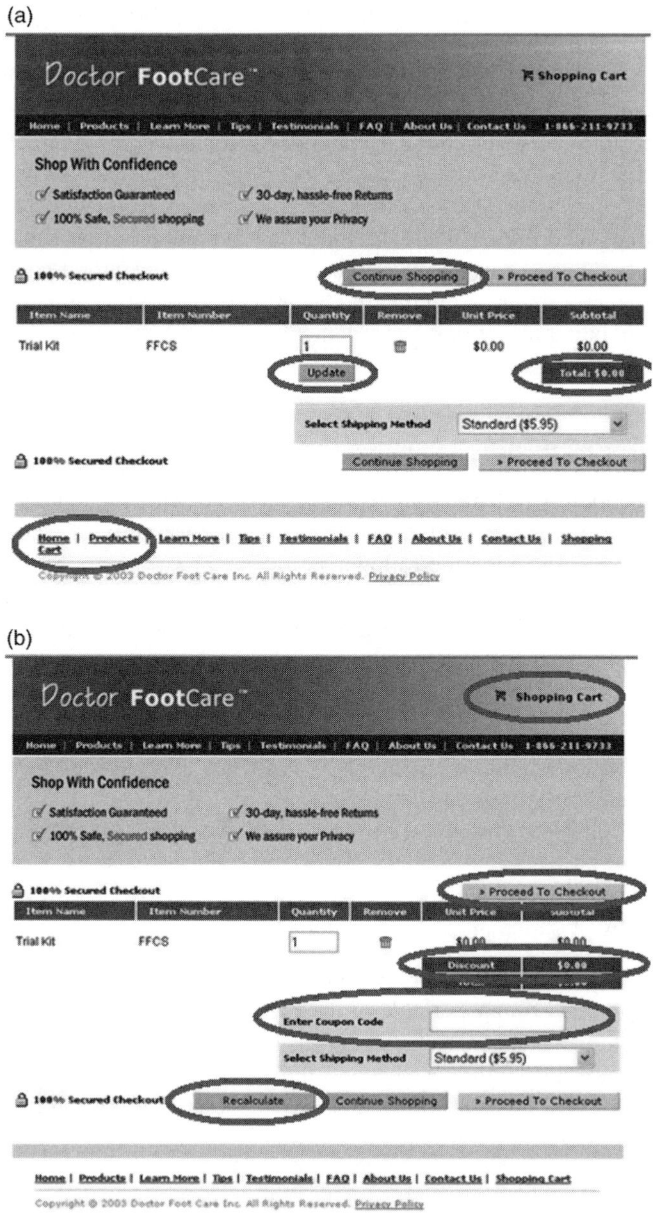

Figure 7.1 (a) Design A—before usability assurance; (b) Design B—after usability assurance. Based on Kohavi (2006).

7. There is a new “Recalculate” button to the left of “Continue Shopping.”
8. The bottom tool bar appears on two lines.
9. The shopping cart icon is one space farther away from the words “Shopping Cart.”

Usability assurance is a key dimension in product marketing success. Marcus (2002) reviewed many studies on usability return on investment (ROI) in user interface (UI) design. These studies show that investment in usability may profit by:

- Saving development time and reducing development, redesign, training, documentation, support, and maintenance costs.
- Increasing success rates: effectiveness, market share, traffic, user productivity, and efficiency, resulting in increased user satisfaction and job satisfaction.

For example, before 1999, IBM’s Web presence traditionally consisted of a difficult-to-navigate labyrinth of disparate subsites. A major website redesign made it more cohesive and user-friendly. According to IBM, the massive redesign effort quickly paid dividends. The company announced, one month after the February 1999 relaunch, that traffic to the Shop IBM online store had increased 120%, and sales went up 400% (Battey 2001).

7.1.5 Methodologies for Usability Assurance

Usability assurance is conducted throughout the development cycle. At the design stage, it is based on methodologies and practices for:

- Anticipating how users may behave when using the product or the service (as opposed to assessment of how users should behave, according to the product and service developer).
- Designing the user interface to ensure seamless interaction with the product.

An example of a methodology for usability assurance by design is presented in Dustin et al. (2001). At the validation stage, it consists of methodologies and practices for verifying that the users behave as intended and that the user interface responds gracefully in cases of deviations from the designers’ intention. An example of a book presenting common testing practices is the one by Duma and Redish (1999).

7.1.6 Usability Research

Good navigation and website design make it easier for users to find what they’re looking for and allow them to buy it once they’ve found it (Donahue 2001). The primary goal of usability research is to define design rules to be applied to requirement specifications in order to ensure that product and service websites are usable. An example of research-based Web design and usability guidelines is the official online booking of the U.S. Department of Health and Human Services. In website design, the goal is to ensure that the sites are easy to navigate. In e-commerce, we also want to ensure that the users will reach particular target pages, fill in correct purchase details, and submit their orders. Examples of Web usability research may be found in reports by User Interface Engineering (2007). For a recent guidelines book, see Nielsen and Loranger (2006).

7.1.7 Predicted Page Usability

Predicted page usability is the designer's impact on page usability. It is the prediction of page usability when specific design rules are applied based on prior research. Predicted usability is an indication of usability applicable to the specification and design stages of website development. The predicted usability attributes of main website pages are as follows:

- *Responsiveness*: The proper time for feedback to the visitor's link or control activation is between 0.1 and 1 second. Response time of less than 0.1 second is too fast, which means that visitors might not notice the feedback (Dabrowski and Munson 2001). Response time greater than 1 second is too slow, which means that visitors might change their focus to a different task.
- *Performance*: The proper download time should normally be 3–5 seconds. Beyond that visitors might change their focus to other tasks, if they are not determined to explore the page, and look for alternative sites.
- *Predicted readability*: Readability indices evaluate the readability of text on the basis of predictive models (Kenett 1996; Kenett and Baker 1999). Such metrics, based on linguistic attributes, can be used instead of running statistical surveys with actual human readers, also known as *readability surveys*, in order to get an indication of webpage readability. Readability scores are based on characteristics such as average word length and sentence length (as a proxy for syntactic complexity). Well-known indices (some included in Microsoft Word) are the Flesch Kincaid metric, the passive sentence index, and the SMOG Index.
- *Predicted relevance*: It is common practice for usability professionals to design websites according to the expected visitor's workflow, which is obtained through task analysis. Predicted page relevance is high if the pages are designed according to the expected visitor's activity, designed according to these workflows. It is low if the pages are designed according to functions or features, regardless of the visitor's workflow.

7.1.8 Limitations of Predicted Usability Attributes

Usability practitioners typically claim that, because they focus on users, they can represent the visitor's needs better than other designers. However, they also admit that users' behavior is often unpredictable. Eventually, after becoming involved in the project, it is difficult even for the best designers to predict how visitors behave on their first visits. Therefore, developers should not rely on user-centered design alone; user testing is also required. The predicted usability attributes are important for the design, but they are insufficient for launching sites with seamless navigation. Typically, the user's experience is not the same as predicted at design time. For example:

- *Responsiveness*: Certain visitors who are eager to see the page content may be willing to wait longer than other page visitors who do not expect much of this page.
- *Performance*: Certain visitors who expect to find the information they need on a particular page may be willing to wait longer than other visitors who do not expect to find any valuable information there.
- *Predicted readability*: Readability scores used by computer scientists use a limited set of rules about usage of language, mainly syntactic, involving dictionaries and grammar testing. Obviously, this is insufficient to obtain a reasonable approximation to the ways a human reader comprehends text. Actual readability depends on knowledge that people gain after years of daily experience. It is impractical to try to find all the rules that could measure the difference in readability of the two sentences "Monkeys like bananas" and "Bananas like monkeys." Yet, in usability testing, it may be obvious that the first sentence is highly readable, while the second sentence might impose a high mental load on the reader.
- *Predicted relevance*: At design time, site traffic is still unknown. It is difficult to anticipate the knowledge and needs of the actual visitors.

7.1.9 Usability Validation

Not surprisingly, theoretical research does not answer all practical design questions. Usability practitioners know quite well that they often fail to predict the user's behavior, and they almost always recommend validating the design before launching the site by testing it with real users. Usability assurance practices that are applicable to the validation stage consist of methodologies and practices for verifying that the users behave as intended and that the user interface responds gracefully in cases of deviations from the designers' intention. Common usability validation methods are based on user reports and testing. User reports capture major usability deficiencies of which the site visitors are aware. However, usability is the outcome of the aggregation of many tiny details, each of them contributing to the goal of seamless navigation. The limitation of user reports is that most users are busy overcoming the many barriers that they encounter; in terms of cognitive science, the amount of detail they need to mentally process exceeds the capacity of working memory. Typically, users are unaware of most of the design deficiencies. Once they are made aware of a minor usability problem, they prefer not to report it in order not to bother management and not to look stupid. The other method for usability validation involves user testing. A typical product user interface (UI) development cycle has three testing phases: integration testing, alpha testing, and beta testing. Validation in e-commerce is similar to UI validation, with special extra needs.

7.1.10 Integration Testing

The first testing phase is during integration of UI with all the product components. The primary goal is to make sure that the product works under normal conditions. Usability validation typically relies on the skills of programmers involved in the product's development and on consulting with usability specialists. This common practice is also typical to website usability validation. Marketing-related issues of e-commerce websites are typically tested by reference to the sites requirement specifications.

7.1.11 Alpha Testing

The second testing phase is alpha testing, primarily intended to verify that the product resists exceptional conditions typical to normal operation, by technical members of the testing group, in the developer's labs, following the predefined operational procedures. Usability testing is typically conducted by usability professionals, who observe users while they try to do predefined tasks. The testing is sometimes conducted in special usability labs at the developer's site, using special equipment (Duma and Redish 1999). Video cameras are used to capture users' behavior, scan converters are used to capture screens, and special setup and software is used to log the testers' observations and to synchronize all the input for failure analysis. Recently, eye tracking equipment has been used in special laboratory setups to learn about the ways users scan the screens. This common UI testing practice is also applicable to website usability validation. Marketing-related issues of e-commerce websites are typically tested by the marketing people involved in the site's requirement analysis and specification.

7.1.12 Beta Testing

The third testing phase is beta testing, intended to verify that the product resists exceptional conditions typical to normal operation by users, in their real operational conditions, doing their real tasks. Usability testing is conducted by usability professionals, who observe the users' behavior either by visiting them at the users' site or remotely, by special video and digital communication means. This common practice is also typical of website usability validation. A few testing companies also invite user representatives to volunteer as panelists, to try the site and comment on usability problems that they encounter while using it. Marketing-related issues of e-commerce websites are validated using special website analyzers.

7.1.13 Ongoing Testing

After a product has been released, the users test it regularly during the interaction, but only severe problems are typically reported back to the developers. With websites the situation is much better, thanks to the server log files that are regularly generated in most websites. Special software programs, called *web analytic tools*, enable site administrators to analyze users' behavior. Special marketing tools enable marketing administrators to view the marketing attributes of the users' behavior. So far, no method has been found for extracting users' expectations from server log files.

7.1.14 Subjective Page Usability

What are the attributes of actual usability, that is, the attributes that describe actual users' behavior? Subjective page usability focuses on the attributes of the visitors' navigation experience. The attributes of subjective page usability answer the following questions:

- *Perceived responsiveness*: To what degree does the page responsiveness suit the visitors' expectations?
- *Perceived performance*: To what degree does the page performance suit the visitors' expectations?
- *Subjective readability*: To what degree do the page visitors succeed in reading and comprehending the page content?
- *Subjective relevance*: To what degree do the page visitors perceive the page as relevant to their needs?

To distinguish between positive and negative attributes of the user experience, we need statistics. This chapter deals with the statistics required to measure, interpret, and classify these and other attributes of the user experience.

7.2 WEB ANALYTICS

Web analytics is the study of the behavior of website visitors. In general terms, web analytics is the process of collecting data about the activities of website visitors and mining those data for information that can be used to improve the website (Peterson 2005). In a commercial context, web analytics refers especially to the use of data collected from a website to determine which aspects of the website promotes of the business—objectives for example, which landing pages encourage people to make a purchase. Before expanding on web analytics in the context of usability studies, we present a brief introduction to measurement systems by focusing on the popular GQM approach.

GQM (goal/question/metric) is a goal-oriented approach that can be used to manage the whole measurement process. It is widely applied in the field of software process and product measurement (see Basili and Weiss 1984; Kenett and Baker 1999; and http://ivs.cs.uni-magdeburg.de/sw-eng/us/java/GQM). GQM aims at evaluating the achievement of goals by making them measurable. Therefore, metrics are needed, which become evident by asking the questions necessary to verify the goal. One starts by making up a list of goals that should be evaluated and asks the relevant questions. Then one collects metrics defined to answer questions used for evaluating the achievement of the goal. The GQM approach can be used as a systematic way to tailor and integrate process measurement's objectives into measurement goals and refine them into measurable values (Figure 7.2). Carried out through observation, interviews with users, and/or workshops, this process is iterative, systematic, and provides rapid identification of the structure for a software improvement program. It creates a foundation of repeatable procedures

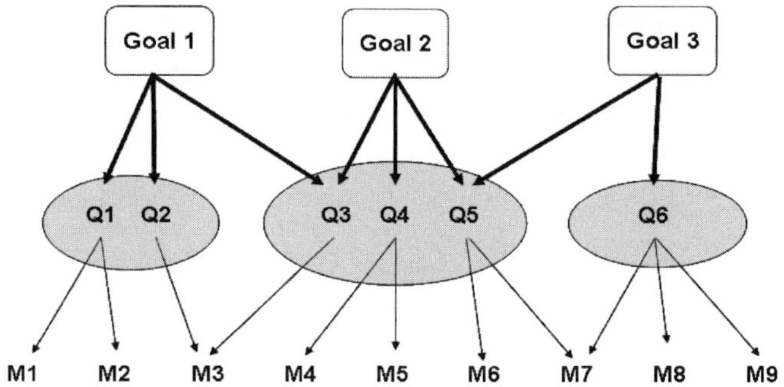

Figure 7.2 The goal questions metrics approach.

for single projects or even for an entire organization. The stages are goal identification, measurement planning, measurement performance, and validation, concluding with analysis and interpretation of the results. It derives the metrics to be used from a thorough analysis of goals and associated questions to be answered quantitatively. By following GQM guidelines, it is possible to:

- Establish the goals of the measurement process.
- Find a proper set of questions to allow one to reach the goals.
- Choose the right set of metrics in order to answer the questions.
- Plan and execute the data collection phase.
- Evaluate the results.

The meaning of the GQM main components—the Gs, the Qs, and the Ms—can be described as follows:

- The Goal describes the measurement's purpose; stating explicit goals gives the measurement program a precise and clear context.
- The set of Questions refines the goal and highlights the quality focus: "What should I know to be able to reach a certain goal?"
- The set of Metrics is used to answer each question. Metric data may result from objective or subjective measurement.

GQM is giving us a context for website evaluation using web analytics. A standard tool used in website management is a recorder, which stores records of the site visits in server log files. Several log analyzers are extensively used to provide usage statistics to website managers, enabling them to learn which pages are visited more frequently than others; how such visits change during the day, the week, or the year; and so on. Certain usage statistics, primarily about rates of site exits and backward navigation, are especially informative. High rates indicate potential usability deficiencies in the page design, which may be very important when they are in the critical path of purchasing procedures. Yet, these measures are weak in diagnostics and do not tell the site managers the reasons for the exceptional rates. High exit rates might indicate dissatisfaction with site navigation but also task completion. Also, the reasons for a high backward navigation rate may be that a page is useless, that the links to this page are wrongly labeled, or that this page is used for additional information to move to another page.

7.2.1 Web Analytics Technologies

There are two main technological approaches to collecting web analytics data. The first method, *log file analysis*, reads the log files in which the web server records all its transactions. The second method, *page tagging*, installs a JavaScript applet on each page view to notify the server about designers' events, such as local activity on the browser.

7.2.2 Server Logs

A straightforward means to collect activity data from website visitors is by server logs. These logs are used routinely by most websites as a main source for backtracking site activity. This chapter offers a method for extracting useful usability information from server logs. Server log files are usually not accessible to general Internet users, only to the webmaster or an other administrator. The two most popular platforms for web hosting, Linux and Windows, support user-defined formats by selection from a set of predefined hit attributes. The most frequently selected attributes are the user IP address, the timestamp (date and time), and the URL. These are also the attributes required for the methodology described here. Additional common attributes, optional in this chapter, include an HTTP code (useful for reducing noise due to technical problems), user agent (browser identification, useful for reducing noise due to robot hits), download size (useful for calculating download speed and measures of textual content), and referrer (useful for better discrimination between internal and external links).

7.2.3 Web Log Analysis Software

Web log analysis software (also called a *web log analyzer*) is software that parses a log file from a web server (like Apache), and, based on the values contained in the log file, derives indicators about who, when, and how a web server is visited. Indicators reported by most web log analyzers include:

- Number of visits and number of unique visitors
- Visit duration and last visit
- Authenticated users and last authenticated visits
- Days of week and rush hours
- Domains/countries of host's visitors
- Host's list
- Most viewed, entry, and exit pages
- Files type

- Operating system used
- Browsers used
- Robots
- Search engines, key phrases, and keywords used to find the analyzed website
- HTTP errors

7.2.4 Web Statistics

Many statistical software packages have been developed to take advantage of the popularity of server logs, providing valuable statistics about different aspects of site navigation, such as traffic analysis, including changes in page popularity during the day or the year, by referrers, or by user agents.

7.2.5 Sales-Oriented Analytics

In a commercial context, web analytics refers especially to the use of data collected from a website to determine which aspects of the website work toward the business objectives; for example, they are often used to calculate trends in page popularity during marketing campaigns and to find out which landing pages encourage people to make a purchase. Two common measures of sales-oriented design are the click-through rate and the conversion rate. The marketing department of any organization that owns a website should be trained to understand these tools.

Section 7.3 describes a theoretical framework for website usability validation, presenting an implementation example. Section 7.4 presents case studies based on real usability data extracted from server logs. Section 7.5 presents conclusions and suggestions for further research.

7.3 MODELING WEBSITE USABILITY

This section presents a theoretical framework for extracting usability diagnostic information from server log files. The framework consists of goal definition and of models used to achieve this goal.

7.3.1 Goal of Website Usability Validation

The theoretical framework is intended to provide two diagnostics about pages and links: page diagnostics and link diagnostics.

Page Diagnostics

- Which pages are likely to be abandoned due to long download time
- Which pages are difficult to read or comprehend
- Which pages are either not interesting or irrelevant to the page visitors
- Which pages are sensitive to performance fluctuation

- Which pages are distracting potential customers from their goals
- Which forms are difficult to fill in or submit

Link Diagnostics

- Which links are likely to be misinterpreted
- Which links are likely to be disregarded
- Which links do not behave according to the visitors' expectations
- Which links are connected to the wrong page

7.3.2 Models Used in Web Usability Diagnostics

This section presents four models used for usability diagnostics. We start with the Markov chain model to represent the overall site view and then go to the page views invoking models of mental activities in page processing. Next, we consider the need for evidence of cause-effect relationships by applying Bayesian networks. Finally, we conclude with statistical analysis of the browsing experience. The models are as follows:

Model 1—Markov Processes: Montgomery et al. (2004) used a dynamic multinomial probit model of Web browsing to show how path information can help to predict visitors' behavior. In analyzing web data, we need to introduce a time dimension so that the dynamics of the navigation sessions are captured. Markov processes model the time dimension in server requests, in which the states represent the webpages and the transitions between states represent the hypertext activity.

Model 2—Mental Activity: Moe (2006b) noted that "though clickstream data do contain a lot of information pertaining to the consumer buying process, it is often difficult to draw cause-and-effect conclusions." To draw such conclusions, we need a model of how certain situations evolve to certain effects. To draw conclusions about the barriers to seamless navigation, we need a model of normal mental activities involved in page handling and possible deviations from normal activities due to design deficiencies. A user-centered model of intrapage mental activities is used to describe the barriers to successful navigation.

Model 3—Bayesian Networks: Bayesian networks map cause and defect relationships between key variables. Here they are used to validate our hypotheses about the relationships between usability design defects and visitors' behavior in response to occurrences of usability barriers.

Model 4—Statistical Analysis: The diagnosis of usability design deficiency involves certain measurements of site navigation, statistical calculations, and rules for statistical decision making. A seven-layer model of statistical manipulations of server log data is described, which enables statistical decisions about possible usability design deficiencies to be made.

7.3.3 Model 1—Markov Processes

In analyzing web data, we need to introduce a time dimension so that the dynamics of the navigation sessions are captured. Markov processes model the time dimension in server requests, in which the states represent the webpages and the transitions between states represent the hypertext activity. Different stochastic models can be introduced to describe and analyze movement within a specific website. We will discuss choices of various Markov processes and illustrate the rationale behind them; then, for illustrative purposes, we will focus on homogeneous Markov chains and present simple estimation techniques, stressing the Bayesian perspective on Markov processes. Each webpage can be identified as a state, and the movement pattern of a visitor within the website can be considered as a sequence of passages from one state to another, with an extra state added to denote the case in which the visitor is outside the website. Markov chains can be used to describe the very simple case of moving from one webpage to another. In this case, from a given webpage, only transitions to another webpage or outside the website are possible. The probabilities of such transitions are the object of the Markov process statistical analysis. A Markov process is a stochastic process *X(t)*, on a probability space with *t* in the parameter space *T*, realizing specific values in the state space *S*, where the Markov property

$$P[X(t) \leq x | X(t_n) = x_n, X(t_{n-1}) = x_{n-1}, \ldots, X(t_0) = x_0] \\ = P[X(t) \leq x | X(t_n) = x_n]$$

holds for all $(t_0, \ldots, t_n)$, $t_0 < \cdots < t_n$, $t_j \in T$, $j = 0, \ldots, n$.

When $T = N$, the Markov process becomes a Markov chain and we denote the process by $\{X_n, n \in T)$; X_n denotes which page (state) the visitor is in at the n-th visited page within the web (see Karlin and Taylor 1975, 1981). Suppose that the website has m pages, and we consider the exit from the website as the state $m + 1$. We consider the home page of the company as state 1 and the webpage with the completion of the action of interest (e.g., order) as state m. The transition probabilities of the Markov chain are represented by a $m + 1$-dimensional square matrix A, whose elements, $p_{ij}(n)$, i, $j = 1$, $m + 1$, are such that $p_{ij}(n) = P(X_{n+1} = j | X_n = i)$ and it holds that $\sum_j p_{ij} = 1$ for all i. The transition probabilities depend on the stage n of the visitor's visit to the website. The assumption can be realistic, since a visitor might spend his or her first stages in the website browsing for all possible offers and details, whereas later he or she will be more likely to make a decision about either leaving the website without taking any action (e.g., ordering or paying) or going quickly through all the webpages needed to complete the action. A Markov process also can be used to account for the time spent in a state. As discussed elsewhere in this chapter, the length of stay at a webpage could be an important index of poor usability. Another possible Markov process model considers the transition probabilities as a function of some covariates often available in clickstream files of e-commerce sites, such as the sex and age of the visitor, number of years doing business with the company, etc. Since this chapter aims to show how usage statistics can be helpful in analyzing usability in practice, we only consider the simple case of a homogeneous Markov chain where the transition probabilities do not depend on stage n. In this case the transition probability becomes p_{ij}, which is the object of our statistical analysis. From the log file of the accesses to the company's website, it is possible to observe N, the total number of transitions, the number n_i of transitions from state i, and the number n_{ij} of transitions from state i to state j, $i, j = 1, m+1$. The likelihood function of what we observe is given by $\prod_{i,j} p_{ij}^{n_{ij}}$ and the maximum likelihood estimates are given by $\hat{p}_{ij} = n_{ij}/n_i$, using a multinomial model for the number of transitions from state i, for all i. Based on the estimated probabilities, it is possible to identify the critical transitions and investigate if usability is the cause of unexpected behaviors. In particular, it is important for the study of the probabilities p_{im} and p_{im+1}, i.e., the transition probabilities to completion of the action (e.g., ordering) and exit from the website, respectively. Experts are therefore asked to interpret a posteriori the results of the statistical analysis. It is also possible to use experts' opinions in making inferences about the probability, following a Bayesian approach. We concentrate on the probabilities $p_i = (p_{i1}, \ldots, p_{im+1})$ for each row i of the transition matrix A. We consider a Dirichlet prior $D(\alpha_{i1}, \ldots, \alpha_{im+1})$ on p_i which is conjugate with respect to the multinomial model, so that the posterior distribution will be a Dirichlet $D(\alpha_{i1} + n_{i1}, \ldots, \alpha_{im+1} + n_{im+1})$. Considering the posterior mean, i.e., the Bayesian (optimal) estimator under the squared loss function, then, we obtain $\hat{p}_{ij} = (\alpha_{ij} + n_{ij})/(\sum_k \alpha_{ik} + n_i)$. A more complex Bayesian model is considered by Di Scala, et al. (2004), who assume a multinomial model for the number of transitions from each state i and a multivariate logit transform of the (nonnull) transition probabilities P_{ij}, defined as $\gamma_{ij} = \alpha_i + \beta_j$. They interpret α_i as a measure of effectiveness of the page i, i.e., of its ability to suggest interesting links, whereas β_j measures the attractiveness of page j. A Bayesian hierarchical model is proposed so that the γ_{ij} are not treated as independent, but their estimation *gets strength* (using the Bayesian jargon) from the observations on other γ_{kl} with $k = i$ or $l = j$. For more on robust Bayesian statistics, see Rios Insua and Ruggeri (2000).

7.3.4 Model 2—User's Mental Activities in Website Navigation

To analyze the barriers to seamless navigation, we need to have a model of normal mental activities involved in page handling and possible deviations from normal activities due to website design deficiencies. A model of website navigation, describing common barriers to successful navigation used in usability diagnostics, is now described. The model assumes that the visitor enters a first page, which is often the home page, or a page referred to by an external hyperlink. Then the visitor repeats the following sequence:

- Evaluates the page download and content.
- Reacts according to the need for additional information.

7.3.4.1 Page Evaluation During Navigation. Page evaluation typically involves the following sequence of activities:

1. Wait for the page to start downloading.
2. Read the page while downloading, looking for information that meets the goal.

3. Realize that the pages have finished downloading.
4. Look for more information related to the goal.
5. Evaluate the information, looking for the best way to proceed.
6. Evaluate the relevance of the page and the overall relevance of the site to the goal.

7.3.4.2 Visitor's Reaction to Evaluation Results. The visitor's reaction to the evaluation typically results in any of the following:

- Return to the previous page if the current page is perceived as less relevant to the goal, than the previous page.
- Link to the next website page, which is perceived as a potential bridge to a goal page.
- Try another hyperlink from a (top or sidebar) main menu if the overall navigation experience is still positive.
- Exit the site if the goal has been reached or if the overall site navigation experience is negative.

7.3.4.3 Barriers to Seamless Navigation. Users may abandon a website after being dissatisfied with its content or behavior. The following is a list of common barriers to page usability:

- The page is difficult to find.
- The time from page request to initial feedback is too long or the initial feedback is unnoticeable, and therefore visitors might suspect that there are technical problems with the requested page.
- The page download time is too long. This is the most widely recognized barrier to a positive navigation experience (Nielsen 1994).
- The page download time is too short. This may happen for extremely fast pages that impose an extremely high mental load (Dabrowski and Munson 2001).
- The page does not indicate clearly when the download is completed. This is a problem in webpages with large amounts of content that download gradually.
- The page is irrelevant to the visitors' goal; the information the visitors need may not be found there.
- The page contains distracting animation, preventing users' focus on their search goals.
- The page is difficult to read or comprehend.
- The data formats in form filling impose unnecessary constrains, or they are not suited to the visitor's profile, and many visitors are unable to provide the required data.
- The rules for form filling are too restrictive, forcing extra work, such as selecting dates from long lists of numbers.

- The links on the page are not bold enough, and visitors often miss them.
- The label describing the hyperlink to this page is misleading; therefore, it is misinterpreted.
- The links may lead visitors to pages that are irrelevant to their needs.

The goal of usability assurance is to make sure that such barriers are removed.

7.3.4.4 Analysis of the Barriers to Seamless Navigation. Common design deficiencies are now listed according to the first phase—page evaluation—of the model above.

- The visitor might wait too long for the page to start downloading. A design weakness may occur when there is no indication that the new page is about to start downloading. The visitor often suspects that there might be a technical problem with this page.
- While downloading, the visitor might first see irrelevant information, banners, or just an indication that the page is downloading. Visitors often suspect that the page does not have the information they are looking for.
- The visitor might be confused if the page finishes downloading very rapidly, behaving like a local application. The visitor may be even more confused if, later, additional information is appended gradually, making the visitor wonder when to decide that the page does not have the needed information.
- The visitor might not see the desired information, even though it is displayed on screen. Or the visitor may see the information only after spending too much time reading all the text on screen.
- The visitor might misinterpret information which is irrelevant his or her needs, consequently proceeding to a wrong page.
- The visitor might conclude that the page is irrelevant to his of her needs, and the desired information might not be found at this site.

7.3.4.5 Attributes of Page Usability. In terms of the barriers to navigation described above, and invoking the GQM approach mentioned in Section 7.1, the attributes or goals of page usability can be derived by answers the following questions:

- *Accessibility*: How easy is it to find the page? Are the links to the page noticeable?
- *Responsiveness*: Does the page provide immediate feedback? Is this feedback noticeable?
- *Performance*: Does it meet the user's expectations? Is it long enough, so that the visitors notice it? Is it longer than the visitors may tolerate?
- *Download completion indication*: Can the visitors be sure that the download is completed? Can they tell when the download is not finished?
- *Relevance*: Is the page irrelevant to the visitors' goal?

- *Focus support*: Can the visitors focus on reading the page content and finding the links they need? Or does the page contain visual or sound elements that distract the visitors from their task?
- *Subjective readability*: Is the page easy to read and comprehend?
- *Compatibility of data formats*: Do the data formats in form filling suit the visitor's profile?
- *Ease of data entry*: Does the page provide support for easy data entry?
- *Link visibility*: Are the links on the page visible and noticeable?
- *Link labeling*: How faithfully do the labels describing the hyperlinks to this page represent the page content?
- *Link information*: Do the hyperlinks refer to other pages according to the descriptions and explanations?

7.3.5 Model 3—Bayesian Networks

Predictions and diagnostics rely on basic structures of cause an defect. Model 3 provides such a structure by invoking Bayesian networks (Ben Gal 2007; Kenett 2007). Bayesian networks (BNs), also known as *belief networks*, belong to the family of probabilistic *graphical models* (GMs). These graphical structures are used to represent knowledge about an uncertain domain. In particular, each node in the graph represents a random variable, while the edges between the nodes represent probabilistic dependencies among the corresponding random variables. These conditional dependencies in the graph are often estimated by using known statistical and computational methods. Hence, BNs combine principles from graph theory, probability theory, computer science, and statistics. BNs correspond to another GM structure known as a *directed acyclic graph* (DAG) that is popular in the statistics, machine learning, and artificial intelligence societies. BNs are both mathematically rigorous and intuitively understandable. They enable effective representation and computation of the joint probability distribution over a set of random variables (Pearl 2000).

The structure of a DAG is defined by two sets: the set of nodes (vertices) and the set of directed edges. The nodes represent random variables and are drawn as circles labeled by variables names. The edges represent direct dependence among the variables and are drawn by arrows between the nodes. In particular, an edge from node X_i to node X_j represents a statistical dependence between the corresponding variables. Thus, the arrow indicates that a value taken by variable X_j depends on the value taken by variable X_i or roughly speaking, that variable X_i "influences" X_j. Node X_i is then referred to as a *parent* of X_j and X_j is referred to as the *child* of X_i. An extension of these genealogical terms is often used to define the sets of *descendants*—the set of nodes that can be reached on a direct path from the node or *ancestors* nodes—the set of nodes from which the node can be reached on a direct path. The structure of the DAG guarantees that no node can be its own ancestor or its own descendant. Such a condition is of vital importance to the factorization of the joint probability of a collection of nodes, as seen below. Note that although the arrows represent a direct causal connection between the variables, the *reasoning process* can operate on a BN by propagating information in any direction.

A BN reflects a simple conditional independence statement, namely, that each variable is independent of its nondescendants in the graph given the state of its parents. This property is used to reduce, sometimes significantly, the number of parameters required to characterize the joint probability distribution (JPD) of the variables. This reduction provides an efficient way to compute the posterior probabilities given the evidence (Lauritzen et al. 1988; Pearl 2000; Jensen 2001).

In addition to the DAG structure, which is often considered the qualitative part of the model, one needs to specify the quantitative parameters. The parameters are described in a manner consistent with a Markovian property, where the conditional probability distribution (CPD) at each node depends only on its parents. For discrete random variables, this conditional probability is often represented by a table listing the local probability that a child node takes on each of the feasible values—for each combination of values of its parents. The joint distribution of a collection of variables can be determined uniquely by these local conditional probability tables (CPTs). Formally, a BN B is an annotated DAG that represents a JPD over a set of random variables $\mathbf{V}$. The network is defined by a pair $B = \langle G, \Theta \rangle$, where G is the DAG whose nodes $X_1, X_2, \ldots, X_n$ represent random variables and whose edges represent the direct dependencies between these variables. The graph G encodes independence assumptions, by which each variable X_i is independent of its nondescendants given its parents in G. The second component Θ denotes the set of parameters of the network. This set contains the parameter $\theta_{x_i|\pi_i} = P_B(x_i|\pi_i)$ for each realization x_i of X_i conditioned on π_i, the set of parents of X_i in G. Accordingly, B defines a unique JPD over V, namely:

$$P_B(X_1, X_2, \ldots, X_n) = \prod_{i=1}^{n} P_B(X_i|\pi_i) = \prod_{i=1}^{n} \theta_{X_i|\pi_i}.$$

For simplicity of representation we will omit the subscript B.

If X_i has no parents, its local probability distribution is said to be *unconditional*; otherwise, it is *conditional*. If the variable represented by a node is *observed*, then the node is said to be an *evidence node*; otherwise, the node is said to *be hidden* or *latent*. The complexity of a domain may be reduced by models and algorithms that describe an approximated reality. When variable interactions are too intricate for application of an analytic model, we may represent current knowledge about the problem, such as a cause generating at least one effect (Pearl 2000), where the final effect is the target of the analysis; for example, in Figure 7.3, the network topology (Lauritzen and Spiegelhalter 1988) of cause and effect is built by choosing a set of variables (e.g., "Visit Africa," "Smoking") that describe the domain (a patient presents some problems, and the physician wants to identify his or her disease and the correct therapy). The domain knowledge allows experts to draw an arc to a variable from each of its direct causes (i.e., visiting Africa may cause tuberculosis). Given a BN that specified the JPD in a factored form, one can evaluate all possible inference queries by marginalization, i.e., summing out over irrelevant variables.

Two types of inference support are often considered: *predictive support* for node X_i, based on evidence nodes connected to X_i through its parent nodes (called also

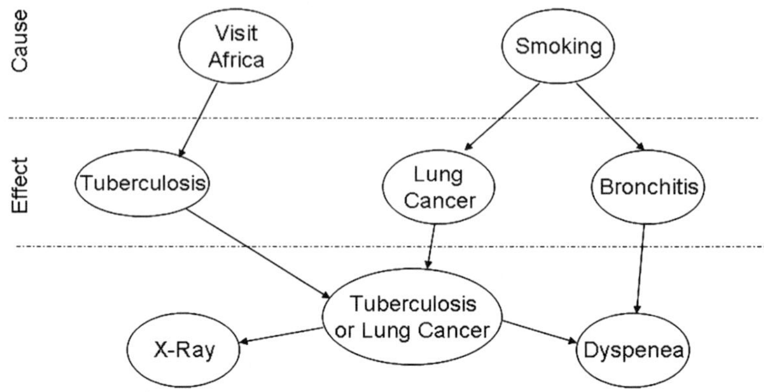

Figure 7.3 An example of a causal network (Lauritzen and Spiegelhalter 1988).

top-down reasoning), and *diagnostic support* for node X_i, based on evidence nodes connected to X_i through its children nodes (called also *bottom-up reasoning*). In general, the full summation (or integration) over discrete (continuous) variables is called *exact inference* and is known to be an NP-hard problem. Some efficient algorithms exist to solve the exact inference problem in restricted classes of networks. In many practical settings, the BN is unknown and one needs to learn it from the data. This problem is known as the *BN learning problem*, which can be stated informally as follows: Given training data and prior information (e.g., expert knowledge, causal relationships), estimate the graph topology (network structure) and the parameters of the JPD in the BN. Learning the BN structure is considered a harder problem than learning the BN parameters.

Moreover, another obstacle arises in situations of *partial observability* when nodes are hidden or when data is missing. In the simplest case of known BN structure and full observability, the goal of learning is to find the values of the BN parameters (in each CPD) that maximize the (log) likelihood of the training dataset. This dataset contains m cases that are often assumed to be independent. Given training dataset $\Sigma = \{\mathbf{x_1}, \ldots, \mathbf{x_m}\}$, where $\boldsymbol{x_l} = (x_{l1}, \ldots, x_{ln})^T$, and the parameter set $\theta = (\theta_1, \ldots, \theta_n)$, where θ_i, is the vector of parameters for the conditional distribution of variable X_i (represented by one node in the graph), the log likelihood of the training dataset is a sum of terms, one for each node:

$$\log L(\Theta|\Sigma) = \sum_m \sum_n \log P(x_{li}|\pi_i,\theta_i).$$

The log-likelihood scoring function *decomposes* according to the graph structure; hence, one can maximize the contribution to the log likelihood of each node independently. Another alternative is to assign a prior probability density function to each parameter vector and use the training data to compute the posterior parameter distribution and the Bayes estimates. To compensate for zero occurrences of some sequences in the training dataset, one can use appropriate (mixtures of) conjugate prior distributions, e.g., the Dirichlet prior for the multinomial case or the Wishart prior for the Gaussian case. Such an approach results in a maximum a posteriori estimate and is also known as the *equivalent sample size* (ESS) method.

BNs are gaining popularity in a wide range of application areas such as risk management (Cornalba et al. 2007) and management science in general (Godfrey et al. 2007; Kenett et al. 2008). Availability of software for analyzing BNs is further expanding their role in decision analysis and decision support systems (Jensen 2001; Bnlearn 2008; Genle 2006; Hugin 2007; SPSS 2008). BNs can be used to validate hypotheses about the relationships between usability design defects and the visitor's behavior in response to occurrences of usability barriers. This will be demonstrated in Section 7.5.

7.3.6 Model 4—Data Analysis in Usability Diagnostics

In e-commerce, we assume that site visitors are task driven, but we do not know if the visitors' goals are related to our website. Also, we have no way to tell if they know anything about the site, if they believe that the site is relevant to their goals, or if they have visited it before; it may be that the visitors are exploring the site or following a procedure to accomplish a task. Yet, their behaviors reflect their perceptions of the site's contents and their estimates of their effort in subsequent site investigation. The diagnosis of usability design deficiency involves certain measurements of site navigation, statistical calculations, and statistical decision. A model of statistical manipulations of server log data is now described which enables statistical decisions about possible usability design deficiencies.

7.3.6.1 Data Sources. Data analysis is possible only if the logs of the visitors' activities include at least three attributes:

- The IP address, used to distinguish between users.
- The page URL, used to specify the particular page.
- The timestamp, used to compute the time interval between hits.

7.3.6.2 Quantifying Website Usability Attributes. How can we tell whether the visitor encounters any difficulty in any of the evaluation stages? Server logs provide timestamps for all hits, including those of page html text files but also those of image files and scripts used for the page display. The timestamps of the additional files enable us to estimate three important time intervals:

- The time the visitors wait until the beginning of file download, used as a measure of page responsiveness.
- The download time, used as a measure of page performance.
- The time from download completion to the visitor's request for the next page, in which the visitor reads the page content but also does other things, some of them unrelated to the page content.

The challenge now is to decide whether the visitors feel comfortable with these time intervals. When do they feel that they wait too much, when is the download time too long or too short, and how do the visitors feel about what they see on screen?

7.3.6.3 Time Analysis of Mental Activities. How can we conclude that visitors consider a time interval acceptable, too short, or too long? For example, consider an average page download time of five seconds. Site visitors may regard it as too long if they expect the page to load rapidly, for example, in response to a search request. However, five seconds may be quite acceptable if the users' goal is to learn or explore specific information that they expect is related to their goal. Diagnostic-oriented time analysis observes the correlation between page download time and page exits. If the visitors care about the download time, then the page exit rate should depend on the page download time. When the page download time is acceptable, visitors may stay at the site, looking for additional information. When the download time is too long, more visitors might abandon the site and go to competitors. Longer download times imply higher exit rates; otherwise, if the visitors are indifferent about the download time, then the page exit rate should be invariant with respect to the page download time. We can now change our perspective on these variables and consider how the download time depends on the exit behavior. We compare the download time of successful visits with that of visits that ended in site exit. If the site visitors care about the download time, we should expect the average download time of those visitors who abandoned the site to be longer than the average of those who continued with site navigation. Otherwise, if the site visitors are indifferent about the download time, we should expect that the download time of the two groups will not be significantly different.

To determine the degree of the usability barrier, we need statistics. A reasonable statistic may be the correlation between download time and exit rate. To determine the significance of the usability barrier, we compare the download time of two samples: one of page views that ended in site exit and the other of all other page views. The null hypothesis is that the two samples are of the same population. If the null hypothesis is rejected, we may conclude that the page visitors' behavior depends on the download time: If the download time of the first sample exceeds that of the second sample and the difference is statistically significant, then we may conclude that the download time of the particular page is significantly too long. A simple two-tailed T-test may be sufficient to decide whether the visitors' behavior is sensitive to the page download time.

7.3.6.4 Analysis of the Visitor's Response Time. In the time interval between the page display on screen and the visitor's action to get the next page, the visitor is busy doing different things. Some of them are mental activities related to the visitor's task, and others are unrelated to the visitor's task (e.g., having a coffee break). It would useful to estimate the task-related and idle (those unrelated to the task) parts of the response time. The problem is that there is no way to distinguish between the two types of activities by manipulations of server logs. However, it makes sense to assume that the variance of the task-related activities is much smaller than that of the idle activities. We can diminish the effect of the idle activities by statistics, such as harmonic average, that weigh more on short time intervals and less on long time intervals.

7.3.6.5 Methodology for Usability Diagnostics. The goals of usability diagnostics are to identify, for each site page, all the design deficiencies that hamper the positive navigation experience at each evaluation stage. To understand the user's experience, we need to know the user's activity compared to the user's expectation. Neither one is available from the server log file, but they can be estimated by appropriate processing. The methodology here involves integration of two types of information:

- Design deficiencies, which are common barriers to seamless navigation, based on the first part of the model described above—visitor's page evaluation.
- Detectors of these design deficiencies, common indicators of possible barriers to seamless navigation, based on the second part of the model—visitor's reaction.

7.3.6.6 Usability Problem Indicators. The way to conclude that the download time of a particular page is too long is by a measure of potentially negative user experience, namely, the site exit rate. Exit rate is a preferred measure for deciding on the effect of download time, but it is irrelevant to the analysis of the visitors' response time. The reason for this is that server logs do not record the event of the visitor's leaving the site, so we cannot measure the time intervals of terminal page views. Therefore, we need to use other indicators of a potentially negative navigation experience. Model 2, the user's mental activities in website navigation described above, listed the most likely visitors' reactions to exceptional situations. Based on this model, we can list the following indicators of the visitor's tolerance of webpage design deficiencies:

- The visitor returning to the previous page may indicate that the current page was perceived as less relevant to the goal than the previous page.
- The visitor linking to the next website page may indicate that the link was perceived as a potential bridge to a goal page.
- The visitor activating the main menu may indicate that he or she is still looking for the information after failing to find it in the current page.
- The visitor exiting the site may indicate that either the goal has been reached or the overall site navigation experience became negative.

So, besides site exit, other indicators of potential usability problem are the rates of navigation back to a previous page and escape from the page to the main menu. These events, as well as the event of site exit, are called here *usability problem indicators* (UPIs).

7.3.6.7 Time Analysis of the Task-Related Mental Activities. Once we have an estimate of the task-related mental activities and a UPI suited to nonterminal activities, we can adapt the method for analyzing problematic performance to analyzing problematic task-related mental activities. How can we conclude that the response time interval is proper for the page, too short, or too long? For example, consider an average page reading time of 20 seconds. Is it too long or too short? Obviously, the reading time of long pages should be longer than that of short pages, so we should not expect absolute measures to be valuable in scoring the value of the reading time. Visitors who feel that the information is irrelevant to their needs are more likely to respond quickly, go backward, or select a new page from the main menu. Therefore, the average time of a page for visitors who navigated backward or retried the main menu should be shorter than that of the average of all visitors. On the other hand, visitors who believe that the information is relevant to their needs, but do not easily understand the page text, are likely to spend more time than average reading the page content, and the average time on the page should be longer. The following assumptions are used for the diagnosis:

- Visitors satisfied with the page display are less likely to exit the site, to return to a previous page, or to escape to the main menu than visitors not satisfied with the page display.
- The exception to the first assumption are terminal pages, at which the users' goals are achieved.

The time that visitors spend reading a page depends on various perceptual attributes, including the relevance of the page to their goals, the ease of reading and comprehending the information on the page, the ease of identifying desired hyperlinks, etc. The method described above for download time analysis may be applicable to these perceptual attributes, provided that we know how usability barriers should affect the visitor's response time. The analysis is based on the following assumptions:

- Task-driven visitors are sensitive to readability deficiencies: Both the response time and the rate of UPIs for pages that are easy to read should be lower than those for pages that are difficult to read. Casual visitors, on the other hand, are less sensitive to readability deficiencies.
- Task-driven visitors are sensitive to the relevance of the information on the page to their needs, but in a different way: The UPI rate of relevant pages should be lower than that of irrelevant pages, but the response time of relevant pages should be higher than that of irrelevant pages Casual visitors, on the other hand, are less sensitive to relevance deficiencies.

7.3.6.8 Interpreting the Time Analysis of Page Readability. The time that visitors spend reading a page depends on various perceptual attributes, including the relevance of the page to their goals, the ease of reading and comprehending the information on the page, the ease of identifying desired hyperlinks, etc. Assume that for a particular page, the average time on screen before backward navigation is significantly longer than the average time over all page hits. Such a case may indicate a readability problem due to design flaws, but it can also be due to good page content, which encouraged users who spent a long time reading the page to go back and reexamine the previous page. It would be nice if we could distinguish between reading the page and other task-related mental activities, such as evaluating a product on sale, comparing prices, etc. Unfortunately, there is no direct way to measure the time it takes for the visitor to accomplish each of the mental activities. Therefore, the convention here is that a problem identified in page readability could be attributed to the other task-related mental activities. This is yet another example of the limits of artificial intelligence and of our need to rely on the human intelligence.

7.3.6.9 Interpreting the Time Analysis of Page Relevance to the Visitors' Needs. Task-driven visitors are likely to respond quickly by a invoking a UPI if they perceive irrelevant information that threatens to lose their focus. Possible sources of such perception are inappropriate content, such as banners distracting visitors from their original tasks; inappropriate layout design; and misleading links to the page due to unethical marketing such as by portals, to poor explanations about the links or to labeling mistakes. The psychological explanation for such perceptions is the limited capacity of human working memory. Too many mental resources required for processing the information on a page might compete with the need to mentally maintain the visitor's goal. Consequently, task-driven users will always prefer navigating pages that do not require too much mental processing. Another possible reason for quick activation of a UPI is that the page is well designed and the page designers wanted visitors who are finished reading the page to navigate backward or try an item from the main menu. Apparently, the latter reason is hypothetical; the main reason for quick UPI activation is poor design, which forces the visitor to take measures in order to stay in focus.

7.3.6.10 Analysis of Misleading Links. One cause of visitors' perception of the page as irrelevant is when a link from a portal or from another page is misleading, either intentionally or by a design mistake. Once a page is identified as being perceived as irrelevant to visitors, the site administrator may want to know why and what links are responsible for the visitors' negative attitude. The problematic links can be found by time analysis similar to that of page response time. This is achieved by sampling the transitions from all pages to the tested page and by comparing the response time of the transitions that were followed by a UPI with that of the other transitions.

7.4 IMPLEMENTATION FRAMEWORK

The implementation framework in this section is based on the structure of reports derived by WebTester, a patented software tool developed by ErgoLight since 1999 to analyze server log files and generate reports about potential usability deficiencies in websites (see www.ergolight-sw.com). This tool demonstrates the feasibility of the method presented in Section 7.2. The usability data extracted using this tool are used in the next section to demonstrate the kinds of diagnostics the method can provide.

7.4.1 WebTester Reports

The reports that WebTester generates include site-level scoring, usage statistics, and diagnostic information as follows:

Page Diagnostics

- Pages that are abandoned due to long download time
- Pages that are difficult to comprehend
- Pages that are not interesting or irrelevant
- Pages that distract potential customers from their goals

Link Diagnostics

- Links that are likely to be misinterpreted
- Links that are likely to be disregarded
- Links that do not behave according to visitors' expectations
- Links that connect to the wrong page

Page Statistics

- Number of page views (view count)
- Number and percentage of site entries from the pages
- Number and percentage of site exits from the pages
- Average download time
- Average reading time
- Average search time for links to these pages

7.4.2 Data Processing

7.4.2.1 Preliminary Activity—Analysis of the Structure of Server Log Files. Server log files consist of records of optional fields. WebTester analyzes the record format and identifies most of the fields automatically according to the two main format conventions (UNIX and MS). Few fields, such as the Size field, are of common integer format and cannot be identified automatically. The operator's control is required to select the proper field and to confirm the automatic field identification.

7.4.2.2 The Lowest Layer—User Activity. WebTester processes the server logs to obtain compressed logs of user activity, in which the records correspond to significant user actions (involving screen changes or server-side processing). The data processing includes algorithms for:

- Visitor identification, including support for variable IP addresses.
- Filtering out traces of robot visits. Robots are identified by the Referrer field, by the User Agent field, or by certain characteristics of the robot sessions.

- Associating embedded objects (images, JavaScripts, etc.) with the page containing them.
- Associating pages with frames by time proximity.

7.4.2.3 The Second Layer—Page Hit Attributes. WebTester computes estimates of record-level usage attributes:

- Page size.
- Page download time.
- Page introduction time, namely, the download time of the first visit in a navigation session.
- Page processing time, calculated by subtracting the statistics of download time of elementary objects (images, JavaScripts, etc.) from the download time.
- User response time (from end-of-page download to beginning of the download of the next page).

The page response time, which may enable diagnostics of lack of or extreme system response time, is not available in server logs, and it was not included in standard usability reports.

7.4.2.4 The Third Layer—Transition Analysis. Using the statistics for page transitions WebTester identifies:

- The site main pages (those accessible through main menus).
- Repeated form submission (indicative of visitors' difficulties in form filling).

7.4.2.5 The Fourth Layer—UPI Identification. Marking certain visits as indicators of possible navigational difficulty:

- Estimates for site exit by the time elapsed until the next user action (as no exit indication is recorded on the server log file).
- Backward navigation.
- Transitions to main pages, interpreted as escaping the current subtask.

7.4.2.6 The Fifth Layer—Usage Statistics. Obtaining statistics of hit attributes over pages, such as:

- Average page size.
- Average introduction time.
- Average download time.
- Average time between repeated form submission.
- Average time on the screen (indicating content-related behavior).
- Average time on a previous screen (indicating ease of link finding).

WebTester computes the statistics over all page views and also over those views that indicate possible difficulties. The average time computations are by the harmonic average, with the intention to weigh out fluctuations due to technical problems and visitors' idle time. A typical e-commerce URI consists of a common page (html, asp, etc.) and parameters defining the deliverables. Typically, the parameters are arranged according to the hierarchy of the deliverables, namely, categories of products or services. In practice, the samples for full URIs are too small to enable significant results. Also, it is desired that the diagnostics are provided for all levels along the URI parameters to support all levels of the categorization. The statistics in WebTester are calculated for all categories, of all levels of detail.

7.4.2.7 The Sixth Layer—Statistical Decision. For each of the page attributes, WebTester compares the statistics over the exceptional page views to those over all the page views. The null hypothesis is that (for each attribute) the statistics of both samples are the same. A simple two-tailed T-test was used to reject it and therefore to conclude that certain page attributes are potentially problematic. The error level was set to 5%.

7.4.2.8 The Top Layer—Interpretation. For each of the page attributes, WebTester provides a list of possible reasons for the difference between the statistics over the exceptional navigation patterns and those over all the page hits. However, the usability analyst must decide which of the potential source of visitors' difficulties is applicable to the particular deficiency.

7.5 CASE STUDIES OF WEBSITE USABILITY DATA ANALYSIS

In this section, we present case studies of Web-based usability data analysis using models 1–4 described above. The data used in this section was obtained by applying ErgoLight WebTester to server logs of the website of www.israeliz.com. The reports generated by applying the tool are available at the ErgoLight website, in the section of report examples: http://www.ergolight-sw.com/pub/Sites/Israeliz/All/First/Reports/wtFrameSet.html. Not all visitors' actions are recorded on the server. Yet, the most significant user actions are the page views, that result in screen changes and in form submission, which have sufficient traces in the server log file. We first present a case study illustrating the application of Markov chain to web statistics and then focus on an implementation of BNs.

7.5.1 Markov Chain Applied to Web Usability Statistics

The apparent UPIs include

- Backward navigation.
- Site exit.
- Escape to the main menu.

A scheme for page event sequences is: The user links to Page -> Page starts download -> Page download complete -> User recognizes page -> User reads page -> User activates a link, about to leave the page. Table 7.1 presents the measured variables and their definitions in the dataset we have analyzed. In that table, key variables are highlighted. The time variables used for the statistics are variables that apparently have an impact on the visit experience. They are:

- The page download time, measured from the first page view until the end of subsequent embedded files (images, JavaScripts, etc.).
- The elapsed time until the next user event, interpreted as reading time.
- The elapsed time since the previous user event, interpreted as the time until the user found the link to this page.

TABLE 7.1 Variables in Case Study (Key Variables Highlighted)

ID	Variable	Definition
1	ActionCount	"Count of page visits as recorded in the log file, including actual visits and faked visits (refreshing and form submission)"
2	NuOfRealVisits	"The actual visits, after removing the faked visits."
3	**TextSize**	Size of html page
4	**PageSize**	TextSize + size of subsequent images recorded individually
5	NuOfEntries	Count of site entries through this page
6	NuOfPreExits	Count of site exit through next page
7	NuOfExits	Count of site exit through this page
8	NuOfPreBacks	Count of backward navigations from this page
9	NuOfBacks	Count of backward navigations to this page
10	NuOfDownloads	Count of page visits with positive download time
11	NuOfGetToPost	"Count of transitions from Get (typically, viewing) to Post (typically, form submission)"
12	TotalDownloadFrequency	Sum of (1/DownloadTime) over all real page visits
13	TotalReadingFrequency	Sum of (1/ReadingTime) over all real page visits
14	TotalSeekingFrequency	Sum of (1/SeekingTime) over all real page visits
15	TotalProcessingFrequency	Sum of (1/ProcessingTime) over all real page visits
16	TotalGetToPostFrequency	Sum of (1/FormFillingTime) over all real page visits
17	AvgDownloadFrequency	Average = Total/Count
18	AvgReadingFrequency	Average = Total/Count
19	AvgSeekingFrequency	Average = Total/Count
20	AvgGetToPostFrequency	Average = Total/Count
21	AvgPostToPostFrequency	Average = Total/Count
22	**AvgDownloadTime**	Harmonic average of page download time over all real page visits
23	**AvgReadingTime**	Harmonic average of page reading time over all real page visits

(*Continued*)

TABLE 7.1 *Continued*

ID	Variable	Definition
24	**AvgSeekingTime**	Harmonic average of page seeking time over all real page visits
25	AvgProcessingTime	Harmonic average of page processing time over all real page visits
26	AvgResponseTime	Harmonic average of page response time over all real page visits
27	PercentExitsSlowDownload	"Number of page visits characterized as slow downloads, followed by site exits/NuOfRealVisits * 100"
28	PercentPbackSlowDownload	"Number of page visits characterized as slow downloads, followed by backward navigation/ NuOfRealVisits * 100"
29	PercentExitsSlowSeeking	"Number of page visits characterized as slow seeking, followed by site exits/NuOfRealVisits * 100"
30	PercentExitsFastSeeking	"Number of page visits characterized as fast seeking, followed by site exits/NuOfRealVisits * 100"
31	PercentPbackSlowSeeking	"Number of page visits characterized as slow seeking, followed by backward navigation/NuOfRealVisits * 100"
32	PercentPbackFastSeeking	"Number of page visits characterized as fast seeking, followed by backward navigation/NuOfRealVisits * 100"
33	PercentPexitSlowReading	"Number of page visits characterized as slow reading, followed by site exit from next page/ NuOfRealVisits * 100"
34	PercentPexitFastReading	"Number of page visits characterized as fast reading, followed by site exit from next page/ NuOfRealVisits * 100"
35	PercentPbackSlowReading	"Number of page visits characterized as slow reading, followed by backward navigation/NuOfRealVisits * 100"
36	PercentPbackFastReading	"Number of page visits characterized as fast reading, followed by backward navigation/NuOfRealVisits * 100"
37	**UsabilityScore**	"Normalized harmonic average of: download time, seeking time, reading time"
38	**UsabilityAlert**	Sum of all UPIs (6–11 above)
39	**AccessibilityScore**	A measure of the speed of finding the links to the current page
40	**PerformanceScore**	A measure of the download speed
41	**ReadabilityScore**	A measure of the reading speed (characters per second)

The page visitor experience is regarded as negative or positive according to whether or not a UPI was indicated afterward. Table 7.2 is a summary table relating content and reading time to visit experience. The expected relationships between the key variables are as follows:

- Page size should directly affect the page download time.
- The amount of text on the page should directly affect the page reading time.
- The ratio TextSize/PageSize may be equal to 1 if the page is pure html (has no graphics, video, etc.) or is very small (since images consume much more space than text).
- Page download time should affect the visit experience. Too long a download time should result in a negative visit experience. Research indicates that too short a download time might also result in a negative visit experience; however, this might be a problem only for extremely good service (server performance).
- Page-seeking time may affect the visit experience. Too much seeking time might have a "last straw" effect if the page is not what visitors expected to find.
- Page reading time may affect the visit experience. Users may stay on the page because it includes a lot of data (in this case study, the data are textual) or because the text they read is not easy to comprehend.

The descriptive statistics of key variables from 122 links are presented in Table 7.3.

Figure 7.4 presents a cluster analysis of the key variables using MINITAB™ version 14 (www.minitab.com). Processing time and response time are naturally closely related. The three score functions form a separate cluster indicating internal consistency.

Figure 7.5 presents histograms of TextSize, PageSize, UsabilityScore, and UsabilityAlert.

We apply the methods about Markov processes illustrated in Section 7.1 to the data at hand. For privacy protection reasons, we modified the names of the webpages. We illustrate the methods when concentrating on one page and analyzing the moves out of it. We consider the case of the webpage *content/man8.htm* and the other ones accessible from it. We identify them as *content/man1.htm*, *content/man6.htm*, *content/man7.htm*, *tool/cont200.htm*, and *tool/cont202.htm*, and we add the state corresponding to the exit from the website. The transition matrix of the Markov chain has a very high dimension since the website has many pages. We concentrate on the estimation of pages accessible

TABLE 7.2 The Effect of Content and Reading Time on Usability

Property of Page Content	Average Reading Time	Visit Experience
Reasonable text size	Short	Positive
Reasonable text size	High	Negative
Large text size	High	Positive
Animation	Short	Negative

TABLE 7.3 Descriptive Statistics of Key Variables in the Case Study Data Set

Variable	Mean	StDev	CoefVar	Minimum	Median	Maximum
TextSize	20109	15817	78.66	0.000000000	17941	72359
PageSize	79577	259545	326.16	0.000000000	40273	2352613
PercentExitLoss	14.25	15.95	111.99	0.000000000	10.00	100.00
AccessibilitySco	15.50	17.67	113.97	0.0178	8.74	76.70
PerformanceScore	46.46	25.13	54.10	3.53	41.70	100.00
ReadabilityScore	2780	4746	170.71	0.000000000	1409	43135
UsabilityScore	73.04	27.15	37.17	14.22	69.61	139.40
UsabilityAlert	58.83	39.54	67.21	0.000000000	52.50	222.00
AvgDownloadTime	2.164	4.243	196.09	0.000000000	1.000	27.000
AvgReadingTime	153	1141	746.28	0.000000000	10.0	12288
AvgSeekingTime	62.2	508.6	818.12	0.000000000	10.0	5624.0

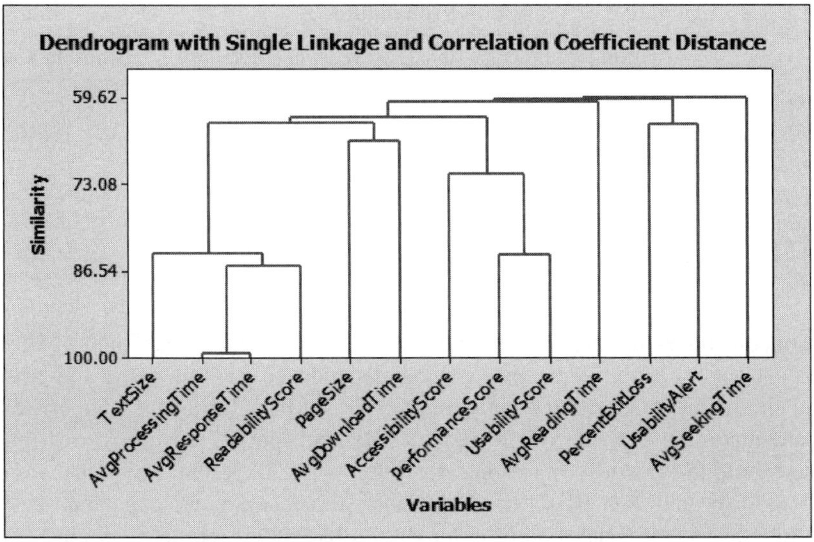

Figure 7.4 Dendrogram from cluster analysis of variables in Table 7.3.

only from *content/man8.htm* and consider a Dirichlet prior with coefficients equal to 1 for the probabilities of transition to accessible states and 0 for the others. This assumption allows only for transitions to the pages linked by *content/man8.htm* and does not take into account a possible move to another page of the website just by typing its address. In this context, we observed 16 transitions to six states. Table 7.4 gives the data and both frequentist and Bayesian estimates.

Both types' estimates have positive and negative properties. The frequentist estimate does not exploit available information, if any, but gives a unique result, not

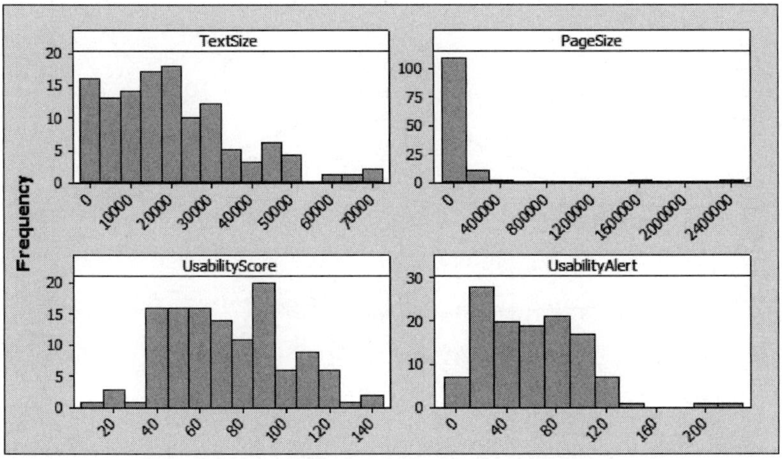

Figure 7.5 Histograms of TextSize, PageSize, UsabilityScore, and UsabilityAlert.

TABLE 7.4 Estimation of Transition Probabilities

Page	Number of Transitions	Frequentist Estimate	Bayesian Estimate
content/man1.htm	2	0.125	0.136
content/man6.htm	2	0.125	0.136
content/man7.htm	4	0.250	0.228
tool/cont200.htm	2	0.125	0.136
tool/cont202.htm	2	0.125	0.136
outside	4	0.250	0.228

questionable once this approach is taken. The Bayesian estimate incorporates the expert's opinion at the price of a strong dependence on this prior and possible wrong elicitation. A sensitivity study, in the spirit of Rios Insua and Ruggeri (2000), can attenuate the impact of a wrong statement about the prior distribution. Furthermore, it is worth observing that the prior Bayesian estimates were all equal to $1/6$ and the posterior Bayesian estimates are, as expected, between the prior Bayesian estimates and the maximum likelihood estimates. Both estimates indicate that transition probabilities to the outside and to *content/man7.htm* are highest.

7.5.2 BNs Applied to the Web Usability Statistics

The data were then reanalyzed using the basic BN presented in Figure 7.6. The network combines background information with a learned network generated using the GeNIe version 2.0 software (http://genie.sis.pitt.edu).

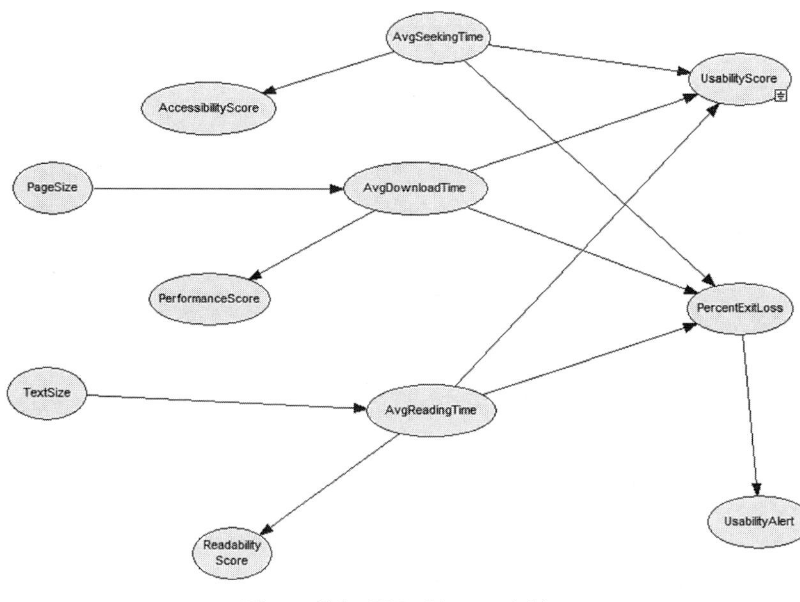

Figure 7.6 BN of key variables.

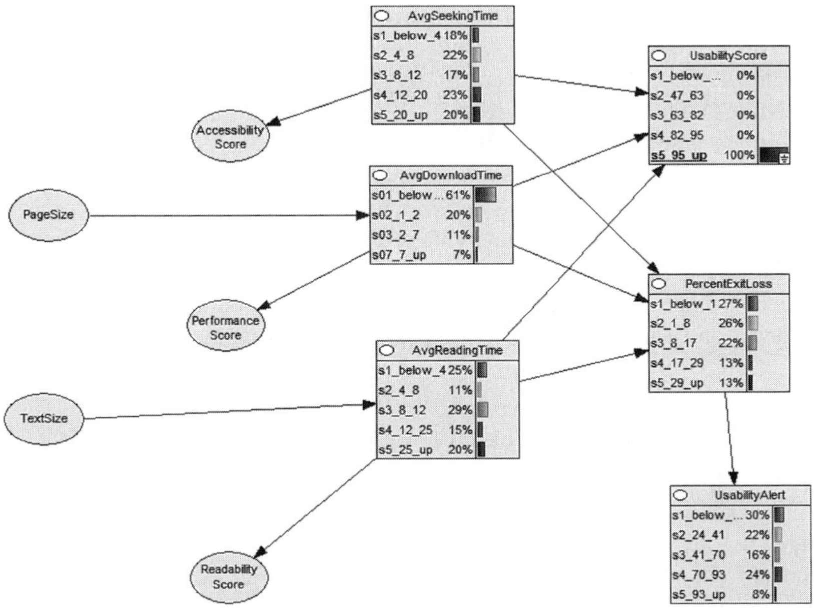

Figure 7.7 Diagnostic distributions conditioned on the UsabilityScore being at its highest level.

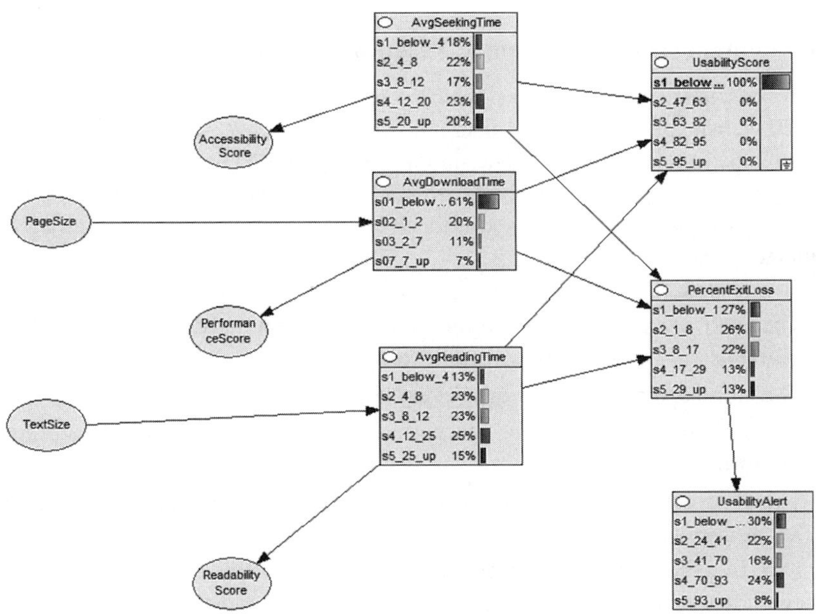

Figure 7.8 Diagnostic distributions conditioned on the UsabilityScore being at its lowest level.

One the basis of the network, we can perform various diagnostic checks. Figure 7.7 presents distributions of various variables conditioned on the UserScore being at its its highest level. Note that the percent of AvgSeekingTime, AvgDownloadTime, and AvgReadingTime at the lowest levels is 18%, 61%, **25%**, and respectively. We can now perform similar diagnostics by conditioning the UsabilityScore to be at its lowest level. Figure 7.8 provides such an analysis.

Note that the percent of AvgSeekingTime, AvgDownloadTime, and AvgReadingTime at the lowest levels is 18%, 61%, and **13%**, respectively. AvgReadingTime has dropped from 25% to 13%. The other variables have not changed. AvgReadingTime is therefore the differentiating factor between high and low values of UsabiliScore. Moreover, conditioning on UsabilityScore has no impact on UsabilityAlert.

7.6 CONCLUSIONS AND AREAS FOR FUTURE RESEARCH

This chapter has presented a method for identifying design deficiencies that result in degraded usability based on web usage data. The method is based on the integration of models from statistics with models of visitors' behavior. It was implemented in a report-generating tool, WebTester. We also introduced several modeling approaches and discussed case studies using Markov chains and BNs. Further analysis linked UPIs to conversion rates. A general approach for such an analysis involves proactive interventions and expert panels.

7.6.1 Limitations of the Example Diagnostic Reports

7.6.1.1 Standard Log Files Are Insufficient. WebTester had produced reports for about 20 websites; five of them concerned small businesses. All of these reports were based on data in server log files. The sample data presented here show the potential of applying this method. Yet, the results are not very impressive due to shortcomings of the source data. Server logs are used extensively in the industry because of their many advantages:

- The web server normally produces log files routinely, so the raw data are available for the analysis with no extra efforts.
- The web server reliably records every transaction it makes.
- The data are on the company's own servers, enabling the company to control the file format.
- The data are in a standard format, ensuring compatibility with commercial log analyzers.
- Log files contain information on visits from search engine spiders, which enables search engine optimization.
- Log files contain information on client requests that the server failed to execute.

However, server logs are ineffective for usability diagnostics. The main limitation is that server logs do not have indications about all necessary client activity. The most important problems are:

- Lack of indication of page reload from the cache, which is essential for employing backward navigation required for subjective readability and relevance analysis. If a person revisits a page, the second request will often be retrieved from the browser's cache, so no request will be received by the web server. This means that the person's path through the site is lost. Caching can be defeated by configuring the web server, but this can result in degraded performance for the visitor to the website.
- Missing records about the visitors' interaction with order forms. The server log does not include records of the visitor's clicks, only of the GET and PUT requests from the client to the server. The problem is that most form-filling activities are processed locally, without notifying the server about them, which means that they are not recorded on the server logs. This limitation is quite significant, because major usability barriers involve the visitors' difficulties in following the flow of data entry and editing and recalling the proper data formats.
- Missing records of pages loaded by third parties, such as gateways (web proxies).
- A convenient way to work around this problem is to install Java script tags in the page views so that the client will deliver a request to the server about any significant visitor event, such as data entry in form filling. This technique is used extensively in e-commerce sites. The cache problem is solved with no extra effort, because the request is sent to the server on each page view. Another limitation of server logs is the high level of noisy data. Main sources of the noisy data are:
- An IP address shared by different people, such as members of the same household. This problem is negligible in most practical cases, as the likelihood that two or more people who share the same IP address will visit the same website simultaneously is very low. Also, visitor ambiguity may be reduced by identifying the visitors not only by the IP address, but also by the User Agent (the browser used for the navigation). This method was implemented in WebTester.
- Distributed IP addressing in a multiserver operational environment, in which the client request arrives through different routing from the client browser. In e-commerce this may be a problem in b2b applications, in which the client is an enterprise operated through a complex system. A distributed IP address system has a common part and a variable part. Records from distributed IP addresses can be aggregated to the proper session by the common part of the IP address. This method was implemented in WebTester.
- Spider visits (by search engine crawlers) are recorded in the server log files. These visits can be detected and filtered out by special characteristics of the

spiders, including a special User Agent (browser identification), known referrer, and certain navigation patterns (e.g., extremely fast access to the server). These methods were implemented in WebTester.

- Obviously, the effect of noise can be reduced by enlarging the sample size. In WebTester, the size of log files used was 10–20 megabytes, and apparently the sample size for main pages was sufficient. To enable support of this capability, a new log analyzer should be developed that will use a proprietary clickstream logger, such as by Java script tagging.

7.6.1.2 Learning Effect. Johnson et al. (2000) confirmed that the power law of practice used in cognitive science also applies to website navigation, implying that users spend less time per session the more they visit the site. However, another study shows that this is not a real limitation. Bucklin and Sismeiro (2001) found that repeat visits by a user lead to fewer pages viewed per session but to no change in average page view duration. This means that the power law of practice applies to learning of the navigation path, but not to the usability attributes of subjective readability and relevance to the visitors' needs.

7.6.2 Further Analysis

7.6.2.1 Proactive Interventions. Proactive interventions such as design changes are, de facto, empirical experiments. The statistical methodology for the design of experiments involves linking variables and parameters using eight steps explained in Table 7.5 that identify factors, specify their levels, and, using appropriate combinations (experimental runs), determine an experimental array. Responses are the measured target variables of an experiment (Kenett and Zacks 1998).

7.6.2.2 Expert Panel. In order to determine the percentage of usability design flaws that can be identified based on log files, one can use an expert panel. In

TABLE 7.5 Designed Experiment Checklist

Design Concern	Design Activity
1. Problem Definition	Plain language description of the problem and how it was identified
2. Response Variables	What will be measured and how the data will be collected
3. Factors	What factors can be controlled by design decisions
4. Factor Levels	What are the current factor levels and what are reasonable alternatives
5. Experimental Array	What are the factor level combinations to be tried and in what order
6. Number of Replications	How long will the experiment run
7. Data Analysis	How the data will be analyzed, with what software
8. Budget and Project Control	Who is in charge of the project, what is the timetable, and what resources are needed

running such a panel, one first lets usability experts analyze specific site usability heuristically (few experts are required for a site, as demonstrated in prior research). Their evaluation can be, for example, recorded using a specially designed questionnaire. Second, one compares the effect of usability problems identified though web surfing usage data with the expert opinions. As an example, consider the dilemma of banners and online advertising. Clearly, advertisement on websites hampers usability. If we want to determine the effect on conversion rates, we can run an experiment with and without ads. This is a simple experiment with one factor at two levels. Measurement of conversion rates and UPIs will help us determine if we have reached the right trade-off between advertisement design and usability. Finally, in considering usability, the temporal dimension should also be analyzed. Usability can be affected by changes in performances over time. In tracking time, many approaches based on sequential statistical analysis can be implemented, such as statistical process control or more advanced procedures such as Shiryayev-Roberts optimal detection (Kenett and Pollak 1986; Kenett and Zacks 2003).

REFERENCES

Basili, V. and Weiss, D. (1984) A methodology for collecting valid software engineering data. *IEEE Transactions on Software Engineering*, 728–738.

Battey, J. (1999). IBM's redesign results in a kinder, simpler web site. Retrieved October 10, 2001, from http://www.infoworld.com/cgi-bin/displayStat.pl?/pageone/opinions/hotsites/hotextr990419.htm.

Ben Gal, I. (2007). Bayesian networks. In *Encyclopaedia of Statistics in Quality and Reliability* (F. Ruggeri, R.S. Kenett, and F. Faltin, editors in chief), Wiley, Chichester, UK.

Bnlearn Package (2008) Available at http://cran.dsmirror.nl/web/packages/bnlearn/index.html.

Bucklin, R.E., Lattin, J.M., Ansari, A., Bell, D., Coupey, E., Gupta, S., Little, J.D.C, Mela, C., Montgomery, A., and Steckel, J. (2002). Choice and the Internet: From clickstream to research Stream. *Marketing Letters*, 13(3): 245–258.

Bucklin, R.E. and Sismeiro, C. (2001). A model of web site browsing behavior estimated on clickstream data. Working Paper, Anderson School at UCLA.

Cornalba, C., Kenett, R., and Giudici, P. (2004). Sensitivity analysis of Bayesian networks with stochastic emulators. *Proceedings of the ENBIS-DEINDE Conference*, University of Torino, Turin, Italy.

Dabrowski, J.R. and Munson, E.V. (2001). Is 100 milliseconds too fast? *Conference on Human Factors in Computing Systems*, Seattle, Washington.

Di Scala, L., La Rocca, L., and Consonni, G. (2004). A Bayesian hierarchical model for the evaluation of web sites. *Journal of Applied Statistics*, 31(1): 15–27.

Donahue, G.M. (2001). Usability and the bottom line. *Software IEEE*, 18(1): 31–37.

Duma, J.S. and Redish, J.C. (1999). *A Practical Guide to Usability Testing*. Intellect Ltd, Bristol, UK.

Dustin, E., Rashka, J., and McDiarmid, D., (2001). *Quality Web Systems: Performance, Security, and Usability*. Addison-Wesley, Reading, MA.

GeNle (2006) Decision Systems Laboratory, University of Pittsburgh, USA. Available at http://genie.sis.pitt.edu.

Godfrey, A.B. and Kenett, R.S., (2007). Joseph M. Juran, a perspective on past contributions and future impact. *Quality and Reliability International*, 23: 653–663.

GVU (1997) GVU's 8th WWW User Survey. Available at http://www.gvu.gatech.edu/user_surveys/survey-1997-10.

Hugin Decision Engine (2007) Hugin Expert, Denmark. Available at http://www.hugin.com.

International Organization for Standardization (1998) ISO 9241-11, *Guidance on Usability*, Geneva, Switzerland.

Jensen, F.V. (2001). *Bayesian Networks and Decision Graphs*. New York: Springer.

Johnson, E.J., Bellman, S., and Lohse, G.L. (2000). what makes a web site "Sticky"? Cognitive lock in and the power law of practice. Working Paper, Graduate School of Business, Columbia University.

Karlin, S. and Taylor, H.M. (1975). *A First Course in Stochastic Processes* (2nd ed.). New York: Academic Press.

Karlin, S. and Taylor, H.M. (1981). *A Second Course in Stochastic Processes*. New York: Academic Press.

Kenett, R. (1996). Software specification metrics: A quantitative approach to assess the quality of documents. *Proceedings of the IEEE Convention of Electrical and Electronics Engineers in Israel*, Tel Aviv, Israel.

Kenett, R.S. (2007). Cause and effect diagrams. In *Encyclopaedia of Statistics in Quality and Reliability* (F. Ruggeri, R.S. Kenett, and F. Faltin, editors in chief), Wiley, Chichester, UK.

Kenett, R. and Baker, E. (1999). *Software Process Quality: Management and Control*. New York: Marcel Dekker.

Kenett, R., De Frenne, A., Tort-Martorell, X., and McCollin, C. (2008). The statistical efficiency conjecture. In *Applying Statistical Methods in Business and Industry—the State of the Art* (S. Coleman T. Greenfield, and D. Montgomery, eds.). Wiley, pp. 61–96.

Kenett, R. and Pollak, M. (1986). A semi-parametric approach to testing for reliability growth with an application to software systems. *IEEE Transactions on Reliability*, R-35(3): 304–311.

Kenett, R. and Zacks, S. (2003). *Modern Industrial Statistics: Design and Control of Quality and Reliability* (2nd ed.). San Francisco: Duxbury Press.

Kohavi, R. (2006). Focus the mining beacon: Lessons and challenges from the world of e-commerce. *San Francisco Bay ACM Data Mining SIG*, June 13.

Lauritzen, S.L. and Spiegelhalter, D.J. (1988). Local computations with probabilities on graphical structures and their application to expert systems. *Journal of the Royal Statistical Society, Series B (Methodological)*, 50(2): 157–224.

Marcus, A. (2002). Return on Investment for usable UI design. *User Experience Magazine*, pp. 25–31.

Moe, W.W. (2006a). An empirical two-stage choice model with varying decision rules applied to Internet clickstream data. *Journal of Marketing Research*, pp. 680–692.

Moe, W.W. (2006b). A field experiment assessing the interruption effect of pop-up promotions. *Journal of Interactive Marketing*, 20(1): 34–44.

Montgomery, A.L., Kannan Srinivasan, S.L., and Liechty, J.C. (2004). Modeling online browsing and path analysis using clickstream data. *Marketing Science*, GSIA Working Paper #2003-E26.

Nielsen, J. (1994). Response times: The three important limits. In *Usability Engineering*. San Francisco: Morgan Kaufmann, San Francisco, CA.

Nielsen, J. (1999). Why people shop on the Web Nielsen alert box. Available at http://www.useit.com/alertbox/990207.html.

Nielsen, J. and Loranger, H. (2006). *Prioritizing Web Usability*. New Riders Press, Indianapolis.

Pearl, J. (2000). *Causality: Models, Reasoning, and Inference*. Cambridge University Press, New York.

Peterson, E. (2005). *Website Measurement Hacks, Tips and Tools to Help Optimize Your Online Business*. O'Reilly, Media Inc.

Rhodes, J.S. (2001). From banners to scumware to usable marketing. Available at http://webword.com/moving/scumware.html.

Rios Insua, D. and Ruggeri, F. (2000). *Robust Bayesian Analysis*. New York: Springer-Verlag.

Speier, C. and Valacich, J.S. (1999). The influence of task interruption on individual decision making: An information overload perspective. *Decision Sciences*, 30(2): 337–360.

Tilson, R., Dong, J., Martin, S., and Kieche, E. (1998). Factors and principles affecting the usability of four e-commerce sites. *Proceedings of the Conference on the Human Factors and the Web*, Barking Ridge, NJ.

User Interface Engineering (UIE) (2007) Research articles. Available at http://www.uie.com.

SPSS (2008) Clementine. Available at http://www.spss.com/clementine/capabilities.htm.

Zijlstra, F.R.H., Roe, R.A., Leonora, A.B., and Krediet, I. (1999). Temporal factors in mental work: Effects of interrupted activities. *Journal of Occupational and Organizational Psychology*, 72: 163–185.

ADDITIONAL BIBLIOGRAPHY

Becher, J. and Kochavi, R. (2001). Tutorial on e-commerce and clickstream mining. *First SIAM International Conference on Data Mining*, April 5.

Bevan, N. (1999). Common industry format usability tests. *Proceedings of UPA'99*, Usability Professionals Association, Scottsdale, Arizona, 29 June–2 July.

Bevan, N. and Macleod, M. (1994). Usability measurement in context. *Behavior and Information Technology*, 13: 132–145.

International Organization for Standardization (1999) ISO 13407 *Human-Centred Design Processes for Interactive Systems*, Geneva, Switzerland.

International Organization for Standardization (2006) ISO 25062 *Common Industry Format (CIF) for Usability Test Reports*, Geneva, Switzerland.

Kwan, T.T., McGrrath, R.E., and Reed, D.A. (1995). NCSA World Wide Web server: Design and performance. University of Illinois, Urbana, IL.

Lin, T.Y. and Siewiorek, D.P. (1990). Error log analysis: Statistical modeling and heuristic trend analysis. *IEEE Transactions on Reliability*, 419–432.

Mason, L. (2002). E-commerce data mining: Lessons learned and insights gained. *Proceedings of the American Statistical Association Annual Meeting*, Chicago.

Park, Y.-H. and Fader P.S. (2004). Modeling browsing behavior at multiple web sites. *Marketing Science*, 23(Summer): 280–303.

Ragavan, N.R. (2005). Data mining in e-commerce: A survey. *Sadhana*, 30 (Parts 2 and 3): 275–289.

US Department of Health and Human Services (2007) Research-based web design and usability guidelines. Available at http://www.usability.gov/pdfs/guidelines.html.

8

DEVELOPING RICH INSIGHTS ON PUBLIC INTERNET FIRM ENTRY AND EXIT BASED ON SURVIVAL ANALYSIS AND DATA VISUALIZATION

ROBERT J. KAUFFMAN

W.P. Carey Chair in Information Systems, W.P. Carey School of Business, Arizona State University, Tempe, AZ 85287

BIN WANG

Assistant Professor, College of Business Administration, University of Texas—Pan American, Edinburg, TX 78539

8.1 INTRODUCTION

More than a decade after the dot-com boom starting in the early 1990s and several years after the stock market downturn in 2000, the Internet sector is gaining traction. However, many Internet firms are still trying to learn lessons from the dark days between 2000 and 2003 after the burst of the Internet bubble, when an estimated 5000 Internet firms either shut down their websites or were acquired by other firms (Webmergers.com 2003). Entrepreneurs came to realize the importance of the bottom line rather than the installed base or market share. Similarly, investors learned to stop blindly going after anything that could be viewed as a dot-com firm and began to focus on those with a sound business model and profit-generating business strategies.

Statistical Methods in e-Commerce Research. Edited by W. Jank and G. Shmueli

Academic researchers have also examined the factors that affect Internet firm survival and failure. Using results from two case studies of failed Internet firms and preliminary analysis based on 31 Internet firms, Rovenpor (2003) finds that Internet firms fail due to a combination of internal and external factors. Faraj et al. (2005) examine Internet firm performance from the social network perspective and find that those with a higher degree of centrality and external ties have more sales. Other research results also suggest that factors such as operating in a market which has been transformed by Internet technology, selling digital products or services, being a late entrant, avoiding stock issuance at around the time of a negative stock market event, and having a smaller firm size can enhance the survival of public Internet firms (Wang and Kauffman 2007).

The above-mentioned studies are primarily aimed at identifying factors that affect Internet firms' performance and survival. The methods employed are traditional ones such as case studies, time series regression, logistic regression, and survival analysis econometrics. Using the case study method, researchers can examine a small number of firms in detail, even though the results may not generalize to the broader population. Using econometric methods such as regression and survival analysis, researchers are able to analyze the performance and survival of a large number of Internet firms. Compared with logistic regression, survival analysis offers a rich set of methods, including nonparametric, semiparametric, and fully parametric estimation model specifications. This family of methods allows researchers to take into account not only the final survival outcome but also the timing of the exit event in terms of firm age or calendar time. (See the Appendix for a brief introduction to these and other relevant related methods.)

In spite of what we have found from this research, there is still much to be learned about the general survival patterns that Internet firms have exhibited since the mid-1990s. In this chapter, we propose the use of a hybrid approach in our analysis of public Internet firm survival, in which we combine traditional statistics and econometrics with nonparametric and data visualization methods. Through our discussion of the advantages associated with each method, we are able to identify the insight that each method is suited to providing. We compare and contrast them, revealing the rich dynamics of Internet firm survival in terms of velocity and acceleration of failure.

8.2 DATA ON PUBLIC INTERNET FIRM SURVIVAL

Next, we illustrate the use of the nonparametric survival analysis and data visualization techniques in examining the dynamics of public Internet firm survival. Our data include 130 publicly traded Internet firms. Similar to Barua et al. (2001), we defined Internet firms as those generating 90% of more of their revenues through the Internet. We obtained our list of Internet firms from multiple data sources, including Mergent Online (formerly FIS Online), the EDGAR Online IPO Express database, corporate filings with the Securities and Exchange Commission (SEC), and COMPUSTAT. We excluded privately held Internet firms from our sample because the entry and exit of

TABLE 8.1 Public Internet Firm IPOs and Exits (N = 130)

Year	No. of IPOs	No. of Exits	No. of Firms	Year	No. of IPOs	No. of Exits	No. of Firms
1996	6	0	6	2002	1	14	47
1997	10	0	16	2003	3	10	40
1998	10	0	26	2004	10	6	44
1999	62	6	82	2005	05	5	44
2000	23	14	91	2006	—	2	42
2001	0	31	60	**Total**	**130**	**88**	—

these firms were not well documented and it was difficult to determine the revenue breakdown of these companies. Our data include companies in different business categories, such as online portals, Internet retailers, online transaction brokers, content sites, online informediaries, and Internet marketers. We recorded the initial public offering (IPO) date for each firm and the corresponding exit date due to bankruptcy, liquidations, merger, or acquisition if such an event was observed. We summarize the number of IPOs and exits from 1996 to the second quarter of 2006 in Table 8.1.

We then discuss our application of the Kaplan-Meier curve and the cumulative hazard function to these data, follow that with additional consideration of the role of data visualization analysis, and then present our synthesis of methods that we believe will yield additional rich managerial insights for the e-commerce research domain.

8.3 LEVERAGING THE KAPLAN-MEIER CURVE AND THE CUMULATIVE HAZARD FUNCTION

Using nonparametric survival analysis techniques such as the Kaplan-Meier estimator, we compare Internet firm survival based on the firms' ages since they first issued stock, irrespective of the calendar time when the IPOs occurred. Results from this age-based comparison allow us to identify survival patterns related to a learning effect as a firm accumulates experience and grows older.

8.3.1 Aggregate Analysis for the Entire Sample

We plot the Kaplan-Meier and cumulative hazard function curves for all firms in Figures 8.1 and 8.2.

The circles in the Kaplan-Meier curve represent censored observations. From the Kaplan-Meier curve, we can see that during the first year after their IPOs were issued, our sample public Internet firms experienced minimal survival pressure and there were not many exits. Starting from the second year after IPOs issuance, however, Internet firms started to exit the marketplace. This trend continued until five years after they first issued stock, after which most firms were able to survive. The

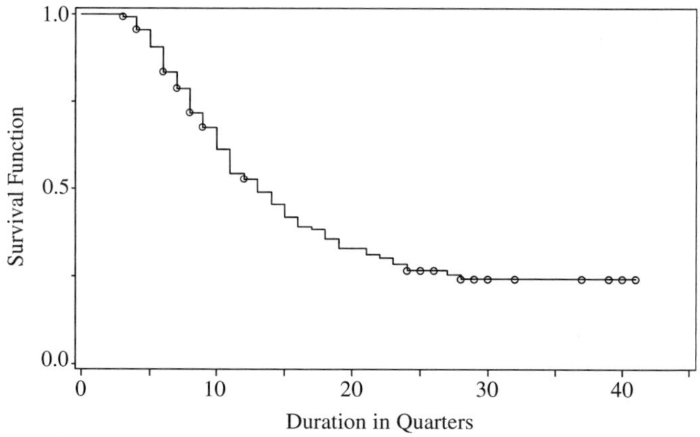

Figure 8.1 The Kaplan-Meier curve for our sample public Internet firms ($N = 130$).

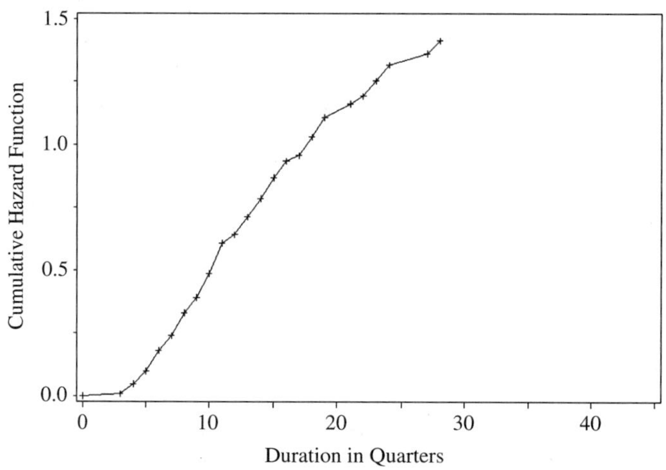

Figure 8.2 Cumulative hazard function for our sample of public Internet firms ($N = 130$).

Kaplan-Meier curve reveals that the period from year 2 to year 5 after IPO issuance was critical to the long-term survival of an Internet firm during the time frame of our data. Our cumulative hazard function curve tells a similar story. The cumulative hazard rate remained low during the first year after IPO issuance but increased steadily from years 2 through 5.

8.3.2 Subgroup Analyses and Comparisons

In addition to the aggregated analysis for our entire sample of Internet firms, we can perform a number of subgroup comparisons. We report the subgroup analysis results in Figure 8.3. Due to space constraints, we only show the Kaplan-Meier curves. The cumulative hazard function curves reveal similar patterns.

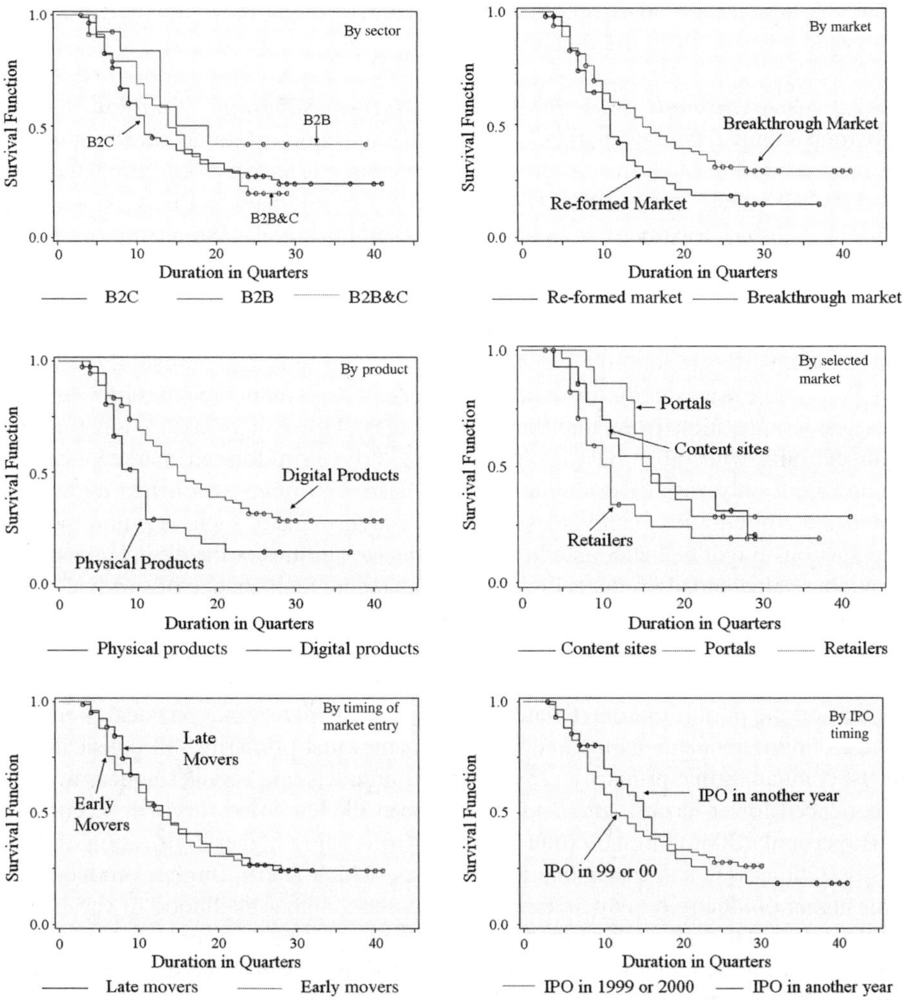

Figure 8.3 Kaplan-Meier curves for subgroup comparisons.

8.3.2.1 Sector Subgroup Analysis. In the subgroup-by-sector comparison, we divided our sample Internet firms into three categories—business-to-business (B2B), business-to-consumer (B2C), and business-to-business-and-consumer (B2B&C). In the last category, a company has both other businesses and consumers as customers. The Kaplan-Meier curves for the three sectors show that B2B companies experienced the lowest hazard rate and the highest survival rate, while B2C companies experienced the highest hazard rate early on after IPO issuance. Our sample of 130 Internet firms includes 82 B2Cs, 13 B2Bs, and 35 B2B&Cs. The large number of B2C firms might have led to more intensive competition, resulting in many early exits. In addition, the B2B marketplace might have a higher barrier to entry. Hence, only those firms with more resources might have been able to enter the marketplace. Once they did, though, they were better able to withstand pressure from the financial markets.

8.3.2.2 Breakthrough and Re-Formed Market Subgroup Analysis. In the subgroup comparison by market, we divided our sample Internet firms into two groups—a breakthrough market group and a re-formed market group. Breakthrough markets are newly emerged markets providing brand new products or services (Day et al. 2003). Examples are search engines, online portals, and auction websites. Re-formed markets are those where previously existing products or services are offered in the digital channel. Examples are online retailers and B2B marketplaces. The Kaplan-Meier curves for the two groups show that companies operating in breakthrough markets experienced lower hazard rates and an enhanced probability of survival. One possible reason for the enhanced chance of survival in breakthrough markets was less competition from traditional companies. Firms in breakthrough markets compete only with other Internet firms, while those in re-formed markets need to compete not only with other online firms but also with more established traditional companies. In addition, companies in breakthrough markets such as online portals and auctions might be better able to leverage the capabilities of the digital channel in providing value-added products or services, enhancing their chance of survival.

8.3.2.3 Digital and Physical Product Subgroup Analysis. Next, we divided our sample based on product into two categories—those primarily selling digital products, such as online portals or content sites, and those primarily selling physical products, such as online retailers. Compared with companies that primarily sell physical products, companies that primarily provide digital products and services appear to have experienced lower hazard rates and an enhanced likelihood of survival. According to Barua et al. (2006), digital product Internet firms enjoy higher digitization of their business processes and lower operational costs, which lead to higher productivity. This higher productivity may, in turn, translate into a higher likelihood of survival.

We selected three markets with the largest numbers of firms—content sites, online portals, and online retailers—and compared the Kaplan-Meier curves of these three groups. Our results indicate that online portals enjoyed the lowest hazard rates early on after issuance of their IPOs and constituted the highest surviving fraction. In contrast, many online retailers exited during the two-year period between years 2 and 3 after issuance of their IPOs, after which their chances of survival improved significantly. The differences in the survival patterns might be due to a number of factors. First, of the three, online portals were in the best position because they provided digital products and competed in breakthrough markets; retailers were in the worst position because they sold physical products and operated in re-formed markets. Content sites were in between, as they provided digital products but competed with traditional media companies. Second, many portal sites were among the first to issue Internet IPOs. So, by the time of the stock market downturn in spring 2000, when a large number of public Internet firms failed, they had established themselves on the stock market. As a result, they might have had a higher chance of survival compared with the other two types of firms.

8.3.2.4 Timing of Market Entry Subgroup Analysis. We further divided our data into two subgroups based on the timing of market entry—the early movers that were founded on or before December 31, 1995, and the late movers that were founded after that date. The Kaplan-Meier curves for the two subgroups are intertwined and do not show any clear difference. As a result, there seems to be no early mover advantage for our sample public Internet firms. While there were successful early movers such as Amazon and Yahoo!, we also witnessed the demise of many early movers such as InfoSeek, Smarterkids.com, and PlanetRX.com, as well the meteoric rise of latecomers such as Google and Orbitz. After spring 2000, the number of Internet firm IPOs decreased dramatically and only the strongest companies dared to issue stock. The strong performance by these latecomers might have permitted them to compensate for their disadvantage as the late movers, resulting in indistinguishable survival patterns between the two groups.

8.3.2.5 IPO Timing Subgroup Analysis. Because of the stock market downturn in spring 2000, we also divided our sample into two subgroups based on IPO timing. One group includes firms that floated IPOs in either 1999 or 2000. The other group includes firms that issued new stock in the other years. The Kaplan-Meier curves for the two groups show that firms that went public in 1999 or 2000 experienced higher hazard rates during years 3 and 4 after the IPO was issued, after which the difference were reversed. Previous research suggests that firms founded right before or after a market crash are more likely to fail (Honjo 2000). For Internet firms that went public in 1999 and 2000, years 3 and 4 after issuance of the IPO correspond roughly to the calendar period between 2001 and 2003, during which time the survival pressure mounted in the stock market and the largest number of exits occurred. Hence, we observed increased exits by firms that went public right around the time of the market downturn in spring 2000.

Overall, our subgroup comparisons provide us with rich insights about the differences in the likelihood of survival among firms with different characteristics. The factors that could have led to the different survival patterns should be further investigated in more rigorous empirical studies. They include the customers of the firm, the market in which the firm competes, the extent of competition it faces, its ability to leverage the digital channel in delivering digital products or services, and the timing of its IPO. We now turn to a discussion of data visualization methods.

8.4 DATA VISUALIZATION ANALYSIS

Using data visualization techniques, we can further examine the number of Internet firms, IPOs, exits, and the changes in these numbers based on calendar time. The calendar time–based analysis allows us to evaluate how the environmental factors and market sentiment might have affected Internet firm survival. First, we perform an analysis on our entire sample of public Internet firms. Next, we present the results from subgroup comparisons. In both cases, we offer interpretive remarks about what insights the analyses offer that relate specifically to the different aspects of the observed behavior in terms of the relevant theory, observations in the business press, and our own field study experience.

8.4.1 Aggregate Analysis for the Entire Sample

We first display the total number of public Internet firms from the beginning of 1996 to the second quarter of 2006 based on our sample of 130 public Internet firms in Figure 8.4.

As the figure reveals, the number of firms increased dramatically in 1999 and peaked at 99 in the second quarter of 2000. During the next three-year period, there was significant consolidation in the marketplace and the number of firms fell to 40 in the second quarter of 2003. The number of firms then stabilized at around 40, even though we saw a mini-boom of new public Internet firms in 2004 and 2005.

8.4.1.1 Interpretations Based on Theory. Even though our sample of Internet firms belonged to many industries, the pattern of the total number of public Internet firms closely mimics that of industry shakeouts that Klepper and Simons (2005) observed. Researchers have used the *product life cycle* (PLC) theory to explain industry evaluation and shakeouts (Gort and Klepper 1982; Klepper 1996). According to the theory, as technologies evolve, many industries will exhibit similar patterns with respect to survival. There is first an increase in the number of firms in the industry, followed by a sharp decline, and then the number of firms levels off. Early on, many firms are entering and innovating to provide different versions of the industry's product. The mortality rate for these small firms is low because prices are high. As more firms enter and industry output increases, prices drop. But as

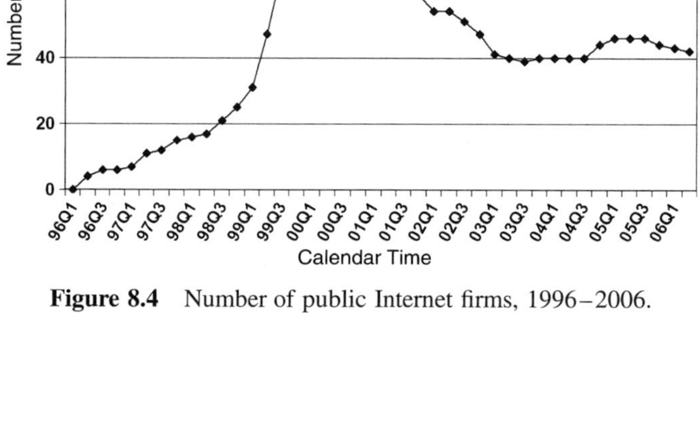

Figure 8.4 Number of public Internet firms, 1996–2006.

the industry grows, chances for innovation diminish and dominant product designs emerge. Firms compete based on the minimum efficient scale, and the less efficient firms begin to fail. Eventually, the entry and exit of firms reach equilibrium, and we see a relatively stable number of firms in the industry.

In the Internet sector since the early 1990s, many firms entered the marketplace and tried out different business models. During the dot-com hype prior to 2000, many investors and venture capitalists followed the Internet firm sector, making it easy for many dot-coms to survive. As more firms entered the marketplace and the Internet bubble burst in spring 2000, survival pressure mounted and Internet firms started to compete based on their performance. The many exits resulted in a sharp decline in the number of public Internet firms. After this period, existing Internet firms were more cautious about going public—possibly because their equity would have been undervalued in the marketplace, making the compensation for loss of control of their companies unattractive—resulting in a relative stable number of public Internet firms as the numbers of entries and exits came into balance.

8.4.2 Probing for Theoretical Explanations of Internet Firm IPO and Exit Patterns

Next, we examine Internet firm IPOs and exits more closely, and display the number of IPOs and exits during the same period in Figure 8.5. Taken together, these numbers depict the first difference changes over time in the total number of Internet firms. We also plot the observed changes in Figure 8.5. From this figure, we can see that the number of Internet firm IPOs increased dramatically starting in 1999, which led to a net increase in the number of Internet firms during this year. After the second quarter of 2000, Internet firm IPOs abruptly declined and the

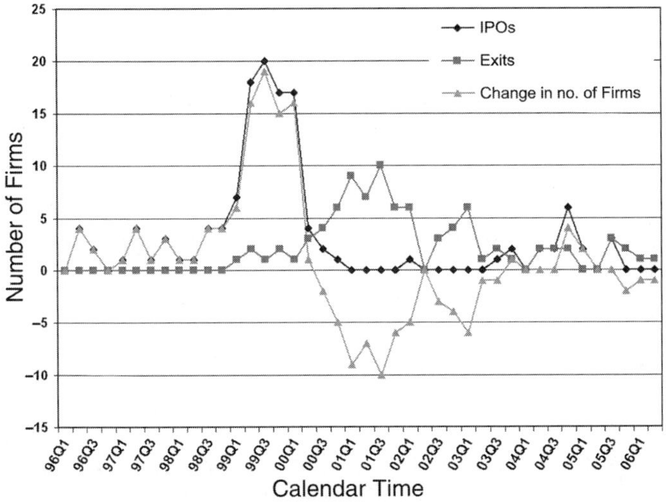

Figure 8.5 Internet firm IPOs and exits and changes in the number of firms, 1996–2006.

number of exits started to increase. This led to a net decrease in the number of Internet firms. After a four-year consolidation period, the number of entries and exits became roughly equal and the number of public Internet firms stabilized.

The entry, exit, and changes that we observed in the number of public Internet firms also conform to the pattern prescribed by the PLC theory. Internet firms started to issue their IPOs in the 1996–1999 period. Because of the small number of public Internet firms and the market hype concerning the dot-coms, in particular, during that halcyon period, we did not observe any public Internet firm exits. Between mid-1999 and early 2000, public Internet firm entry reached its peak; the number of firms tripled by 2000 in comparison with earlier years. As the number of firms increased, though, the competition intensified and firm exits started to occur. After spring 2000, the stock market took a dramatic turn for the worse—as investors and analysts recalibrated their beliefs about the future prospects of the dot-coms—and the pattern of entries and exits began to change dramatically. We started to see more exits and the number of entries quickly declined, resulting in an overall decrease in the number of firms. We observed the most exits occurring between 2001 and 2003, after which the numbers of entries and exits became roughly equal. The public Internet firm exit data that we have studied matched the overall pattern for all Internet firm failure. According to Webmbergers.com, a former dot-com failure statistic aggregator, the number of dot-com shutdowns and bankruptcies (i.e., those that were public or had received at least US$ 1 million in investment) increased from 225 in 2000 to 537 in 2001 and then decreased to about 200 by the end of 2002 (Webmergers.com 2003).

We also plot the changes in the number of IPOs and exits in Figure 8.6. These are like the second derivatives of the underlying function that gives rise to the number of Internet firms. From this figure, we can see that the number of IPOs increased very quickly in 1999 but was followed by a sharp decline in 2000. Changes in the

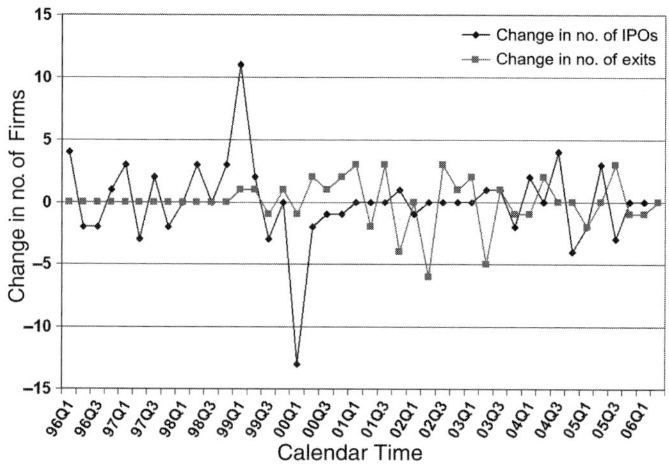

Figure 8.6 Changes in the number of Internet firm IPOs and exits, 1996–2006.

number of exits in 2001 and 2002 varied during those two years. In the other years, the numbers of IPOs and exits did not show dramatic changes.

The sharp peak in the change in IPOs issued in the first quarter of 1999 reveals not only that were there Internet firm IPOs during that quarter, but also that the number of IPOs increased dramatically from the previous quarter. In contrast, one year later in the first quarter of 2000, the number of IPOs took a dive and far fewer firms went public. The changes in the number of exits point to two separate periods of Internet firm failure increases—one from mid-2000 to early 2001 and the other from late 2002 to early 2003. Our subgroup comparison based on sector will show that a first wave of increasing exits was characterized by failing B2C and B2B&C companies, while the second wave was based mainly on the failure of B2Bs and B2B&Cs.

8.4.3 Subgroup Comparisons for Internet Firm IPO and Exit Patterns

Next, we compare Internet firm IPOs and exits based on calendar time by sector, market, product, timing of market entry, IPO timing, and selected markets, and offer additional interpretive insights as before, based on the different information that these analyses offer us.

8.4.3.1 Dot-Com IPOs in the B2C, B2B, and B2B&C Sectors. The subgroup comparison of the number of IPOs based on sectors suggests that IPOs for all three types of Internet firms—B2C, B2B, and B2B&C—significantly increased in 1999 and early 2000 as the capital markets made funding available in historically large amounts. (See Figure 8.7a.) After the spring 2000 stock market downturn, though, IPO issuance halted. The only IPO among dot-com firms in the next two years, 2001 and 2002, was that of PayPal, which had more than proven its capabilities by dominating the Internet payment services market. After the third quarter of 2003, though, the capital markets found renewed interest in dot-com stocks, and better and more IPOs started to emerge for Internet companies.

8.4.3.2 Two Waves: Temporal Differences in B2C and B2B Exits. During that time, we saw more B2C than B2B or B2B&C Internet firms exit after spring 2000. Figure 8.7b shows the two waves of Internet firm exits. B2C and B2B&C firm exits started to increase immediately following the spring 2000 market downturn—the first wave. For public B2B firms, this trend was delayed until late 2002 and early 2003, when the second wave occurred. Most observers believe that the market was first able to see through the ineffective business models of the B2C firms, resulting in their early vulnerability to failure in the capital markets (Lorek 2000; Willoughby 2000). During the same period, sentiments were still fairly positive regarding the B2B firms' likelihood for success, even though the capital markets were no longer funding new ventures (Trombly 2000; Petersen 2001). The "profitability vigil" was on—marketwide.

During this time, however, it became plain to many observers that there was simply too much duplication of innovation effort in the B2B space (Laseter et al. 2001). Observers began to realize that the market could not bear the competition

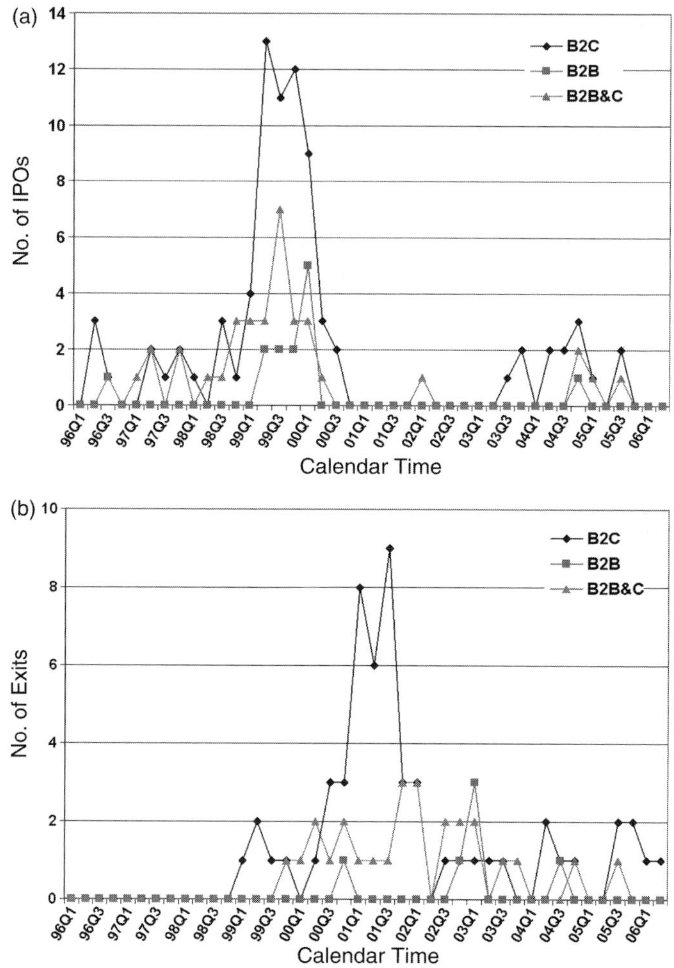

Figure 8.7 (a) B2C, B2B, and B2B&C firm IPOs, 1996–2006; (b) B2C, B2B, and B2B&C firm exits, 1996–2006.

among several B2B procurement market intermediaries which sought profits within single industry bounds (e.g., the metals industry, the electrical parts supply industry, the paper products industry). Slowly, the market came to realize that it would be appropriate for B2B e-market intermediaries to span multiple industries with their procurement services. Once this realization became widespread, the adjustment of the expectations present in the stock market became plain, leading to the exit of many B2B e-market and other B2B services intermediaries (Day et al. 2003).

8.4.3.3 Dot-Com IPOs and Exits in Re-Formed and Breakthrough Markets. The IPOs and exits of Internet firms in breakthrough and re-formed markets do not show any significant differences, based on what is shown in Figures 8.8a and 8.8b. Companies in breakthrough markets, such as search engines

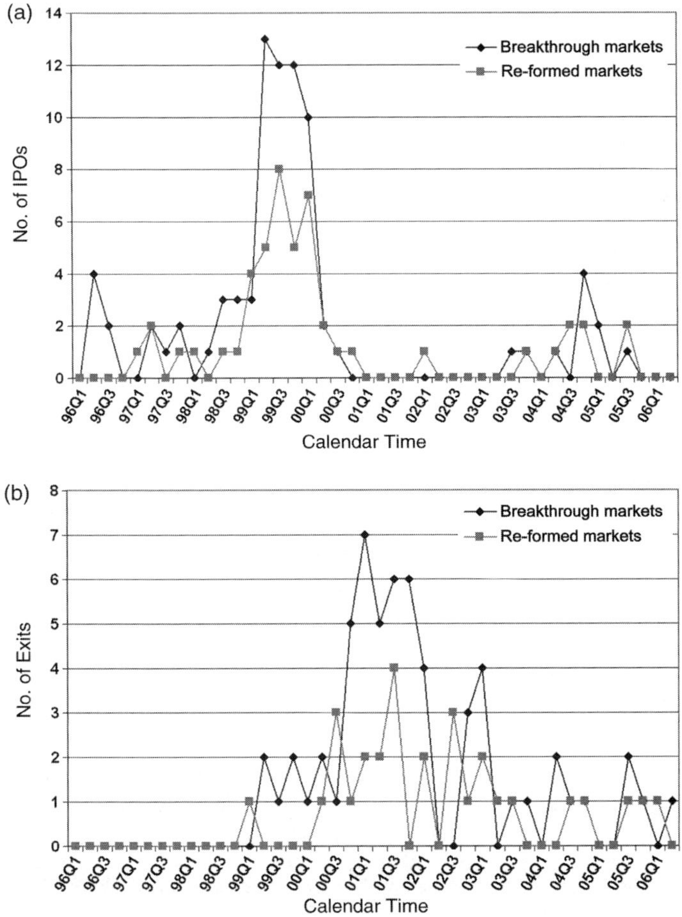

Figure 8.8 (a) Breakthrough and re-formed market firm IPOs, 1996–2006; (b) Breakthrough and re-formed market firm exits, 1996–2006.

and online portals, issued stock as early as early 1996, but the dot-com firms that served re-formed markets caught up quickly after 1997. Overall, we observed more IPOs being issued in breakthrough markets than in re-formed markets. This is an interesting observation because it suggests the lesser ability of financial analysts and fund managers to understand the limits of technology-based entrepreneurship as the digital economy unfolded, in much the same way that other observers have exhibited *irrational exuberance* related to other, more general matters of the economy and the value of financial assets within them (Shiller 2005).[1] In terms of exits, firms in breakthrough markets began to leave even before spring 2000,

[1]For additional interesting commentary about the background of the term *irrational exuberance*, Alan Greenspan's intentions with respect to his initial use of the term, and the subsequent ways that the business press and the investing public have understood its use, see www.irrationalexuberance.com/definition.htm.

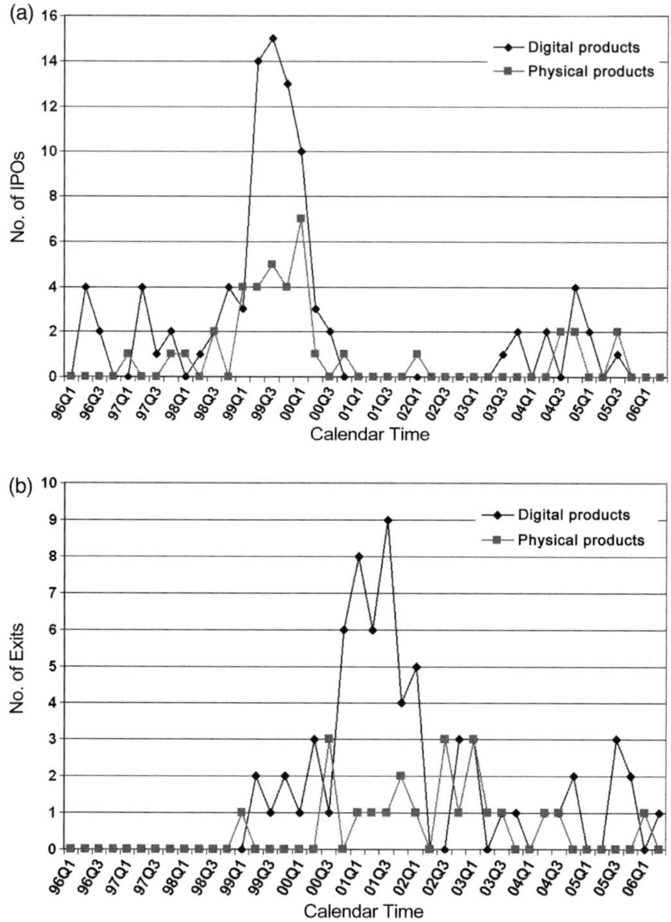

Figure 8.9 (a) Digital versus physical product firm IPOs, 1996–2006; (b) Digital versus physical product firm exits, 1996–2006.

a vanguard group whose exits apparently were due to merger and acquisition activity (Figure 8.8b). This may have suggested the highly positive valuations of the present value of their future growth opportunities. After spring 2000, however, both reformed market and breakthrough market dot-com firms experienced large numbers of exits (Figure 8.8b). This was due to intensified competition, rapidly changing expectations about their likely return on investment, and heightened pressure from the marketplace and the capital markets (Kauffman et al. 2006).

8.4.3.4 Digital and Physical Goods Dot-Com IPOs and Exits. Internet firms that offered digital versus physical products during the 1996–2002 period experienced similar but not identical IPO and exit patterns. (See Figures 8.9a and 8.9b.) Digital product firms' business models were among the earliest business models to be funded by the capital markets—in 1996, in fact—including firms that offered

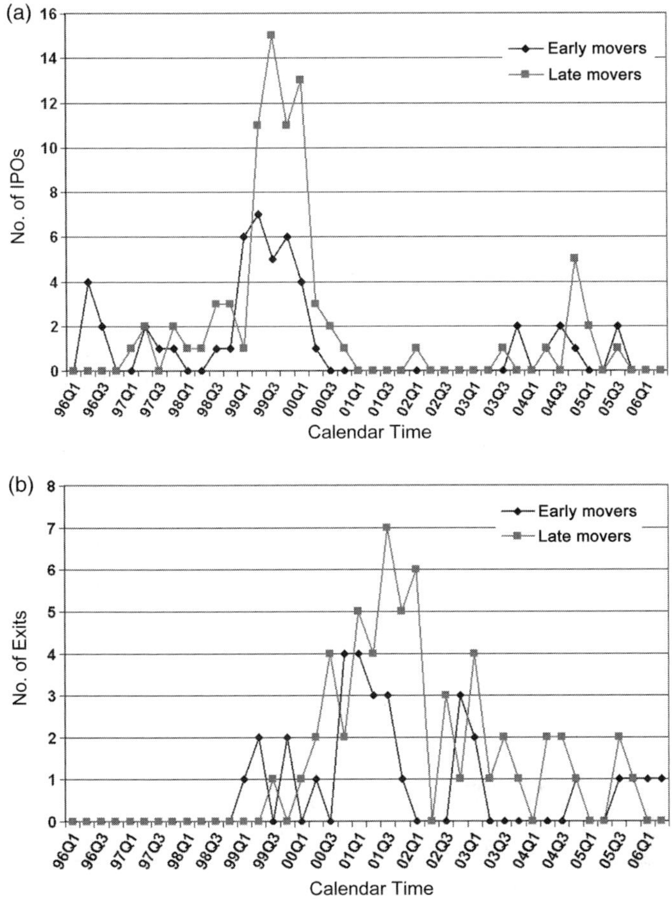

Figure 8.10 (a) Early and late mover dot-com IPOs, 1996–2006; (b) Early and late mover dot-com exits, 1996–2006.

search engine and online portal services and auction sites that sold digital products. Beginning in 1997, other firms whose greater emphasis was on selling physical products started to issue IPOs, though their number was less than that of firms selling digital products or services. Digital product firm exits were more spread out in the period from mid-1999 to mid-2003 (Figure 8.9a), while physical product firm exits were concentrated across two periods: around the fourth quarter in 2000 and from the fourth quarter of 2001 to the fourth quarter of 2003.

Internet firms that were early movers, i.e., those that were founded on or before December 31, 1995, started to go public in 1996, one year before the late movers. (See Figure 8.10a.) During the Internet firm IPO peak in 1999 and early 2000, more late movers went public than early movers. Early movers also started to exit the marketplace earlier. As soon as the market turned down in 2000, however, late mover exits increased quickly and more late movers exited the marketplace. (See Figure 8.10b.)

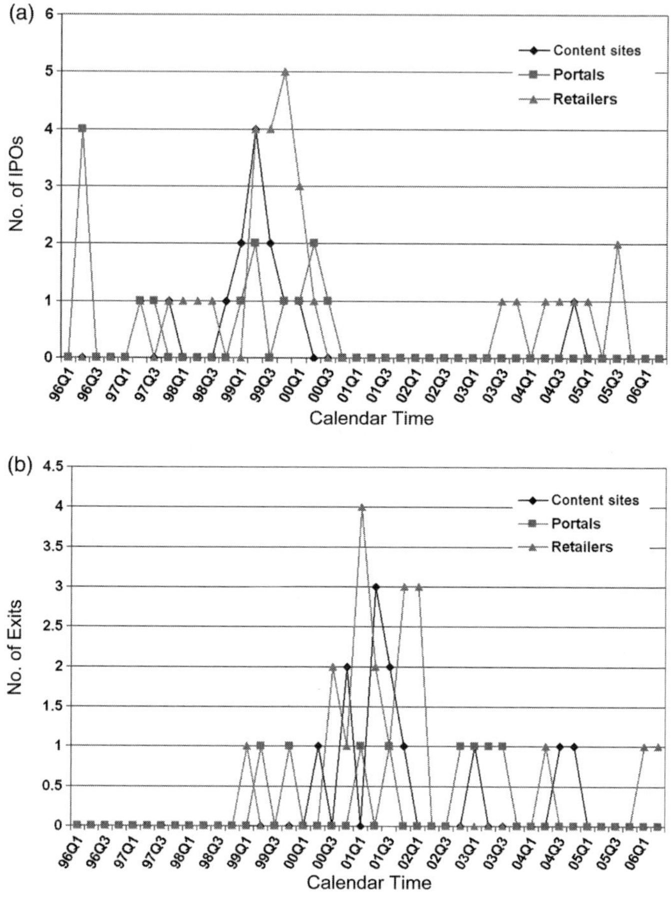

Figure 8.11 (a) Content sites, portals, and retailers: IPOs, 1996–2006; (b) Content sites, portals, and retailers: exits, 1996–2006.

The IPO figures for three selected markets—content sites, online portals, and online retailers—show that online portals were among the first Internet firms to go public in 1996. (See Figures 8.11a and 8.11b.) Online retailer and content site IPOs started to appear in 1998. During the IPO peak period in 1999 and early 2000, more content sites and online retailers went public, while fewer portal sites issued IPOs. During the second IPO peak between the third quarter of 2003 and the fourth quarter of 2005, more online retailers went public, while only one content site issued stock. At that time, no online portals went public.

These results suggest that the online portal market had the highest barriers to entry in the IPO market, while the online retail market had the lowest. Four portal sites—Yahoo!, Excite, Lycos, and Infoseek—were founded and went public early. As a result, they were able to enjoy a first-mover advantage and large installed bases of customers; this surely presented significant barriers to entry for later movers.

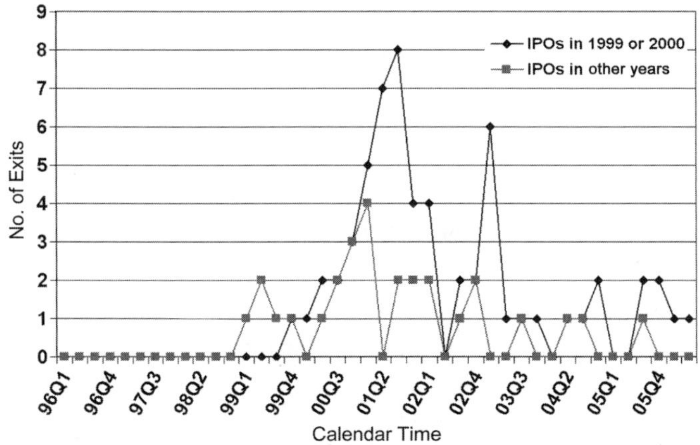

Figure 8.12 Firm exits with IPOs in 1999 and 2000 versus other years.

Despite some early IPO issuers, such as Amazon and N2K Inc., in the online retail market, the late movers were still able to identify market niches and issue their IPOs—quite surprisingly, even as late as 2005. Thus, the exits in these markets show a mixed pattern. Overall, the portal and content site exits seemed to be concentrated in shorter periods compared with the retailer exits. This suggests that market consolidation for portals and content sites occurred quickly—almost as though the market came to a clear understanding of their business models and the present value of their growth opportunities before they figured out the other e-commerce business models. The exits of online retailers show a much slower failure pattern.

The final figure for this group compares Internet firm exits based on IPO timing and shows an interesting aspect: More companies that went public in 1999 or 2000 exited right after spring 2000. (See Figure 8.12.) Companies that went public in 1999 or 2000 had limited histories on the stock market at the time of the market downturn, and so many of them—as might be expected for the "newest of the new" market entrants—were still operating on losses, and might have been viewed as having limited prospects for achieving positive future cash flows. As a result, they were the least able to withstand the pressure from the stock market and more likely to exit, which led to additional pressures that ultimately resulted in the large number of observed failures.

8.5 SYNTHESIS: DEVELOPING RICH INSIGHTS FROM HYBRID METHODS

The boom and bust that Internet firms went through has fascinated financial market analysts and academic researchers alike. The latter have used traditional statistical methods, such as logistic regression, time-series regression, and survival analysis, to examine the factors that seem to have most affected the Internet firms' performance and survival. In this chapter, we have discussed and illustrated the use of more explanatory methods such as nonparametric survival analysis, data visualization techniques, and functional data analysis in analyzing the survival patterns of 130 publicly traded Internet firms. The use of the exploratory methods is complementary to the traditional methods, and it provides researchers and practitioners with the insights described in Sections 8.5.1 and 8.5.2.

8.5.1 Analytical Richness from Traditional Statistics and Data Visualization Methods

The combination of data visualization methods with traditional statistics provides the following insights compared with the use of traditional statistics alone. First, data visualization methods allow analysts to visually display and identify trends in the data. The purpose of data visualization methods is to illustrate the patterns in the data and provide a basis for the formation of analytical intuition. Results from data visualization analysis may also reveal directions for additional statistical analysis.

Second, data visualization analysis allows analysts to examine the patterns of velocity and acceleration in the data. These kinds of information are not so easily revealed through traditional statistical analysis, which is better at conveying information about other aspects of the analysis—especially the marginal effects of individual variables on the dependent variable, levels of significance, and the extent to which the variance of the dependent variable can be explained by the model. Third, data visualization methods provide an easy way to identify outliers and check modeling assumptions. Traditional statistical analysis relies heavily on assumptions, and the results are sensitive to influential outliers. Using data visualization methods, researchers can identify and isolate these outliers—irrespective of the functional form that is adopted for estimation—and, in this way, provide more accurate and robust estimates that are more useful for managers and policymakers. Fourth, data visualization methods provide greater flexibility in slicing and dicing the data. They are very useful tools for revealing unseen relationships across cuts of the data that are tactically stratified by the analyst to focus on the role of specific variables in the overall setting that is being considered in the research.

A fifth benefit of the use of data visualization approaches comes in the study of aspects of observed behavior in complex systems (Cilliers 1998; Colander 2000) that seem to defy the human capacity to write down simple models and representations of likely outcomes in the presence of the dynamics of such systems. Studies of the movement of leading indicators in the macroeconomy are difficult to predict due to the underlying complexity and dynamics of the economic system (Arthur 1990; Arthur et al. 1997). Similar to this is the difficulty of predicting future prices of financial assets, given the inherent complexity of processing relevant information in the financial markets to yield specific asset price outcomes. In such settings, data visualization techniques offer researchers, investors, and public policymakers the equivalent of three-dimensional images that simplify the true-form n-dimensional images of the phenomena that we wish to understand. This can be said, for example, for the vast number of pharmacology studies that have been conducted to ascertain the effects of new pharmaceuticals on the human body, where the complexity of the underlying biochemical and biophysical processes can only be understood in greatly simplified form—as if the analysts were only studying several of many possible dimensions of a problem.

8.5.2 Analytical Richness with Hybrid Methods for the Internet Firm IPO and Exit Data

Our use of the traditional statistical methodology of nonparametric survival analysis, coupled with data visualization techniques on public Internet firm survival, provided a number of useful insights. First, the Kaplan-Meier curves illustrate the overall survival patterns for public Internet firms, as well as the differences in survival patterns among firms with different characteristics. Their differences are based on their sectors of operation, the nature of the competition in the markets they serve, the kinds of products they offer, the timing of their market entry, the timing of their IPOs, etc. Although it is possible with statistical analysis to include modeling dummies and categorical variables to tease out the relative effects by coding subgroups, the information provided by parameter estimates and significance levels doesn't give the analyst the same intuition provided by the visual nature of the various graphs.

Second, plotting the number of public Internet firms over time reveals the extent and timing of the shakeout in the Internet sector. The fact that the pattern is similar to other patterns that have been recognized for a variety of industry shakeouts in the past suggests that the industry evolution literature (e.g., Sutton 1991, 1998; Jovanovich and McDonald 1994; Klepper 1996) may be useful in informing researchers on the survival and failure dynamics for firms in the digital marketplace. Another potentially valuable literature examines patterns of competition in "rugged landscapes," which are competitive settings that require different kinds of "fitness" on the part of competitors to successful in the marketplace (Levinthal 1997; Levinthal and Warglien 1999).

In empirical research, we advocate the development of as complete an understanding of the data as possible via multiple methods—including data visualization, descriptive statistics, multiple estimation models, post-estimation examination of the error terms, the influence of different observations, and other analysis of the underlying assumptions of the estimation approaches used. With this perspective in mind, we often remind our students and our colleagues to be cautious in drawing conclusions about a theory or an empirical phenomenon without fully understanding the role of proxy variables, modeling control variables, and the overall appropriateness of the data that are used as a basis for the chosen research design and statistical tests.[2]

[2]Although we have not attempted to develop the arguments more fully here, in our doctoral teaching related to empirical research on technology, IS and e-commerce-related issues, we have stressed the importance of giving careful consideration to these other aspects of the empirical research design process. Choices that an analyst makes about the operationalization of key theoretical constructs as study variables need to be considered in terms of issues like *proxy distance* (i.e., the difference between an ideal measure of a construct and what the constraints of a real-world setting permit), the appropriate use of *control variables* (i.e., to ensure that the variance of the dependent variable relative to important but not theory-bearing variables is being absorbed), and the extent to which a given research design and an empirical test permit a *true test of theory* (as opposed to knowledge about associations between variables). For additional useful

Third, examining different statistical views of the same setting often offers useful informationn. In this case, we use one figure to focus on the entries and exits of public Internet firms over the 10-year period from 1996 to 2006. We use another figure to depict the changes in the number of entries and exits to help illustrate the velocity and acceleration of dot-com firm entries and exits. Through these curves, we also are able to identify specific years in which the number of entries or exits increased or decreased. We should note, however, that in most empirical research settings it is inappropriate for the researcher to put too much faith in prejudging how things work—unless the emphasis is on testing theory. If that is the case, then even partial representations of the full spectrum of variables will be analytically appropriate, as well as relatively small degrees of overall variance in the dependent variable explained. Still, as Sims (1980) and others (e.g., Bajari and Hortacsu 2003; Kauffman and Wood 2007a) have pointed out, there are many situations in which it is better to permit the data to speak for themselves, and there are other statistical and nonstatistical methods that suit this purpose very well.

Fourth, our plots on the subgroup comparisons of IPOs and exits provide us with another reading and the basis for additional insights into the velocity of market entries and exits. Such analysis illustrates the flexibility of data visualization methods in examining the data from different perspectives.

8.6 CONCLUSION

In this chapter, we discussed and illustrated the use of nonparametric survival analysis and data visualization techniques as complementary to traditional statistical methods in analyzing Internet firm survival. We argued that, through the use of such a hybrid approach, researchers can obtain a more comprehensive understanding of the dynamics of dot-com firm entry and exit in the digital marketplace during the 1996–2006 period (and beyond). We believe that these blended analytical capabilities provide a basis for achieving unique insights in to the issues under study.

8.6.1 Contributions to Research and Practice

8.6.1.1 Higher-Level Contributions. Our research has made the following high-level contributions to research and practice. We have argued in favor of the use of a combination of graphical and statistical methods, drawing researchers' attention to the use of such methods in analyzing Internet firm survival in particular and e-commerce phenomena where many data exist more generally. Our research also provides academic scholars and business practitioners with a more visual understanding of a large chunk of the 10-year evolution from 1996 to 2006 that Internet firms have experienced.

insights into these issues, the interested reader should read Kauffman and Wood (2007b), Kauffman and Tallon (2007), and Mithas et al. (2007).

8.6.1.2 Lower-Level Contributions. We also note several additional lower-level contributions in our work. First, using the Kaplan-Meier estimator, we plotted the Kaplan-Meier curve and the cumulative hazard function for our entire sample. Second, our subgroup comparisons of the Kaplan-Meier curves for different Internet firms based on sector, market, product, timing of market entry, IPO timing, and firms in selected markets revealed how the survival patterns differed for firms with different characteristics. In addition, our Kaplan-Meier curves gave us far more information on how the survival patterns differed at different ages of the firms, something that is not easily revealed in an analysis using traditional methods. For example, empirical analysis using the Cox proportional hazards model only shows that Internet firms in breakthrough markets are more likely to survive, as noted by Wang and Kauffman (2007). In our subgroup comparison based on market type, the Kaplan-Meier curves for the two groups of firms showed that the survival patterns were similar during the first two years after the IPOs occurred. The two Kaplan-Meier curves started to diverge during the third year, though, after which firms in breakthrough markets experienced a lower hazard rate and a higher survival percentage.

Third, our data visualization analysis of the entire sample in terms of the total number of firms, the number of IPOs and exits, and the changes in the number of IPOs and exits showed the underlying dynamics in the digital marketplace. Fourth, our subgroup comparisons on the number of IPOs and exits in different periods based on sector, market, product, timing of market entry, IPO timing, and firms in selected markets show how the IPOs and exits differ for Internet firms with different characteristics.

8.6.2 Limitations

Although exploratory methods using nonparametric survival analysis and data visualization techniques provide us with insight into the dynamics of Internet firm survival, they have their limitations. First, such methods do not allow researchers to quantify the extent of a factor's impact on an Internet firm's likelihood of survival. Parametric models are better suited for that purpose. Second, when performing subgroup comparisons, the analyst should note that the differences in the groups are based on a single factor only, and the confounding effects of the other factors are not considered simultaneously. Confirmatory methods such as semiparametric or fully parametric survival analysis allow researchers to answer some of these questions. This is also our basis for recommending the use of a hybrid approach in examining Internet firm survival to obtain a more holistic picture.

Even though we discussed the possible use of functional data analysis in analyzing Internet firm survival, we did not apply it to our data. Future research can collect such data as revenues, financial capital, number of employees, and cash on hand for publicly traded Internet firms and identify the functional object for Internet firms, as well as the velocity and acceleration in the changes. Researchers can also compare the functional objects for firms based on sector, market, product, timing of market entry, IPO timing, etc.

ACKNOWLEDGMENTS

We would like to thank our faculty and doctoral colleagues at the University of Texas–Pan American, the University of Minnesota, and Arizona State University for their comments and encouragement in this area of our research work. Rob Kauffman also thanks the participants of a doctoral seminar at Arizona State for related discussions and sharing of ideas during the spring 2007 semester. He appreciated ongoing discussions with other research colleagues and coauthors; our joint work informs some of the more general interpretations of the role of hybrid methods and developing a "deep" understanding of empirical data. They include Paul Tallon, Chuck Wood, Ajay Kumar, Nelson Granados, and Alok Gupta. We also appreciated input given by other colleagues at our university seminar visits in the United States, Hong Kong, Taiwan, and China, as well as conference presentations in the United States and China. We further benefited from the comments and criticisms we obtained on related research papers that have been reviewed at various journals. Some of the work on this chapter was supported by the MIS Research Center at the Carlson School of Management, University of Minnesota; the Center for Advancing Business through Information Technology (CABIT) at the W.P. Carey School of Business, Arizona State University; and the W.P. Carey Chair in Information Systems, which are all due our thanks.

REFERENCES

Arthur, W.B. (1990). Positive feedbacks in the economy. *Scientific American*, 262(2): 92–97.

Arthur, W.B., Durlauf, S.N., and Lane, D.A. (eds.) (1997). *The Economy as a Complex Evolving System II*. Reading, MA: Addison Wesley.

Bajari, P. and Hortacsu, A. (2003). The winner's curse, reserve prices and endogenous entry: Empirical insights from eBay auctions. *RAND Journal of Economics*, 34(2): 329–355.

Barua, A., Pinnell, J., Shutter, J., Whinston, A.B., and Wilson, B. (2001). Measuring the Internet economy. Working Paper, Center for Research on E-Commerce, McCombs School of Business, University of Texas at Austin.

Barua, A., Whinston, A.B., and Yin, F. (2006). Not all dot coms are created equal: An exploratory investigation of the productivity of Internet based companies. Working Paper, Center for Research on E-Commerce, McCombs School of Business, University of Texas at Austin.

Cilliers, P. (1998). *Complexity and Postmodernism: Understanding Complex Systems*. London: Routledge.

Cleveland, W.S. (1993). *Visualizing Data*. Summit, NJ: Hobart Press.

Colander, D. (2000). *The Complexity of Vision and the Teaching of Economics*. Northampton, MA: Edward Elgar.

Congdon, P. (2003). *Applied Bayesian Modelling*. Chichester, England: Wiley.

Cox, D. (1975). Partial likelihood. *Biometrika*, 62(2): 269–275.

Day, G.S., Fein, A.J., and Ruppersberger, G. (2003). Shakeouts in digital markets: Lessons from B2B exchanges. *California Management Review*, 45(2): 131–150.

Faraj, S., Gosain, S., and Yeow, A. (2005). Survival and performance in the Internet industry: Examining network characteristics of high performing internet firms. *Proceedings of the INFORMS Conference on Information Systems and Technology*, San Francisco, Available on CD-ROM.

Gort, M. and Klepper, S. (1982). Time paths in the diffusion of product innovations. *Economic Journal*, 92(367): 630–653.

Honjo, Y. (2000). Business failure of new firms: An empirical analysis using a multiplicative hazards model. *International Journal of Industrial Organization*, 18(4): 557–574.

Hosmer, D.W.J. and Lemeshow, S. (1999). *Applied Survival Analysis: Regression Model of Time to Event Data*. New York: Wiley.

Ibrahim, J.G., Chen, M.-H., and Sinha, D. (2001). *Bayesian Survival Analysis*. New York: Springer-Verlag.

Jovanovich, B. and McDonald, G.M. (1994). The life cycle of a competitive industry. *Journal of Political Economy*, 102(2): 322–347.

Kauffman, R.J., Miller, T., and Wang, B. (2006). Reflections on: When Internet companies morph. *First Monday*, 11(Special Issue 6). Available at http://www.firstmonday.org/issues/special11_7/kaufman/index.html. Accessed April 22, 2007. This article is a reprint of an earlier article by the authors with additional commentary.

Kauffman, R.J. and Tallon, P. (2007). Beyond the bounds of statistical inference: Opportunities and challenges for economics, is and e-commerce research. In *Economics, Information Systems, and Electronic Commerce: Empirical Methods* (R.J. Kauffman and P. Tallon, eds.). Armonk, NY: M.E. Sharpe.

Kauffman, R.J. and Wood, C.A. (2007a). Follow-the-leader: Price change timing in Internet-based selling. *Managerial Decisions and Economics*, 28: 1–22.

Kauffman, R.J. and Wood, C.A. (2007b). Revolutionary research strategies for e-business: A philosophy of science view of research design and data collection in the age of the Internet. In *Economics, Information Systems, and Electronic Commerce: Empirical Methods* (R.J. Kauffman and P. Tallon, eds.). Armonk, NY: M.E. Sharpe.

Klepper, S. (1996). Entry, exit, growth, and innovation over the product life cycle. *American Economic Review*, 86(3): 562–583.

Klepper, S. and Simons, K.L. (2005). Industry shakeouts and technological change. *International Journal of Industrial Organization*, 23(1–2): 23–43.

Laseter, T., Long, B., and Capters, C. (2001). B2B benchmark: The state of electronic exchanges. *Strategy + Business*, 25: 33–42.

Le, C.T. (1997). *Applied Survival Analysis*. New York: Wiley.

Levinthal, D.A. (1997). Adaptation on rugged landscapes. *Management Science*, 43(7): 934–950.

Levinthal, D.A. and Warglien, M. (1999). Landscape design: Designing for local action in complex worlds. *Organization Science*, 10(3): 342–357.

Lorek, L. (2000). Dot com bubble bursts; layoffs begin. *Inter@ctive Week*, 7(20): 28.

Mithas, S., Almiral, D., and Krishnan, M.S. (2007). A potential outcomes approach to assess causality in information systems research. In *Economics, Information Systems and Electronic Commerce: Advanced Empirical Methodologies* (R.J. Kauffman and P. Tallon, eds.). Armonk, NY: M.E. Sharpe.

Petersen, S. (June 4, 2001). B2B's silver lining emerges. *eWeek*, 18(22): 45.

Powers, D.A. and Xie, Y. (2000). *Statistical Methods for Categorical Data Analysis*. San Diego, CA: Academic Press.

Ramsey, J.O. and Silverman, B.W. (1997). *Functional Data Analysis*. New York: Springer-Verlag.

Rovenpor, J. (2003). Explaining the e-commerce shakeout. *e-Service Journal*, 3(1): 53–76.

Shiller, R.J. (2005). *Irrational Exuberance* (2nd ed.). Princeton, NJ: Princeton University Press.

Shmueli, G. and Jank, W. (2005). Visualizing online auctions. *Journal of Computational and Graphical Statistics*, 14(2): 1–21.

Shmueli, G. and Jank, W. (2007). Modeling dynamics in online auctions: A modern statistical approach. In *Economics, Information Systems, and Electronic Commerce: Advanced Empirical Methodologies* (R.J. Kauffman and P. Tallon, eds.). Armonk, NY: M.E. Sharpe.

Simonoff, J.S. (1996). *Smoothing Methods in Statistics*. New York: Springer-Verlag.

Sims, C. (1980). Macroeconomics and reality. *Econometrica*, 48(1): 1–48.

Smith, A.F.M. and Roberts, G.O. (1993). Bayesian computation via the Gibbs sampler and related Markov chain Monte Carlo methods. *Journal of the Royal Statistical Society, Series, B* 55(1): 3–23.

Sutton, J. (1991). *Sunk Costs and Market Structure: Price Competition, Advertising, and the Evolution of Concentration*. Cambridge, MA: MIT Press.

Sutton, J. (1998). *Technology and Market Structure: Theory and History*. Cambridge, MA: MIT Press.

Trombly, R. (2000). E-business models. *Computerworld*, 34(49): 61.

Tufte, E.R. (1990). *Envisioning Information*. Cheshire, CT: Graphics Press.

Tufte, E.R. (2001). *The Visual Display of Quantitative Information* (2nd ed.). Cheshire, CT: Graphics Press.

Wang, B. and Kauffman, R.J. (2007). The duration of Internet firms in breakthrough and re-formed markets: A semiparametric Cox survival analysis of their bankruptcies, mergers and acquisitions. Working Paper, Center for Advancing Business through Information Technology, W.P. Carey School of Business, Arizona State University.

Webmergers.com. (August 2003). Internet companies three years after the height of the bubble. Research report. San Francisco: Webmergers.com.

Willoughby, J. (March 20 2000). Burning up-warning: Internet companies are running out of cash-fast. *Barron's*, 80(12): 29–32.

APPENDIX: STATISTICAL METHODS FOR ANALYZING INTERNET FIRM SURVIVAL

We briefly discuss the statistical methods that can be used to examine Internet firm survival, including traditional statistics, Bayesian data analysis, data visualization methods, and functional data analysis (FDA). Even though we do not illustrate the use of FDA and Bayesian methods in this chapter, we point out these methods and discuss their potential applications to Internet firm survival.

8.A.1 TRADITIONAL METHODS AS AN APPROACH TO CREATING MANAGERIAL INSIGHTS

We divide the traditional statistical methods that can be used in our research context into two categories: those that *describe* Internet firm survival patterns and those that try to *explain* the survival patterns. For the first, we discuss survival analysis and the Kaplan-Meier estimator, two well-established approaches that permit us to evaluate many different kinds of failures processes.

8.A.1.1 Survival Analysis Concepts

Survival analysis has been widely used in biostatistics to examine the effectiveness of drugs in treating diseases and in economics to analyze business survival. Researchers view the occurrence of an event such as death or disease relapse, the reincarceration of a previously released criminal, or a firm's bankruptcy as the result of a *failure process* that begins at a certain point in time, such as the birth of an individual, the initial release of a prisoner, or the inception of a firm. There are four basic concepts in survival analysis. *Duration* refers to the time elapsed from the time the failure process starts until the occurrence of the event or the end of the observation period, whichever occurs sooner. In the case where the study period ends before an event is observed, the observation is *right-censored*. The *hazard rate* refers to the instantaneous likelihood of observing an event for an observation when it is still at risk (i.e., or still alive in terms of people with diseases) right before time t.

Finally, the *survival function* depicts the probability that an observation will have a duration longer than t. It can also be interpreted as the proportion of the population that is still alive after time t (Le 1997). Even though most applications of survival analysis compare observations based on duration or age, it is also possible to perform calendar time–based analysis, which is especially relevant to Internet firm survival. This enables us to answer questions about why certain things happened at given points in time.

The nonparametric techniques that survival analysis offers allow researchers to describe the survival patterns in terms of survival function and hazard rate without making any assumptions about the underlying distribution. As a result, these techniques have been very useful in obtaining visual plots of the survival pattern. Next, we discuss one widely used nonparametric statistic called the *Kaplan-Meier estimator*, which is especially well suited for this purpose.

8.A.1.2 The Kaplan-Meier Estimator

This statistic calculates a survival function for the sample data using $\hat{S}_t = \prod_{t(q)\leq t} \frac{n_{t(q)} - d_{t(q)}}{n_{t(q)}}(q = 1, 2, 3, \ldots, Q)$, where Q is the number of distinct event times in the sample, $t(q)$ is the time (in terms of the number of quarters after an IPO occurred) when a firm exit was observed, and $n_{t(q)}$ is the number of firms that are still in operation up to that time $t(q)$. As a result, these firms are still at risk of failure at time $t(q)$. $d_{t(q)}$ is the number of firms that exited at time $t(q)$ (Le 1997). The survival rates at different times can be plotted against firm duration, resulting in the *Kaplan-Meier curve*. In addition to plotting the survival function using the Kaplan-Meier curve, we can plot the hazard rate using the formula $h\hat{c}_{t(q)} = d_{t(q)}/n_{t(q)}$, where $h\hat{c}_{t(q)}$ represents the hazard rate at time $t(q)$. Using the Kaplan-Meier estimator, we can calculate the survival function for an Internet firm at age $t(q)$ after issuance of its IPO and plot it as a Kaplan-Meier curve. In addition, we can plot the cumulative hazard function using the function $H_t = \int_0^t h(x)dx = -\ln(S_t)$.

In contrast to the methods that describe survival patterns, we can employ methods that are helpful to explain survival phenomena. When we only consider the final outcome of failure or survival, we can use logistic regression to examine how a set of explanatory variables leads to the observed exits. However, when we want to take into consideration both the final outcome and the timing of the outcome, survival analysis offers semiparametric and fully parametric techniques that are very useful.

8.A.1.3 Logistic Regression

Logistic regression is a member of a family of *discrete choice models* that are widely used in economics, social sciences, and epidemiology to handle dependent variables that are not continuous. The probability of the dependent variable being a certain choice is $Prob(y = 1) = \frac{e^{\beta' x}}{1 + e^{\beta' x}} = \Lambda(\boldsymbol{\beta}'\boldsymbol{x})$, a logistic cumulative distribution function. With choices of $y = (y_1, y_2, \ldots, y_n)$ for n observations in the sample, the joint probability is $Prob\,(y = (y_1, y_2, \ldots, y_n)) = \prod_{i=1}^{n} [F(\boldsymbol{\beta}'\boldsymbol{x}_i)]^{y_i}[1 - F(\boldsymbol{\beta}'\boldsymbol{x}_i)]^{1-y_i}$ (Powers and Xie 2000). Taking the logarithm, the coefficients can be estimated with maximum likelihood methods, and evaluated with the usual χ^2 goodness-of-fit statistic.

The marginal effects in a logit model are given by $\frac{\partial E[y|x]}{\partial x} = \Lambda(\boldsymbol{\beta}'\boldsymbol{x})[1 - \Lambda(\boldsymbol{\beta}'\boldsymbol{x})]\boldsymbol{\beta}$, and they are calculated at the means of the independent variables. The *odds ratio* is often used to interpret the estimated coefficients. The *odds of an outcome* is the ratio of the probability that the outcome will occur over the probability that it will not. For binary dependent variables, the odds of $y = 1$ is $\frac{Prob(y=1)}{Prob(y=0)} = \frac{Prob(y=1)}{1 - Prob(y=1)}$. The *odds ratio* is the impact of a variable on the odds of an outcome. A coefficient estimate of β_i for the ith independent variable is associated with an odds ratio of $exp(\beta_i)$. Using logistic regression, we can examine how different factors may affect the survival outcomes of Internet firms.

8.A.1.4 The Cox Model

One widely used semiparametric survival analysis technique is the Cox proportional hazards model (Cox 1975). For a firm at time t, its hazard function has a nonparametric baseline hazard $h_0(t)$ that depends only on time t and a parametric part (with βs to estimate the strength of the effects) that reflects the impact of market, firm, and e-commerce variables on the hazard rate, $h(\boldsymbol{t}, \boldsymbol{x}, \boldsymbol{\beta}) = h_0(t)\exp(\boldsymbol{\beta}'\boldsymbol{x})$. The independent variables x differ across firms over time. We can obtain the *cumulative baseline hazard function* (Le 1997) as $H_0(t) = \int_0^t h_0(y)dy$. The *partial likelihood function* is $PL(\beta) = \prod_{i=1}^{I} \left[\frac{\exp(\boldsymbol{\beta}'\boldsymbol{x}_{i,t(i)})}{\sum_{j \in R(t(i))} \exp(\boldsymbol{\beta}'\boldsymbol{x}_{j,t(i)})} \right]^{c_i}$, where I is the sample size with each firm denoted by i. The notation $t(i)$ here is firm i's duration up to its failure (Hosmer and Lemeshow 1999).[3] $R(t(i))$ is the set of firms at risk of failing at $t(i)$, when firm i fails. This includes all firms with durations that are the same as or longer than the duration of firm i. $x_{i,t(i)}$ is the vector of independent variables for firm i at the time it fails, and $x_{j,t(i)}$ is the vector of time-varying covariates for firm j that is at risk of failure. c_i is 0 if observation i is censored and 1 otherwise. The baseline hazard cancels out.[4] In addition to the Cox proportional hazards model, researchers can use parametric models when they make assumptions about the distribution of the baseline hazard function. Two frequently used distributions are the exponential

[3]We distinguish between $t(i)$ used in the partial likelihood (PL) function and $t(q)$ used in the Kaplan-Meier estimator. $t(i)$ denotes firm i's duration. $t(q)$ refers to points in time before t and when Internet firm exits were observed.

[4]An intercept is unnecessary because variations in the hazard rate that are the same for all firms at the same age are absorbed into the baseline hazard. The PL function assumes no *tied durations*: No two firms exited at the same duration after issuing an IPO. Depending on the number of ties, an adjustment to the PL function is necessary to account for the conjoint probability of observing two or more events at the same time.

and Weibull distributions. Using semiparametric and parametric survival analysis, we can examine how a set of explanatory factors may affect Internet firm survival and the timing of the exits.

8.A.2 DATA VISUALIZATION METHODS

Data visualization methods use graphical techniques such as points, lines, and maps to represent quantitative, spatial, or temporal information (Tufte 1990, 2001). Data visualization involves graphing the data and fitting mathematical models to them (Cleveland 1993). Some of the simple techniques include the use of histograms, scatter plots, multiway dot plots, quantile plots, quantile-quantile plots, and box plots to plot the data directly, data after transformation, or the residuals. Function-fitting involves estimating linear or smooth curves for the data. Information systems (IS) researchers are also starting to use data visualization methods. For example, Shmueli and Jank (2005) construct profile plots of willingness-to-pay data over the course of one or many online auctions. They also use box plots to represent the bidding intensity in auctions.

We can visually describe the underlying dynamics of Internet firm survival from three perspectives. First, we can display the number of Internet firms in total and by sector at different points in time, which gives us an overall picture of how the number of Internet firms has been changing over time. Second, we can display the number of Internet firm entries and exits. This is equivalent to the first derivative of the number of firms, and it shows us the *velocity* of entry or exit. Third, we can plot the *acceleration* in Internet firm entry and exit.

8.A.3 FUNCTIONAL DATA ANALYSIS

FDA is another method that can be used to describe Internet firm survival patterns. FDA views curves or images as representing functions and aims to analyze such functional data (Ramsey and Silverman 1997). When a process can be observed multiple times and hundreds of observations per unit are available, FDA can be used to plot the trajectory of value changes of a variable. In addition to plotting the curves, researchers can use data-smoothing methods to estimate the functional object that best describes the curves, and also plot the first and second derivatives to examine the velocity and acceleration of the variable. Some of the most often used smoothing methods include kernel smoothing, local polynomial smoothing, and spline smoothing (Simonoff 1996).

In IS research, Shmueli and Jank (2007) use FDA to analyze how bid prices change with the progression of online auctions. In this case, FDA is helpful because researchers can repeatedly observe the prices over a seven-day auction period for hundreds of items and establish meaningful empirical price trajectories.

For Internet firm survival, we have observed similar processes for multiple firms with revenues, financial capital, number of employees, cash on hand, etc. We can then use FDA to visually plot such data, see how the curve varies for different firms, and use data-smoothing methods to estimate the functional object describing the overall pattern. In addition, we can plot the first and second derivatives in Internet firm revenues, financial capital, number of employees, cash on hand, etc., to show how these variables change and the acceleration or deceleration of the changes.

8.A.4 BAYESIAN DATA ANALYSIS

In addition to the above-mentioned methods, researchers can use Bayesian statistics in examining Internet firm survival. Bayesian data analysis is based on the Bayes theorem, which allows updating of the distribution of a parameter given some observed data and prior knowledge about its distribution (Congdon 2003). It is especially helpful in research contexts where historical data are available as a good starting point for parameter estimation. More formally, Bayesian analysis in this context specifies the posterior distribution of a parameter, θ, given the observed data, D, and a prior distribution for $\pi(\theta)$: $\pi(\theta|D) = \frac{L(\theta|D)\pi(\theta)}{\int_{\Theta} L(\theta|D)\pi(\theta)d\theta}$. The θ here is the parameter we want to estimate. $\pi(\theta)$ is the known prior distribution for θ, and $\pi(\theta|D)$ is the updated posterior distribution for θ. $L(\theta|D)$ is the likelihood function for θ given observed data D, with Θ the parameter space of θ (Ibrahim et al. 2001). Previous data are incorporated into the posterior distribution through $\pi(\theta)$, and the current data contribute to the posterior distribution through $L(\theta|D)$. A Markov chain Monte Carlo (MCMC) simulation technique called the *Gibbs sampler algorithm* is frequently used to generate the parameter estimates (Smith and Roberts 1993).

9

MODELING TIME-VARYING COEFFICIENTS IN POOLED CROSS-SECTIONAL E-COMMERCE DATA: AN INTRODUCTION

ERIC OVERBY

Georgia Institute of Technology, College of Management, Atlanta, GA

BENN KONSYNSKI

Emory University, Goizueta Business School, Atlanta, GA

9.1 INTRODUCTION

E-commerce is a relatively new phenomenon, and the electronic marketplace continues to evolve at a rapid pace. New technologies, new business models, new legislation, and an ever-expanding Internet user base are some of the reasons why phenomena related to e-commerce are dynamic. Just as the electronic marketplace changes over time, it is likely that the empirical relationships that describe e-commerce change over time as well. For example, the influence of an eBay reputation score on the winning price of an auction may change over time as users become more experienced with online trading or as new transaction assurance mechanisms are developed.

The purpose of this chapter is to describe several statistical methods available for testing whether relationships uncovered in e-commerce research evolve over time. In particular, we focus on methods to assess whether coefficients in regression models vary over time. The literature on time-varying regression coefficients is dispersed, and the motivation for this chapter is to consolidate and summarize this literature and

Statistical Methods in e-Commerce Research. Edited by W. Jank and G. Shmueli

discuss its application to e-commerce research. Our discussion is of an introductory nature and is designed to be accessible to researchers with varied backgrounds. For example, we have provided graphical illustrations, as well as the more typical textual and notation-based depictions, to illustrate many of the methods. More technical treatments are available elsewhere (e.g., Brown et al. 1975; Hastie and Tibshirani 1993; Fan and Zhang 1999; Orbe et al. 2005).

The purpose of this chapter is not to critique existing research practice but rather to influence future practice. It is not that e-commerce researchers are misapplying techniques to investigate how empirical relationships might evolve over time; it is that too often they are not applying them at all! Many datasets collected for e-commerce research span time (e.g., eBay or Amazon.com data gathered over the course of several months), but analysis is often done on only the pooled data, while potential dynamism over time is left unexplored. Although the data provide the opportunity to investigate how relationships among variables evolve over time, this opportunity is too often left to lie fallow or else shunted into the purgatory referred to as "future research opportunities." This may be a function of data limitations (such as overly short time spans), but it may also be due to a lack of awareness of available methodology. As we will discuss in this chapter, there are multiple statistical methods that may be used to test for changes in empirical relationships over time, many of which do not require particularly advanced statistical training. We hope that this chapter stimulates e-commerce researchers to analyze how relationships within their data may be evolving. After all, due to its relative youth, e-commerce is a particularly dynamic research field.

The chapter is structured as follows. First, we discuss how the method by which data are collected and how the resulting structure of the dataset affects the types of statistical methods that are appropriate, focusing on a particular dataset structure found frequently in e-commerce research: pooled cross sections gathered over time. Second, we present a dataset from the wholesale automotive industry that we will use for illustrative purposes throughout the chapter. Third, we discuss available statistical methods that are appropriate for testing for time-varying relationships in the types of pooled cross-sectional data often found in e-commerce research. We discuss methods to test for both continuous and discontinuous change in individual coefficients as well as sets of coefficients, including (1) tests for structural change, such as the Chow test and the CUSUM/MOSUM test; (2) rolling regression; and (3) varying coefficient models in which the coefficients are modeled as functions of time. We close with a discussion of how researchers might use these methods in concert to analyze how relationships in e-commerce research vary over time.

9.2 DATA STRUCTURES: POOLED CROSS SECTIONS, PANEL DATA, AND TIME SERIES DATA

Much of the data used in e-commerce research are either (1) downloaded from the Internet (either manually or via software agents) or (2) given to a researcher by a firm engaged in e-commerce. It is common for these data to span time. For example, some of the observations may have occurred in January, others in May, and others in September. Thus, it is possible to examine how relationships within these datasets evolve over time. For example, a researcher can determine if a given relationship is the same at the end of the time span covered by the dataset as at the beginning.

Collecting observations from multiple points in time can cause the resulting data to be structured as pooled cross sections, a time series, or a panel. How the researcher gathers data and the resulting data structure have implications for which statistical methods are appropriate. In this section, we discuss each of these data structures, along with the assumptions underlying the statistical methods used to analyze them. In the balance of the chapter, we discuss methods appropriate for analysis of pooled cross-sectional data, which is common in e-commerce research.

Prior to discussing pooled cross-sectional data, we present a definition of these data. *Cross-sectional data* consist of multiple observations at a given point in time (Wooldridge 2002). For example, if a researcher records eBay Motors transactions conducted on May 1, these data represent a cross section. If the researcher augments this data by recording eBay Motors transactions conducted on other days, the combined dataset represents pooled cross sections. Thus, *pooled cross-sectional data* result when cross sections from different points in time are combined in the same dataset (Wooldridge 2002).

When gathering pooled cross-sectional data, the researcher makes no explicit attempt to gather data on the same observational units (e.g., people, firms, products) in each cross section. Continuing the eBay Motors example, this means that the researcher is not tracking the same buyer, seller, or vehicle over time. Instead, he may observe a given set of buyers, sellers, and vehicles in the first cross section (taken at time 1), then observe a different set in the second cross section (taken at time 2), and then observe a different set in the third cross section (taken at time 3), and so on. If he observes the same buyer, seller, or vehicle in multiple cross sections, this is taken as coincidental rather than by design. This represents the key distinction between pooled cross-sectional data and time series or panel data. Time series and panel data are constructed by measuring the *same* observational units, such as the same people, the same firms, the same websites, the same products, etc., over time.

Briefly, a *time series* consists of observations of the same observational unit made sequentially in time (Chatfield 1996). For example, the number of visitors to eBay Motors measured on a weekly basis represents a time series. A *univariate* time series is one in which only a single variable is measured. A *multivariate* time series is one in which multiple variables are measured, such as the number of website visitors and the number of page views. Whereas a time series consists of observations of a single observational unit over time, a *panel dataset* consists of observations of a set of observational units over time (Wooldridge 2002). For example, monthly website traffic data for 10 Internet auction sites between 2000 and 2005 would constitute a panel dataset. If data for each month are available for all 10 firms, the panel is said to be *balanced*. If data are missing for some firms in certain months, the panel is said to be *unbalanced*. The term *longitudinal data* is sometimes used to describe data structured as a panel (Wooldridge 2002), as is the term *time-series cross-section* data (Beck and Katz 1995). Observations within time series and panel data are usually recorded at regular intervals, such as daily, weekly, monthly, or annually.

Note that multiple time series measured over the same periods result in the same structure as panel data. For example, if individual time series measuring the number of website visitors on a weekly basis were collected for 10 firms, the combined data would have a panel structure. A representation of the relationship between cross-sectional, pooled cross-sectional, time series, and panel data is shown in Figure 9.1.

Many of the datasets used in e-commerce research are pooled cross sections. First, consider e-commerce transaction data, such as those scraped from eBay. A common data collection design is to download all transactions meeting a certain criterion, such as those for a certain product category, over a period of time ranging from a few days to several months. Often there is no explicit attempt to restrict the data collection to those buyers, sellers, or specific products that appear in each time period. Thus, instead of the data consisting of the same observational units measured at multiple times (as would be the case for panel data), the data consist of different units measured at multiple times (as is the case for pooled cross-sectional data). Second, consider clickstream data, which tend to be structured similarly to e-commerce transaction data in that they contain observations of different units over time. In addition, clickstream data are often de-identified, so that even if the same unit appears multiple times in the data, it may be impossible to distinguish that unit from the others. Similar issues may arise in the analysis of other types of usage logs, such as those produced by transaction systems.

Whether the same or different observational units are measured over time has important implications for statistical methodology. Methods designed for the analysis of time series and panel data are typically designed to leverage the fact that the same

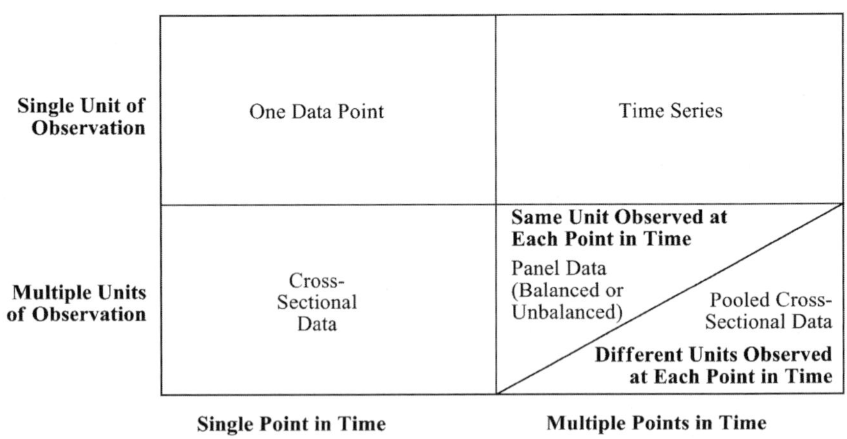

Figure 9.1 Graphical representation of cross-sectional, pooled cross-sectional, time series, and panel data. In panel data, the same observational units (e.g., people, households, firms, products) are measured repeatedly over time. In pooled cross-sectional data, different observational units are measured over time. If the same unit appears in multiple time periods in pooled cross-sectional data, this is taken as coincidental rather than by design.

observational units are measured over time. In particular, these methods build upon the common assumption that the observations for a given unit in a dataset are correlated. In a time series, it is commonly assumed that the observation at time t is correlated with the observation at time $t + 1$. For example, the number of visitors to a website this week is frequently assumed to be a good predictor of the number of visitors next week. The corresponding assumption frequently made within panel data is that observations for each unit observed through time are correlated. For example, in a panel dataset consisting of web traffic data for 10 firms over 52 weeks, it is frequently assumed that the observations for each firm are correlated, as there may be firm-specific factors such as industry membership and website design that influence the amount of traffic that firm receives on its website.

Because these methods build upon the assumption that observations are correlated, they are typically not appropriate for the types of pooled cross-sectional data often used in e-commerce research. This is because there is no a priori expectation in pooled cross-sectional data that the observations are correlated, which is reasonable given that the observations are of different units. In the remainder of this chapter, we focus on a set of statistical methods that are appropriate for analyzing time-varying relationships (represented as regression coefficients) within pooled cross-sectional data. These methods can be used to detect time-based change in an entire set of coefficients as well as in a specific coefficient. We limit our inquiry to regression models. We hope that this discussion will complement the more widely available discussions on investigating time-varying relationships in time series and panel data (Hamilton 1994; Enders 2004).

9.3 MODELING TIME-VARYING COEFFICIENTS IN POOLED CROSS-SECTIONAL DATA

Before discussing the methods available to investigate time-varying coefficients in pooled cross-sectional data, we introduce an empirical example used to illustrate the methods.

9.3.1 The Used-Automobile Wholesale Market

The empirical context is the used-automobile wholesale market. This market facilitates the exchange of used vehicles between institutional sellers and buyers. Sellers include rental car companies, the financial affiliates of automotive manufacturers, and other fleet operators who wish to sell large quantities of used vehicles in the wholesale market. Buyers are typically licensed automobile dealers who seek to purchase used vehicles in the wholesale market for resale to the consumer public. For example, in order to dispose of vehicles no longer appropriate for rental, a rental car firm may use the wholesale automotive market to sell hundreds of late-model vehicles from its fleet. Automobile dealers use the market to purchase the vehicles at wholesale prices, which they then resell to the consumer public at retail prices. There are several intermediaries in this market that provide a range of market-making services, such as arranging for an ample supply of vehicles, attracting a mass of potential buyers, certifying vehicle quality, facilitating price discovery, and providing transaction support. As this is a wholesale market, the consumer public is rarely allowed to participate.

The wholesale automotive market has traditionally operated as a physical market in which buyers, sellers, and vehicles are all collocated at a physical facility where *sales events* are held. In a typical sales event, hundreds of vehicles are driven, one at a time, into the midst of a group of buyers, who then bid on and purchase the vehicles via an ascending auction process. Recently, a new market mechanism has been introduced whereby some vehicles are displayed electronically via a flat-panel monitor rather than physically driven through the facility. An unusual feature of this market is that both the physical and electronic vehicle presentation mechanisms are used in the same sales events. For example, if a sales event involves 200 vehicles, the first 5 might be presented physically, the next 5 electronically, and so on. Figure 9.2 depicts the physical and electronic vehicle presentation mechanisms.

We have collected a dataset of over 100,000 vehicle transactions that occurred in this market over a three-year span. In this chapter, we use a subset of these data to illustrate the methods to test for time-varying coefficients. This subset consists of 10,211 vehicle transactions that occurred in 70 discrete sales events between August 2004 and January 2005. We chose this subset because it represents a portion of the larger dataset in which the relationships appeared particularly dynamic. The data presented here are for illustrative purposes only. A comprehensive analysis of the full dataset, including a discussion of the multiple theoretical mechanisms that might cause the relationships within the data to change over time, is beyond the scope of this chapter due to our focus on statistical methods. That analysis is reported elsewhere (Overby and Jap 2007).

In approximately 17% of the transactions in the example data, the vehicle was presented electronically; in the other 83%, it was presented physically. We use a dummy variable (ELECTRONICVEHICLE) to represent how the vehicle was presented. This variable is set to 1 for vehicles presented electronically and 0 for vehicles presented physically. The dependent variable in our examples is the price of the vehicle (PRICE). In addition to ELECTRONICVEHICLE, the other independent variables in our examples are as follows. VALUATION measures the wholesale valuation for each vehicle in the dataset on the day it was sold. VALUATION is established by the intermediary

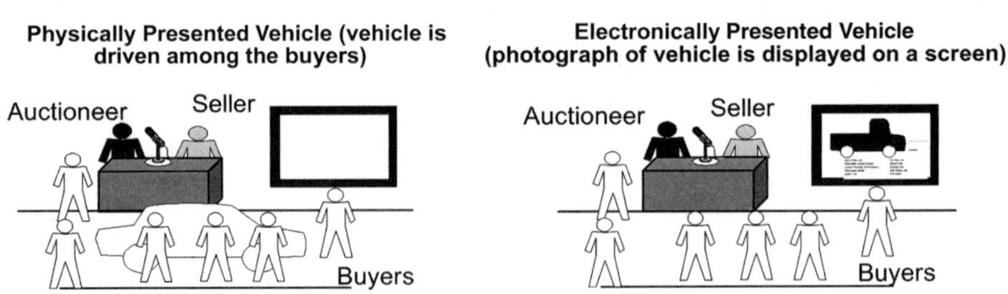

Figure 9.2 The physical and electronic vehicle presentation methods.

based on a vehicle's year, make, model, and mileage. It is based on historical sales prices and is calculated on a 30-day rolling basis, which helps to control for possible seasonal variation. VALUATION accounts for most of the variance in price attributable to vehicle characteristics and is a powerful control variable for our purposes. VALUATION does not take into account vehicle condition, however, so we account for this separately by including the condition grade assigned to each vehicle by the intermediary. CONDITIONNUMBER ranges from 0 to 5, where 0 represents a vehicle suitable only for scrap parts and 5 a near-pristine vehicle. We also include a control variable for the number of buyers who participated in a sales event (NUMBERBUYERS), as this is known to affect price in ascending auctions. Last, we include dummy variables for each seller (SELLERDUMMIES) to control for the possibility that the price of a vehicle depends on a characteristic of the seller, such as trustworthiness or reputation. Table 9.1 and Figure 9.3 provide summaries of the data.

These data are an instance of pooled cross-sectional data. The unit of analysis is the vehicle (identified by its Vehicle Identification Number, or VIN). Each VIN is observed in the dataset only once. Thus, the dataset consists of observations of different units over time. The cross sections are taken at multiple points in time between August 2004 and January 2005.

Note that it would be possible to convert these data into a panel if we changed the unit of analysis from individual vehicles identified by their VINs to vehicle groups based on make/model combinations (e.g., Ford Taurus, Honda Accord). This is because, although we do not observe the same VINs in the dataset over time, we do observe the same make/model combinations over time. Table 9.2 provides an illustration of how this dataset could be "panelized." The main drawback to structuring the data in this manner is that it throws away data by assuming that all vehicles of a given make/model are homogeneous. This is an unrealistic assumption for used vehicles, as there is considerable heterogeneity among used vehicles of the same make and model.

TABLE 9.1 Summary Statistics

Variable	Summary Statistic
Price	Mean: 12914.61 St. Dev.: 10546.74
ElectronicVehicle	# Presented Physically: 8478 (83.0%)
	# Presented Electronically: 1733 (17.0%)
Valuation	Mean: 13464.89 St. Dev.: 10578.28
ConditionNumber	Mean: 3.06 St. Dev.: 0.92
	Count of Vehicles: Condition 0 = 132
	Count of Vehicles: Condition 1 = 318
	Count of Vehicles: Condition 2 = 1887
	Count of Vehicles: Condition 3 = 4769
	Count of Vehicles: Condition 4 = 2668
	Count of Vehicles: Condition 5 = 437
NumberBuyers	Mean: 76.78 St. Dev.: 24.68
SellerID	42 sellers in the dataset (largest seller = 2396 observations, smallest seller = 1 observation)

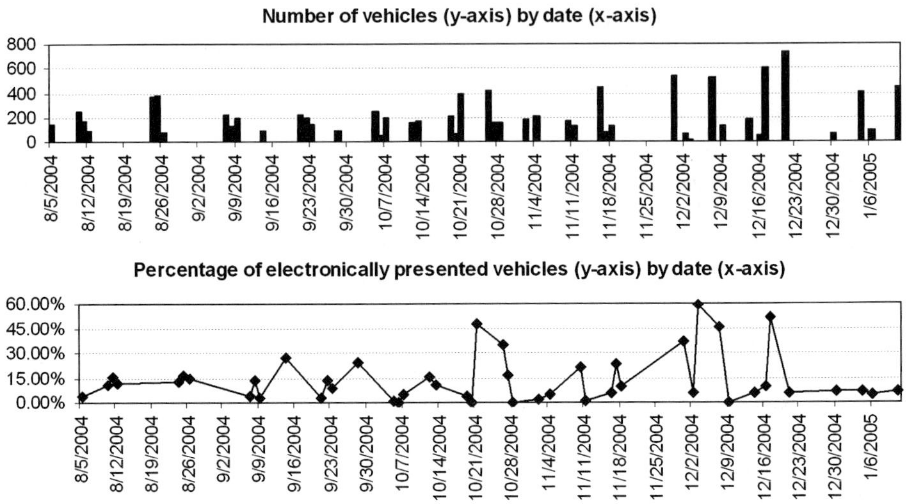

Figure 9.3 Plots of the number of vehicles and the percentage of vehicles presented electronically over time.

In the next section, we discuss multiple methods for testing for time-varying coefficients within the type of pooled cross-sectional data described above. We use the example automotive data to illustrate the methods. In particular, we consider whether the coefficient for the ElectronicVehicle variable changes over time.

TABLE 9.2 Two Ways to Structure the Wholesale Automotive Data

Native Format (Pooled Cross-Sectional)				Transformed Format (Restructured as a Panel)			
Unit of analysis is the vehicle as specified by the VIN.				Unit of analysis is the make/model combination.			
Each VIN is observed only once, at a specific time period (e.g., T1, T2, or T3).				Each make/model combination is observed multiple times in multiple time periods.			
	T1	T2	T3		T1	T2	T3
VIN: 1FAFP52U9WA165605 (Ford Taurus)	✓			Ford Taurus	✓	✓	✓
VIN: 1FAHP56S02A121156 (Ford Taurus)			✓	Honda Accord	✓	✓	✓
VIN: 1FAHP56SX4A219705 (Ford Taurus)		✓					
VIN: 1HGCG56611A112869 (Honda Accord)		✓					
VIN: JHMCG5665YC038981 (Honda Accord)	✓						
VIN: 1HGCG5641YA127413 (Honda Accord)			✓				

9.3.2 A Set of Methods

We present several methods to test for time-varying coefficients. These methods apply to testing both continuous and discrete changes in coefficients, and can be used to test changes in entire coefficient sets or in a specific coefficient. The methods we discuss are

- tests for structural change, including the Chow test and the CUSUM/MOSUM test
- rolling regression
- varying coefficient models

9.3.2.1 Tests for Structural Change. Tests for structural change are useful to diagnose whether coefficients are stable within a dataset or if they shift, either as a function of time or some other variable. Two common methods to test for structural change are the Chow test and the CUSUM/MOSUM test. (CUSUM is short for "Cumulative Sum" and MOSUM for "Moving Sum.") In the context of time-varying coefficients, the Chow test is useful if the coefficients are believed to have shifted after some specific event, while the CUSUM/MOSUM test is useful to diagnose when a shift might have occurred.

9.3.2.1.1 The Chow Test. The Chow test is commonly used to test for "structural breaks" within a dataset (Chow 1960). It can be used to compare coefficients associated with different subsets of observations. The Chow test can be useful in testing for time-varying coefficients, as it allows a researcher to test whether coefficients at one point in time are different from those at another point in time. Figure 9.4 provides a pictorial representation of how a Chow test can be used to estimate coefficients from different subsets within a dataset (referred to as *regimes*), which can then be compared to one another.

The first step in conducting a Chow test is to identify the event after which the coefficients are hypothesized to have changed. Examples of such events in e-commerce research include the introduction of a new website feature and the passage of an important piece of Internet-related legislation. Once the event has

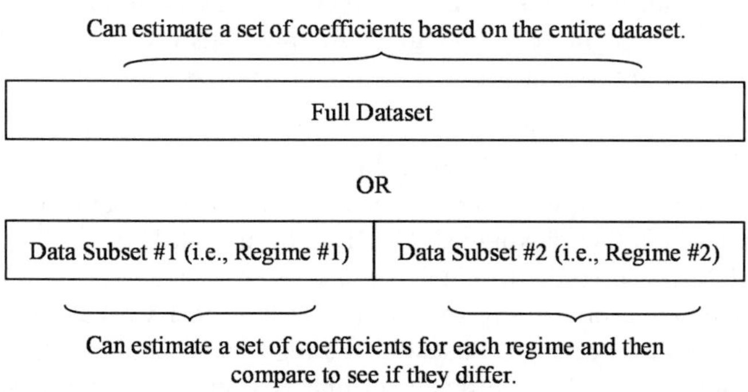

Figure 9.4 Pictorial overview of a Chow test for structural breaks.

been identified, the next step is to divide the data into two regimes. The first regime contains all observations before the event, and the second regime contains all observations after the event. The next step consists of fitting a regression model multiple times: once using the entire sample and once for each individual regime.

The Chow test statistic is given by

$$\frac{SSR_{\text{pooled}} - (SSR_1 + SSR_2)/k}{SSR_1 + SSR_2/n - 2k},$$

where SSR_{pooled} is the sum of the squared residuals from the regression using the entire sample, SSR_1 and SSR_2 are the sum of squared residuals from the regressions using each individual regime, k is the number of variables in the regression model, and n is the total number of observations in the overall sample. The SSR terms can be computed by squaring and summing the residuals generated by each regression. After calculating the Chow test statistic, the next step is to compare it to a critical value to assess significance. The critical value is drawn from the F-distribution with k and $n-2k$ degrees of freedom at the appropriate significance level (usually 95%). If the Chow test statistic exceeds the critical value, then the coefficients are believed to vary between the two regimes.

A drawback to the Chow test is that it is sensitive to the way the regimes are delineated. In some cases, a specific event may suggest a natural breakpoint. However, in many cases, it is difficult to determine a priori what constitutes the appropriate breakpoint. For example, there is no event that represents a natural breakpoint in our automotive example. Despite this, we can still illustrate the Chow test using the example data. We first divide the data into two regimes. The "Early" regime consists of the observations during the first three months of the dataset, and the "Late" regime consists of the observations during the last three months of the dataset. We then regress PRICE on an intercept, ELECTRONICVEHICLE, VALUATION, CONDITIONNUMBER, NUMBERBUYERS, and SELLERDUMMIES using the entire sample and then using each of the two regimes. A Chow test indicates that the coefficients in the Late regime are significantly different than those in the Early regime. The Chow test statistic = 2.72 with 44 and 10,123 degrees of freedom, which is significant at the 1% level. This provides some evidence that the coefficients are evolving, but it doesn't provide much insight into the process of the evolution or indicate which specific coefficients might be evolving. These questions can be analyzed via other methods presented in this chapter.

9.3.2.1.2 The CUSUM/MOSUM Test. In constrast to the Chow test, the CUSUM/MOSUM test does not require a priori specification of a breakpoint. We first discuss the CUSUM test and then distinguish CUSUM from MOSUM.

The CUSUM test uses the residuals from a series of regressions to determine if coefficients are stable or varying over time. Assume that we have n observations ordered by when they occurred. First, we fit a regression model to the first k observations, $k < n$. We use the coefficient estimates from the first k observations to

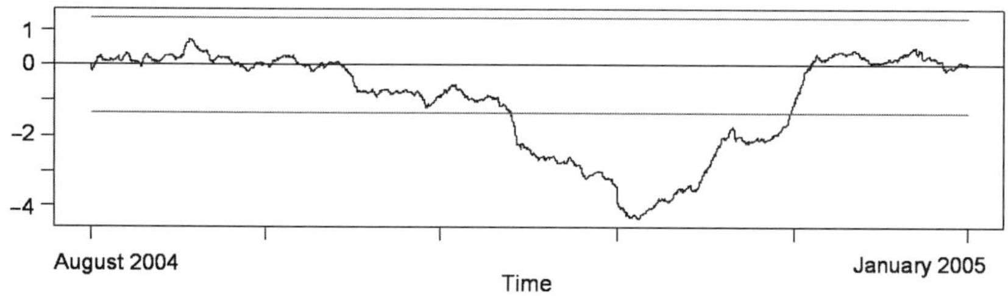

Figure 9.5 Normalized residuals obtained from the CUSUM procedure.

predict the value of observation $k + 1$. We then measure the accuracy of the prediction by examining the residual, which is the difference between the predicted value and the actual value. This process is repeated as k grows. As long as the coefficients are stable, each prediction will be approximately as accurate as the previous prediction. However, if the coefficients change, then predictive accuracy will decrease. This is reflected in an increase in the absolute value of the residuals.

Figure 9.5 shows a plot of the normalized residuals obtained from a CUSUM procedure on the example data. If the coefficients are stable, then the CUSUM (cumulative sum) of the normalized residuals should be approximately 0. The upper and lower lines in Figure 9.5 represent a 95% confidence interval around 0. Predictions appear relatively stable in the first 20% of the observations, after which predictive accuracy begins to drop. Predictions stabilize again toward the end of the dataset. This can be interpreted as follows. The regression fitted using the first 20% of the observations is predictive of the next observations, but the regression fitted using the first 50% of the observations is not. This suggests that the coefficients were relatively constant through the first 20% of the observations but then shifted between the 20% and 50% marks. Note that the regression fitted using the first 80% of the observations is predictive of the next observations. This is because including the additional 30% of the observations in this window effectively smooths out the dynamism present in the middle part of the data. Taken as a whole, the CUSUM test suggests that the coefficients change over time. If they did not, the CUSUM residuals would remain close to 0 throughout the entire time span.

The CUSUM procedure uses all previous observations to predict the next observation. By contrast, the MOSUM procedure uses only the previous g observations. In other words, MOSUM predicts each observation based on a rolling window of prior observations. The results of the MOSUM procedure fitted to the example data, using a window set to 15% of the total observations, are displayed in Figure 9.6. Figures 9.5 and 9.6 show a consistent pattern. Predictions are relatively accurate at the beginning and end of the example data but not in the middle. The "dip" at the midpoint of the MOSUM plot and the "peak" beyond it correspond to the "valley" in the CUSUM plot. To see this, note that the valley in the CUSUM plot is decreasing during the dip portion of the MOSUM plot and then increasing during the peak portion of the MOSUM plot. This is because the CUSUM uses all prior observations, whereas the MOSUM uses only the g prior observations.

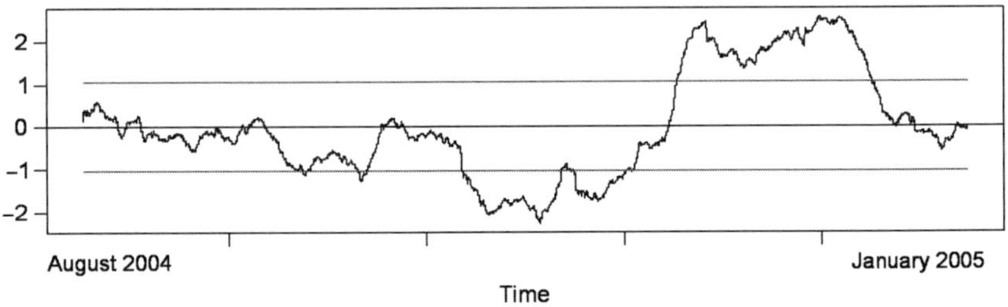

Figure 9.6 Normalized residuals obtained from the MOSUM procedure.

The original CUSUM procedure is attributed to Brown et al. (1975). A more recent discussion of models to detect structural change, including CUSUM/MOSUM procedures and the Chow test, is provided by Andrews (1993). Both the CUSUM and MOSUM procedures are available in statistical packages. For example, they are both implemented in the strucchange package for the R program (http://www.r-project.org).

These types of models are useful to detect whether entire sets of coefficients change over time, but they are not helpful if a researcher is interested in whether a specific coefficient changes over time unless the regression model contains only a single coefficient. We will now discuss models designed to detect changes over time in a specific coefficient.

9.3.2.2 Rolling Regression. Rolling regression, also known as *moving regression* or *moving window regression*, is one method used to investigate time-based effects in a specific coefficient. Rolling regression fits a regression model to a dataset multiple times by moving forward through the dataset in a rolling fashion (Brown et al. 1975). For example, assume that a dataset consists of 1500 observations at different points in time. In a rolling regression procedure, the researcher first orders the observations from earliest to latest and then fit a regression model to the first k observations. For example, let $k = 100$. This means that the first regression is fitted using observations 1 through 100. The researcher records this initial set of coefficient estimates and then fits the same regression model using observations 2 through 101. After recording this set of coefficient estimates, the researcher fits the model using observations 3 through 102, and so on until the end of the dataset. The parameter k is referred to as the *window size*. Rolling regression produces multiple sets of coefficient estimates, depending on the window size, the number of observations in the dataset (denoted n), and the *step size* (denoted s), which is the increment by which the window is moved each iteration.[1] The formula for the number of regressions and the resulting sets of coefficients produced by a rolling regression procedure is $[(n - k)/s] + 1$. The resulting coefficient estimates can be plotted against time to assess whether

[1]Rolling regression is similar to the MOSUM procedure in that each fits the same regression model to a series of windows of observations. They differ in that MOSUM provides a statistical test of whether the sum of the residuals at each window is distinguishable from 0; rolling regression does not.

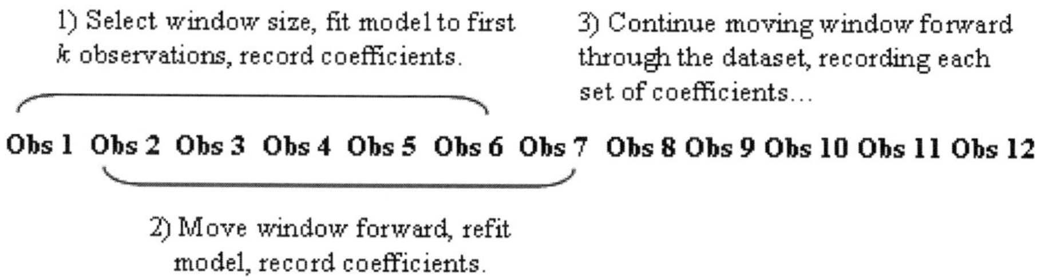

Figure 9.7 Pictorial overview of the rolling regression procedure.

they appear to change over time. Mean-comparison tests can be used to determine if one set of coefficients produced by the rolling regression (e.g., those from the first half of the windows) differs from a different set (e.g., those from the second half). Rolling regression procedures are available in many statistical packages. For example, in STATA the procedure is called *rolling*. Figure 9.7 provides a pictorial overview of the rolling regression procedure.

We illustrate rolling regression using the wholesale automotive data by first ordering the 10,211 transactions according to which occurred first. We then select the window size. Selecting the window size is something of an art, as the window should be large enough for each regression to have adequate power but small enough to reflect changes in the coefficients over time (otherwise, the estimates will too closely resemble the estimates for the sample as a whole). Larger window sizes will cause the plot of coefficients over time to appear smoother. This is because influential observations will have less impact when included in larger windows. Smaller step sizes will also have a smoothing effect on the plot. This is because each window will have more observations in common with the previous window, thereby causing the coefficient estimates to be similar. (Conversely, setting the step size equal to the window size will ensure that adjacent windows contain no shared observations.) Performing sensitivity analyses based on different window and step sizes is good practice. Figure 9.8 displays a plot of the ElectronicVehicle coefficient for different window sizes and a step size = 20 (plots using a step size = 100 are similar). Not all coefficients are plotted in order to maintain resolution in the graphic (we show the coefficients sampled at evenly spaced intervals).

Figure 9.8 suggests that the ElectronicVehicle coefficient changes over time, and also illustrates that the smoothness of the plots produced by rolling regression varies based on the chosen window size. In all of the plots, the ElectronicVehicle coefficient is smaller in absolute value for the second half of the dataset than for the first. The pattern of the change appears smoother for the larger window sizes.[2]

[2]The plots shown in Figure 9.8 are not directly comparable to the CUSUM and MOSUM plots shown in Figures 9.6 and 9.7. This is because the plots in Figure 9.8 are based on only a single coefficient in the model (ElectronicVehicle), while the CUSUM/MOSUM plots are based on all the coefficients in the model (because they show the residuals).

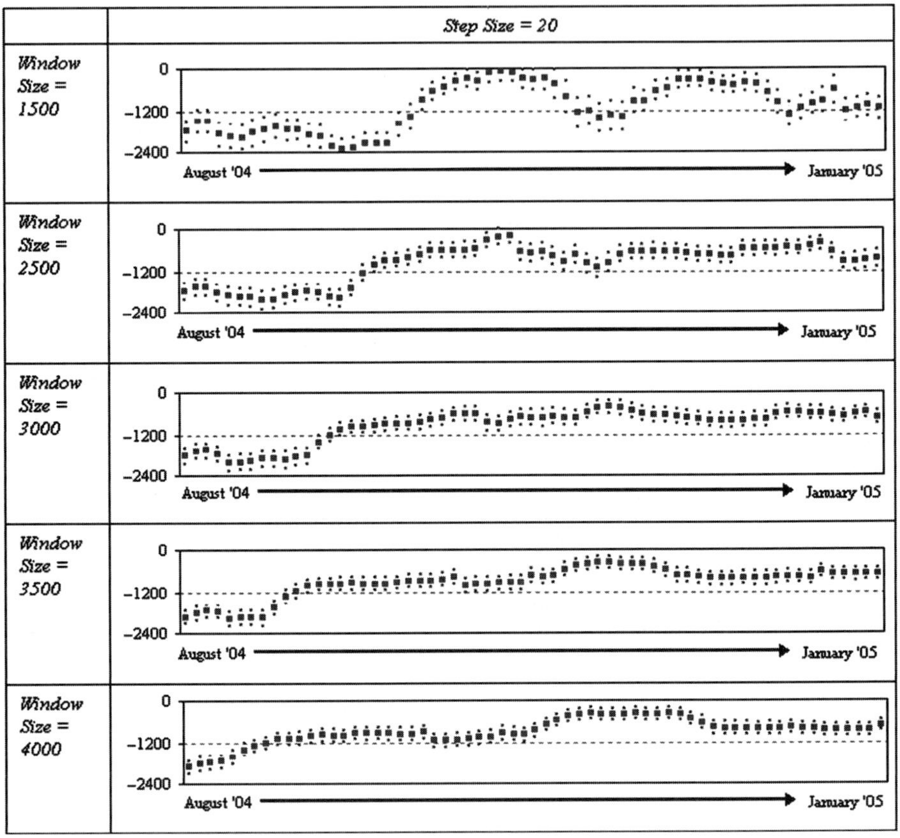

Figure 9.8 Plot of the ELECTRONICVEHICLE coefficient over time, using different window sizes. The large dots represent the ELECTRONICVEHICLE coefficient for a given data window, and the small dots represent the upper and lower bounds of the 95% confidence interval around the ELECTRONICVEHICLE coefficient estimate.

Rolling regression overcomes one of the key limitations of the Chow test in that it does not require the researcher to specify a priori any breakpoints or regimes. Rolling regression permits a view of how coefficients evolve over time on a continuous basis. Rolling regression is particularly useful if the observations are spread relatively evenly through time. In other words, the rolling regression procedure is best leveraged when used with a dataset whose observations are spread more or less continuously through time. If, on the other hand, most of the observations are grouped at specific periods in time, rolling regression may inaccurately suggest continuity in a coefficient's evolution, especially if there are only a few groups. A disadvantage of rolling regression is that it requires a relatively large dataset, as the number of observations in the dataset should be several times larger than the window size to capture the dynamism in a coefficient.

Rolling regression is one method to investigate evolution in coefficient estimates over time. Another method is to parameterize the coefficients using a varying coefficient model, which we discuss next.

9.3.2.3 Parameterizing the Coefficients via Varying Coefficients Models. This section presents methods of parameterizing coefficients to test for both continuous and discrete changes in a coefficient over time. Multiple labels have been used to describe the techniques discussed in this section, including *process functions*, *parameter functions*, *transition equations*, and *evolution equations*. These methods can be categorized as varying coefficient models (Hastie and Tibshirani 1993; Fan and Zhang 1999).

9.3.2.3.1 Modeling a Continuous Change in a Coefficient. The standard regression model assumes that the value of a dependent variable is a function of independent variables, whose influence on the dependent variable is weighted according to regression coefficients. In other words,

$$y = \beta_0 + \beta_1 x_1 + \beta_2 x_2 + \cdots + \beta_k x_k + \varepsilon, \tag{9.1}$$

where y is the dependent variable, x_1 to x_k are independent variables, the β's are regression coefficients, and ϵ is an error term. In the standard regression model, the β's are not considered to be functions of the covariates or other variables.

But consider the possibility of a time-varying beta coefficient. It is possible to represent this beta coefficient as a function that depends on time (Farley and Hinich 1970; Hinich and Roll 1981). Such a function is often referred to by terms such as *process function* (Wildt and Winer 1983; Naik and Raman 2003) or *evolution equation* (Hastie and Tibshirani 1993). For example, we can model β_1 as being a function of time (denoted as t) in the following manner:

$$\beta_{1,t} = \alpha_0 + \alpha_1 t + \alpha_2 t^2 + \cdots + \alpha_k t^k. \tag{9.2}$$

In equation (9.2), $\beta_{1,t}$ is modeled as a polynomial of degree k. α_0 represents the baseline (time-invariant) component of β_1, while the other terms capture how β_1 varies with time. The null hypothesis of $\alpha_1 = \alpha_2 = \cdots = \alpha_k = 0$ would suggest that the coefficient is unaffected by time and is therefore time-invariant. However, rejecting this null hypothesis would suggest that the coefficient is a function of time, and the values of the α coefficients would suggest how. For example, if α_1 is significant and positive, then the β_1 coefficient would increase linearly with time. A positive and significant α_2 coefficient would suggest that β_1 is increasing at a growing pace over time. Alternative evolution equations for β_1 can be constructed based on the researcher's theory of how the coefficient might vary with time. For example, equation (9.3) could be used if the researcher expects the coefficient to increase with time, but at a declining rate:

$$\beta_{1,t} = \alpha_0 + \alpha_1 \ln(t). \tag{9.3}$$

Assume that we have constructed the regression equation shown in equation (9.1), but we believe that the β_1 coefficient varies over time according to the evolution equation shown in equation (9.2). We limit equation (9.2) to be a second-degree polynomial for simplicity. We write down equation (9.1), substituting equation (9.2) in place of the β_1 term, which yields

$$y = \beta_0 + [\alpha_0 + \alpha_1 t + \alpha_2 t^2]x_1 + \beta_2 x_2 + \cdots + \beta_k x_k + \varepsilon. \tag{9.4}$$

By multiplying x_1 by the expression for β_1, we get

$$y = \beta_0 + \alpha_0 x_1 + \alpha_1 t x_1 + \alpha_2 t^2 x_1 + \beta_2 x_2 + \cdots + \beta_k x_k + \varepsilon. \tag{9.5}$$

Effectively, we have introduced interaction terms into our regression and can estimate equation (9.5) in a straightforward fashion. α_0 is the coefficient for the x_1 variable, α_1 is the coefficient for the interaction between the x_1 variable and the time variable, and α_2 is the coefficient for the interaction between the x_1 variable and the time variable squared. Inspection of the α_1 and α_2 variables indicates whether the β_1 coefficient varies over time in accordance with the evolution equation we have specified.

To illustrate this method using the automotive example, we first construct an evolution equation for the ELECTRONICVEHICLE coefficient, which we refer to as β_1. Suppose we believe the evolution equation for β_1 to be best represented as a logarithmic function such that β_1 increases over time but at a declining rate. (The rolling regression procedure gave us a clue about this.) Let

$$\beta_{1,t} = \alpha_0 + \alpha_1 \ln(t). \tag{9.6}$$

As described above, this introduces an interaction term into our model: the interaction between the ELECTRONICVEHICLE variable and the natural log of the TIME variable. (TIME ranges from 1 to 160 and represents which day within the time span a transaction occurred.) Thus, after substituting equation (9.6) for β_1, our model becomes

$$y = \beta_0 + \alpha_0 x_1 + \alpha_1 \ln(t)^* x_1 + \beta_2 x_2 + \cdots + \beta_k x_k + \varepsilon. \tag{9.7}$$

Table 9.3 lists the coefficients for this regression model, along with those in the regression model excluding the time interaction variable.

Table 9.3 indicates that the ELECTRONICVEHICLE* LN(TIME) coefficient, which is our estimate of α_1, is positive and significant. This suggests that the coefficient is changing over time.

An advantage of this type of varying coefficient model is that the researcher can explicitly model how the coefficient varies with time. Also, it is straightforward to

TABLE 9.3 Coefficients Estimates for the Base Model (without the Time Interaction Variable) and the Time-Varying Coefficient Model Depicted in Equation (9.7)

	Base Model (Does Not Include Time Interaction Term)		Time-Varying Model Depicted by Eq. (9.7) (Includes Time Interaction Term)	
	Coefficient Estimate (Std. Error)	*p*-Value	Coefficient Estimate (Std. Error)	*p*-Value
β_0: Intercept	−2200.96 (168.44)	<0.01	−2168.18 (167.96)	<0.01
α_0: ElectronicVehicle	−983.58 (54.22)	<0.01	−2738.72 (226.12)	<0.01
α_1: ElectronicVehicle* Ln(Time)	n/a	n/a	407.03 (50.92)	<0.01
β_2: ConditionNumber	723.97 (20.92)	<0.01	720.56 (34.54)	<0.01
β_3: Valuation	0.95 (0.01)	<0.01	0.95 (0.01)	<0.01
β_4: NumberBuyers	3.44 (0.66)	<0.01	3.17 (0.66)	<0.01

Coefficient estimates for SellerDummies available from the authors.
$R^2 = 0.98$ for both models. (Note that the Valuation variable is highly predictive, as the intermediary is known for its precision in estimating wholesale vehicle values.)

test whether evolution equation coefficients are statistically significant. In the model above, this can be done via a *t*-test to determine whether α_1 is statistically significant. A disadvantage of this method is that the researcher must choose the appropriate evolution equation for the coefficient. It is not always clear which equation is best, although theory and model-fitting techniques can provide guidance. Another drawback is the number of interaction terms introduced into the model, particularly if the researcher is interested in testing whether more than one coefficient varies over time. This can lead to problems with multicollinearity and statistical power. If the researcher wants to examine the possibility of multiple time-varying coefficients, then rolling regression may be an attractive alternative, as it will provide a picture of the evolution of multiple coefficients without requiring the inclusion of multiple interaction terms in the model.

It is possible for the evolution equation to be nonparametric and represented via smoothing techniques (Hastie and Tibshirani 1993; Orbe et al. 2005). It is worthwhile to note, however, that rolling regression also produces a nonparametric estimate of an evolution equation. As discussed earlier, smoothing can be achieved by adjusting the window and step sizes.[3]

[3]It is also worth mentioning that varying coefficient models are similar to hierarchical models. In both types of models, the coefficients in the focal model are modeled as functions of the covariates or other variables. We hesitate to use the term *hierarchical model* in our context, however, as these models are often prescribed when observations are nested (e.g., observations on students nested by school or observations on houses nested by neighborhood). Although this type of nesting is certainly possible in pooled cross-sectional data, it is not characteristic of them.

9.3.2.3.2 Modeling a Discrete Change in a Coefficient. The above description applies if a given coefficient is believed to evolve continuously over time. Other methods are useful if the coefficient is thought to have changed as a result of a discrete event, such as the introduction of a new website feature or the passage of new Internet-related legislation. These methods are closely related to the Chow test described above, as they involve examining whether a coefficient differs across regimes.

To illustrate using the automotive example, we first divide the dataset into regimes. The Early regime consists of the observations during the first three months of the dataset, and the Late regime consists of the observations during the last three months of the dataset. Let z represent a dummy variable equal to 0 for observations in the Early regime and 1 for observations in the Late regime.

Assume that our base regression model is

$$y = \beta_0 + \beta_1 x_1 + \cdots + \beta_k x_k + \varepsilon. \tag{9.8}$$

We define the following evolution equations for β_0 and β_1:

$$\beta_0 = \alpha_0 + \alpha_1 z \tag{9.9}$$

$$\beta_1 = \alpha_2 + \alpha_3 z. \tag{9.10}$$

Substituting equations (9.9) and (9.10) into equation (9.8) yields

$$y = \alpha_0 + \alpha_1 z + \alpha_2 x_1 + \alpha_3 (z^* x_1) \cdots + \beta_k x_k + \varepsilon. \tag{9.11}$$

For observations in the Early regime, $\alpha_1 z$ and $\alpha_3(z^* x_1)$ will drop out, as $z = 0$ for these observations. This means that the coefficient for the x_1 variable in the Early regime is simply α_2. However, for observations in the Late regime (where $z = 1$), the $\alpha_1 z$ and $\alpha_3(z^* x_1)$ terms reduce to α_1 and $\alpha_3 x_1$. This leaves two terms involving x_1: $\alpha_2 x_1$ and $\alpha_3 x_1$. By collecting terms, we can see that the coefficient for the x_1 variable in the Late regime is $\alpha_2 + \alpha_3$. Whether the coefficient in the Early regime is different from the coefficient in the Late regime—in other words, whether the coefficient has changed over time—can be examined by testing whether α_3 is statistically different from 0. We can test for a similar change in the intercept across the two regimes by examining α_1. If there is no need to test for a change in the intercept, then the β_0 term can be left in the model rather than be replaced by equation (9.9).

Table 9.4 shows the coefficient estimates for the regression model specified in equation (9.11).

Note that the estimate for α_3 is positive and significant, suggesting that the ElectronicVehicle coefficient differs between the Early and Late regimes. The estimate for α_1 is not significant, suggesting that there was no shift in the intercepts between the two regimes.

A limitation of the method discussed above is that it only permits analysis of whether the intercept and the ElectronicVehicle coefficient changed over time. The other coefficients in the model are assumed to be invariant across the two

TABLE 9.4 Coefficients Estimates Assuming a Discrete Change in the Electronic Vehicle Coefficient and the Intercept

	Coefficient Estimate	Standard Error	p-Value
α_0: Intercept	−2113.76	168.87	0.00
α_1: Z	−61.03	37.08	0.10
α_2: ElectronicVehicle	−1337.57	71.49	0.00
α_3: Z*ElectronicVehicle	609.09	81.26	0.00
β_2: ConditionNumber	722.95	20.89	0.00
β_3: Valuation	0.95	0.01	0.00
β_4: NumberBuyers	2.98	0.68	0.07

Coefficient estimates for SellerDummies available from the authors.
$R^2 = 0.98$.

regimes. This assumption may be overly restrictive. Consider that each regime might have its own regression model, as shown in equations (9.12) and (9.13):

$$y_{\text{early}} = \beta_{0,\text{early}} + \beta_{1,\text{early}} x_{1,\text{early}} + \cdots + \beta_{k,\text{early}} x_{k,\text{early}} + \varepsilon \tag{9.12}$$

$$y_{\text{late}} = \beta_{0,\text{late}} + \beta_{1,\text{late}} x_{1,\text{late}} + \cdots + \beta_{k,\text{late}} x_{k,\text{late}} + \varepsilon. \tag{9.13}$$

The coefficients for each of the independent variables might differ across regimes. To facilitate testing this, we can combine these two models as shown in equation (9.14):

$$\begin{aligned} y = {} & d_{\text{early}} * (\beta_{0,\text{early}} + \beta_{1,\text{early}} x_{1,\text{early}} + \beta_{2,\text{early}} x_{2,\text{early}} + \cdots + \beta_{k,\text{early}} x_{k,\text{early}} + \varepsilon) \\ & + d_{\text{late}} * (\beta_{0,\text{late}} + \beta_{1,\text{late}} x_{1,\text{late}} + \beta_{2,\text{late}} x_{2,\text{late}} + \cdots + \beta_{k,\text{late}} x_{k,\text{late}} + \varepsilon). \end{aligned} \tag{9.14}$$

The d_{early} and d_{late} terms in equation (9.14) represent dummy variables. d_{early} is set to 1 for observations in the Early regime and 0 otherwise. d_{late} is set up analogously. If an observation belongs to the Early regime (i.e., $d_{\text{early}} = 1$ and $d_{\text{late}} = 0$), then equation (9.14) reduces to equation (9.12), and coefficient estimates are given by $\beta_{1,\text{early}}$, $\beta_{2,\text{early}}$, etc. The parallel is true for observations in the Late regime. The advantage of estimating the coefficients for both regimes simultaneously is that linear hypotheses can be used to determine if coefficients differ across regimes. For example, it is straightforward to test whether $\beta_{1,\text{early}} = \beta_{1,\text{late}}$ using common statistical software.

Table 9.5 shows the coefficients that result from applying this type of model to the automotive example.

A linear hypothesis test indicates that the ElectronicVehicle coefficient in the Early regime is statistically different from the ElectronicVehicle coefficient in the Late regime, which is consistent with the results of other tests.

A test for a discrete change in a coefficient is appropriate when the researcher has data from two or more discrete points in time. For example, if a researcher has a dataset consisting of a group of observations from June 2003, another group from

TABLE 9.5 Coefficient Estimates for the Early and Late Regimes

	Early Regime		Late Regime	
	Coefficient Estimate (Std Error)	*p*-Value	Coefficient Estimate (Std Error)	*p*-Value
β_0: Intercept	−9381.80 (249.21)	0.00	−1227.33 (238.84)	0.00
β_1: ElectronicVehicle	−1270.96 (80.06)	0.00	−862.71 (76.13)	0.00
β_2: ConditionNumber	739.13 (29.10)	0.00	756.81 (30.40)	0.00
β_3: Valuation	0.94 (0.01)	0.00	0.94 (0.01)	0.00
β_4: NumberBuyers	1.53 (1.20)	0.20	2.83 (0.93)	0.00

Coefficient estimates for SellerDummies available from the authors.
$R^2 = 0.99$.
Linear hypothesis test to determine if ElectronicVehicle (Early) − ElectronicVehicle (Late) = 0: $F(1, 10\,054) = 13.66$, p-value < 0.01.

March 2004, and a third group from January 2005, a discrete change test can be useful to determine if coefficient estimates vary across these time periods. If the researcher has data that span a block of time more or less continuously (as opposed to being bunched at specific points), then methods designed for continuous coefficient evolution can be used.

9.4 CONCLUSION

This chapter has discussed several statistical methods available for investigating time-varying relationships in pooled cross-sectional data, which is a data structure commonly found in e-commerce research. We have presented and illustrated methods that are relatively simple to implement, yet can provide significant insight into whether and how empirical relationships in e-commerce evolve over time. Each of the methods has strengths and weaknesses, and researchers applying these methods should consider applying them in concert. For example, an initial rolling regression procedure can provide a general picture of how each coefficient might vary. The researcher can then use the plot of a coefficient against time to help determine the functional form of an evolution equation for that coefficient. As another example, the researcher might perform a rolling regression using a relatively large step size. The resulting plot of coefficients against time can provide clues as to whether there are any structural breaks within the dataset, which can then be further examined via a test for a discrete change in the coefficients. This could potentially alert the researcher to external factors that might have triggered a change in the coefficients, such as new legislation or a system enhancement. Using multiple methods in concert allows a researcher to capitalize on each of their strengths and should increase confidence in any effects observed.

We have illustrated each of these methods using example data drawn from the automotive industry. Results are consistent and suggest that the coefficients vary over time. In particular, tests suggest that the ElectronicVehicle coefficient becomes smaller in absolute value over time. As a result, conclusions based on analyzing only the pooled data (without considering variation over time) might be misleading. For example, the ElectronicVehicle coefficient for the pooled data is −983.58 (see Table 9.2). However, by the end of the time span, the coefficient appears closer to −700. Thus, a forecast based on the coefficient from the pooled estimation might overestimate the discount associated with electronic presentation by approximately $300. This underscores the importance of testing for changes in the coefficients over time. There are multiple theoretical mechanisms that might explain why the ElectronicVehicle coefficient changes over time, including increasing comfort with the electronic mechanism on the part of the buyers, changes in the mix of buyers who participated in the market, changes in the mix of vehicles presented electronically, etc. Due to our focus on expositing and illustrating the statistical methods, we are unable to examine and disentangle these theoretical mechanisms in this chapter. Readers interested in that analysis are referred to Overby and Jap (2007).

The statistical literature on modeling and testing for time-varying coefficients continues to grow. We have designed this chapter to be an introduction to this literature accessible to researchers with a variety of backgrounds; those interested in more technical treatments and methods for specialized situations are referred elsewhere. For example, Hastie and Tibshirani (1993) provide formal notation and present multiple examples of varying coefficient models; Cai et al. (2000) suggest goodness-of-fit tests for varying coefficient models; Orbe et al. (2005) discuss nonparametric evolution equations; and Kim (2007) discusses quantile regression models with varying coefficients.

We hope that this chapter will stimulate e-commerce researchers to investigate how the relationships they uncover evolve over time. E-commerce will continue to be a dynamic research area. For example, developments with the semantic web and the increasing sophistication of electronic software agents will affect how we use the Internet. This will likely cause the empirical relationships that we have documented related to electronic market design, effects on online advertising, etc. to change. Understanding the speed and trajectory of these changes will help us, as e-commerce researchers, better understand and predict how new technologies are affecting commerce.

REFERENCES

Andrews, D.W.K. (1993). Tests for parameter instability and structural change with unknown change point. *Econometrica*, 61(4): 821–856.

Beck, N. and Katz, J.N. (1995). What to do (and not to do) with time-series cross-section data. *The American Political Science Review*, 89(3): 634–647.

Brown, R.L., Durbin, J., and Evans, J.M. (1975). Techniques for testing the constancy of regression relationships over time. *Journal of the Royal Statistical Society (Series B)*, 37(2): 149–192.

Cai, Z., Fan, J., and Li, R. (2000). Efficient estimation and inferences for varying-coefficient models. *Journal of the American Statistical Association*, 95(451): 888–902.

Chatfield, C. (1996). *The Analysis of Time Series: An Introduction*. Boca Raton, FL: Chapman & Hall/CRC.

Chow, G.C. (1960). Tests of equality between sets of coefficients in two linear regressions. *Econometrica*, 28(3): 591–605.

Enders, W. (2004). *Applied Econometric Time Series*. Hoboken, NJ: Wiley.

Fan, J. and Zhang, W. (1999). Statistical estimation in varying coefficient models. *The Annals of Statistics*, 27(5): 1491–1518.

Farley, J. and Hinich, M. (1970). Testing for a shifting slope coefficient in a linear model. *Journal of the American Statistical Association*, 65(331): 1320–1329.

Hamilton, J.D. (1994). *Time Series Analysis*. Princeton, NJ: Princeton University Press.

Hastie, T. and Tibshirani, R. (1993). Varying-coefficient models. *Journal of the Royal Statistical Society. Series B (Methodological)*, 55(4): 757–796.

Hinich, M. and Roll, R. (1981). Measuring nonstationarity in the parameters of the market model. *Research in Finance*, 3: 1–51.

Kim, M.-O. (2007). Quantile regression with varying coefficients. *Annals of Statistics*, 35(1): 92–108.

Naik, P.A. and Raman, K. (2003). Understanding the impact of synergy in multimedia communications. *Journal of Marketing Research*, 40(4): 375–388.

Orbe, S., Ferreira, E., and Rodriguez-Poo, J. (2005). Nonparametric estimation of time varying parameters under shape restrictions. *Journal of Econometrics*, 126(1): 53–77.

Overby, E. and Jap, S. (2007). Electronic vs. physical market mechanisms: Evaluating multiple theories in the wholesale automotive market. Working Paper, Georgia Institute of Technology.

Wildt, A.R. and Winer, R.S. (1983). Modeling and estimation in changing market environments. *Journal of Business*, 56(3): 365–388.

Wooldridge, J.M. (2002). *Econometric Analysis of Cross Section and Panel Data*. Cambridge, MA: MIT Press.

10

OPTIMIZATION OF SEARCH ENGINE MARKETING BIDDING STRATEGIES USING STATISTICAL TECHNIQUES

Alon Matas and Yoni Schamroth
Media Boost Ltd., Ohr Yehuda, Israel

10.1 INTRODUCTION

In the rapidly expanding sector of search engine marketing, analytical methodologies are strongly sought after to support complex decision-making strategies. In this chapter, we present some of the quantitative solutions we have applied to and created specifically for the domain of e-commerce, which allow for effective inference of bidding results. In particular, having introduced the concepts of search engine marketing, we will demonstrate how the bid-to-profit relationship may be estimated through a series of interdependent subrelationships. These include modeling the effect of a particular bid on cost-per-click, position, click-through rate, impressions, conversions, and ultimately profit. Once modeled, these relationships are used to optimize profit at the keyword level. We describe the challenges that we encountered, particularly those stemming from the rapidly changing environment of competitor behavior. Finally, we present sample case studies of keywords we have encountered.

10.1.1 Pay-Per-Click

Search engine marketing is often spoken of in terms of running a pay-per-click campaign, which, simply stated, means that you pay for your website to appear in

Statistical Methods in e-Commerce Research. Edited by W. Jank and G. Shmueli

the advertisements usually found at the top and in the right-hand column of a search engine results page for a specific predefined list of keywords. The order in which the ads appear is auction based—with competitors bidding on the maximum amount they are willing to pay the particular search engine provider each time their ad is clicked. Table 10.1 below contains some of the commonly used terminology.

The campaign manager maintains control of:

- *Keyword selection*: Those keywords for which his ad should appear. A pay-per-click campaign might include 10 to 100,000 different keywords.
- *Geo-targeting*: The specific geographical areas where he wishes his ad to appear.
- *Creatives*: The text of the ad (several different creatives may be used).
- *Day parting*: The time of day or day of the week when the ad should run.
- *Budget*: A budget cap ensuring that the total cost will not exceed his financial limitations.
- *The bid*: The maximum amount he is prepared to pay per click to the search engine provider for each keyword in the campaign.

The search engine provider determines the position of the ad based on a combination of the ad's relevance and the proposed bid relative to those of advertisers competing on the same keyword. The mechanism used, an adaptation of the *generalized second-price* auction, ensures that the fee charged per click will be one cent more than the minimum necessary to keep the same position on the page.

TABLE 10.1 Terms and Definitions

Adgroup	A collection of keywords under a particular campaign, all using the same creatives.
Campaign	A collection of AdGroups. Geo-targeting and budget are generally set on the campaign level.
Average CPC	The actual average cost-per-click charged.
Impressions	The number of times a particular add was shown; a basic measure of Internet traffic.
Average Position	The average position in which the ad appeared, position 1 being the top position.
Clicks	The number of times a particular ad was clicked.
CTR	Click-through rate. The proportion of impressions that are clicked: $\text{CTR} = \dfrac{\text{Clicks}}{\text{Impressions}}$
Conversion	A click that ends in a desired action (purchase, sign-up, etc.).
Conversion Value	The dollar value a conversion is worth to the advertiser.
Conversion Rate	The proportion of clicks that convert (*CRate*): $\text{CRate} = \dfrac{\text{Conversions}}{\text{Clicks}}$

Edelman et al. (2006) demonstrate the advantages of such auction-based sponsored advertisement, now favored by search engines, over the *generalized first-price* auction-based structures previously used. They conclude that in such a setting, empirical data analysis is highly appropriate.

10.1.2 Online versus Offline Marketing

Search engine marketing has numerous advantages over traditional offline marketing:

- *Targeted marketing*: The ability to pinpoint relevant market sectors through carefully selected keywords and geo-targeting.
- *Unlimited radius of impact*: The radius of online marketing is not limited by physical, political, and time barriers.
- *Setup ease*: Within a few hours, one's ads can appear worldwide.
- *Real time*: Data can be analyzed and strategies developed in real time.
- *Measurability*: Availability of highly detailed data allows the effectiveness of each keyword, ad, and bidding strategy to be measured.
- *Data reliability*: The data received by the search engine are exact, with no measurement errors.

In short, the quantative rigor in online advertising easily lends itself to optimization through analytic techniques. However, despite its advantages to consumers and advertisers alike, the general population has been slow to adopt online commercial channels (Forman and Goldfarb, this volume; Langer et al. 2007). Moreover, recent studies (iProspect 2007) have indicated that up to two-thirds of the online search population are in fact driven to search via offline channels. This has prompted many leading authorities to recommend a combination of the two channels.

Historic data reported by the search engine provider are generally attainable per keyword, per creative, on 24-hour resolution.

10.2 PRIMARY MODELS

Our strategy is to allow the campaign manager to gain insight into the expected profit of each keyword as a function of the bid proposed. Profit, of course, depends on cost and revenue, which are ultimately controlled by the bid placed. Higher bids will result in better positions, producing a larger number of clicks, but at a higher expense. Lower bids will result in cheaper clicks but worse positioning on the webpage, which will typically reduce the number of clicks. Figure 10.1 shows how revenue, cost, and profit might depend on the bid for a keyword. The challenge is to model these relationships, allowing for easy identification of the optimum bid.

The generic relationships set up by search engines connecting bid to profit can often be portrayed as in Figure 10.2. Indeed, this was confirmed empirically by

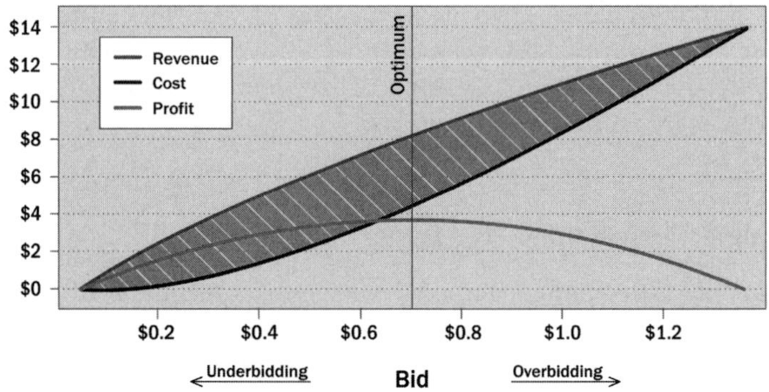

Figure 10.1 Revenue, cost, and profit as a function of bid.

numerous case studies. The proposed bid controls the position of the ad as well as the cost per click (CPC). Impressions, click-through rate, conversion rate, and, on occasion, even conversion value are all influenced by the position attained. These parameters lead to the calculation of cost, revenue, and ultimately profit. Since each of these parameters are changing dynamically at their own pace and because of the interdependent nature among the explanatory variables, we found it necessary to model each of them individually. We now provide a brief overview of the various challenges we encountered in modeling these parameters. In Section 10.3, we present sample case studies for a few of the dependencies mentioned. These particular data

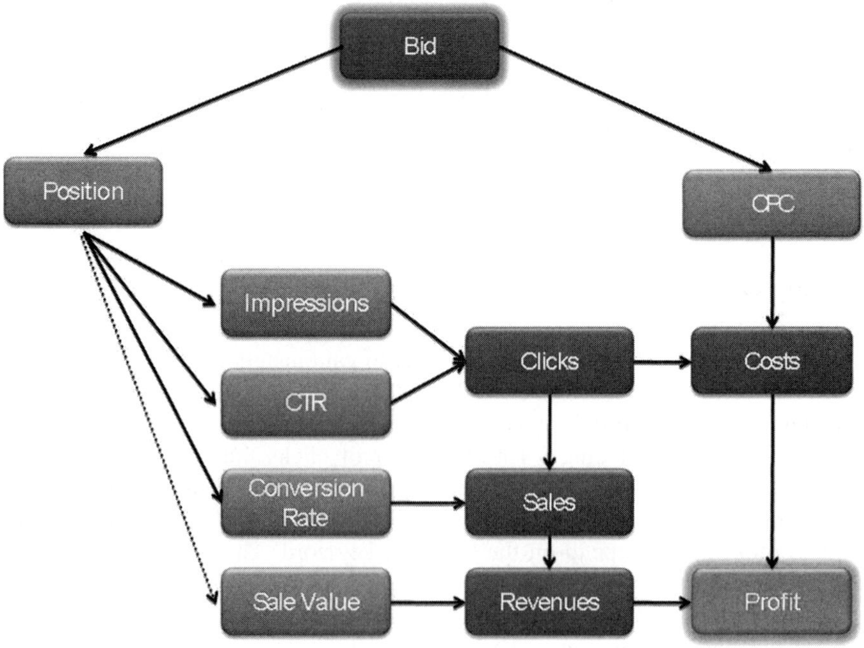

Figure 10.2 Factors linking profit to bid.

were supplied in daily increments over a period of 90 days. Modifying models according to hourly data is easily achieved.

10.2.1 Challenges

We encountered many interesting and unforeseen challenges during our initial exploratory data analysis, which affected numerous areas of our research. We present some of the highlights below together with a brief description of the solutions we provided.

10.2.2 Outliers

Identifying influential outliers and outlying clusters played a significant role in "cleaning" the data before any inference could be performed. In conjunction with the standard definition of *outliers* as extreme observations, the definition of *extreme* had to be redefined to fit within the logical boundaries of the model concerned—namely, monotonicity. Monotonicity is a compelling assumption when modeling impressions, average CPC, and average position. As the bid increases, the average position is not expected to worsen or CPC to decrease. Similarly, an improvement in position should not result in a lower volume of impressions being displayed. Empirical data do not always follow the logical constraints of monotonicity.

For example, in the hypothetical scenario in Figure 10.3, the number of impressions observed at an average position of 1.3 is expected to be high and is

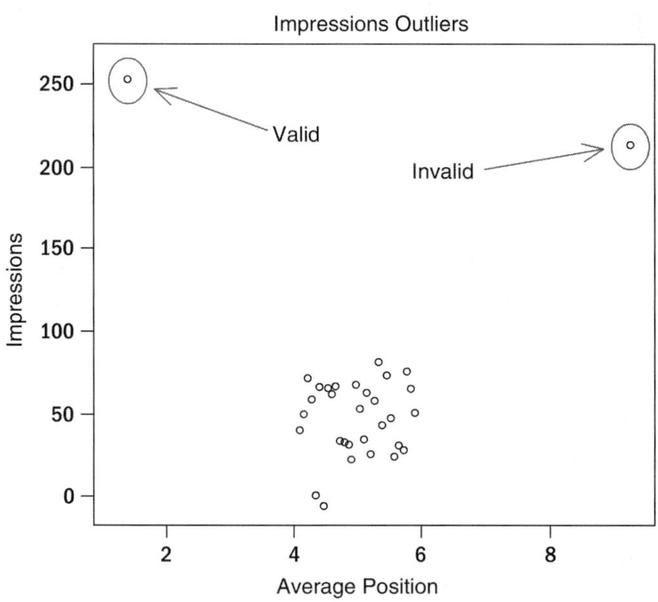

Figure 10.3 Identification of outliers.

valid, whereas that at an average position of 9.3 should be low and is therefore marked as an invalid outlier.

Apart from being caused by an extreme random event, outlying clusters often result from external changes to the campaign. These include

1. Change in budget on a campaign level.
2. Splitting or replicating a campaign over several geo-targeted areas.
3. Day parting—the user preselecting bidding strategies for different hours of the day or days of the week.
4. Adding similar keywords that share the same total Internet traffic.
5. Changing an adgroup's creative text.

Changes 1–4 all result in fluctuating volumes of impressions with no corresponding change in average position or bid. Changes 4 and 5 affect the click-through rate, which may influence ad relevance and thus position. Moreover, in the case of multiple creatives, the search engine provider generally maintains control over the intraday frequency with which each ad is displayed, and these frequencies can change at random, adding noise to the click-through rate inference. These scenarios demanded a solution to the following questions: How much time is required for our training dataset? Can dirty data be detected automatically? Can such data be salvaged?

Concerning detection of a structural change in the overall campaign, this information is generally provided by the search engine provider or the campaign manager. This inside information could be used to mark data as irrelevant for a particular analysis. In the absence of such information, outliers or outlying clusters need to be identified analytically. Owing to the assumptions mentioned above, standard methods were not always successful and we were compelled to develop new techniques.

10.2.3 Changes in Conversion Rate

A change in website or landing page, offline advertisement, and other factors may cause a dramatic change in observed conversion rate and needs to be detected when estimating the current of conversions to clicks proportion. Such a change could be tested on an adgroup or another aggregated level. Once a significant change is detected and the time of the change identified, prior data can be marked as invalid.

10.2.4 Conversion Lag

Conversions sometimes occur much later than the original click. Search engine providers generally track such lagged conversions for up to 30 days. As a result, same-day conversions do not reflect all conversions for that day's clicks. An estimate of this lag effect is required to accurately assess the conversion rate. Historic conversion data were therefore saved daily and aggregated on an adgroup level,

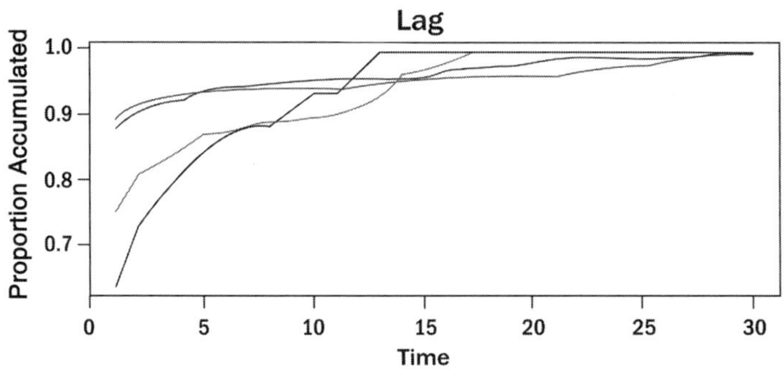

Figure 10.4 Time lag in observed conversions.

enabling this lag to be calculated. Figure 10.4 shows a few examples of these cumulative conversion distributions.

10.2.5 Simultaneous Regression Bias and Asymptotic Behavior

Because average position is used simultaneously as a response variable in one model and as an explanatory variable when modeling impressions, and because average position asymptotes at 1, wrong conclusions could result, as can be seen in Figures 10.5 and 10.6.

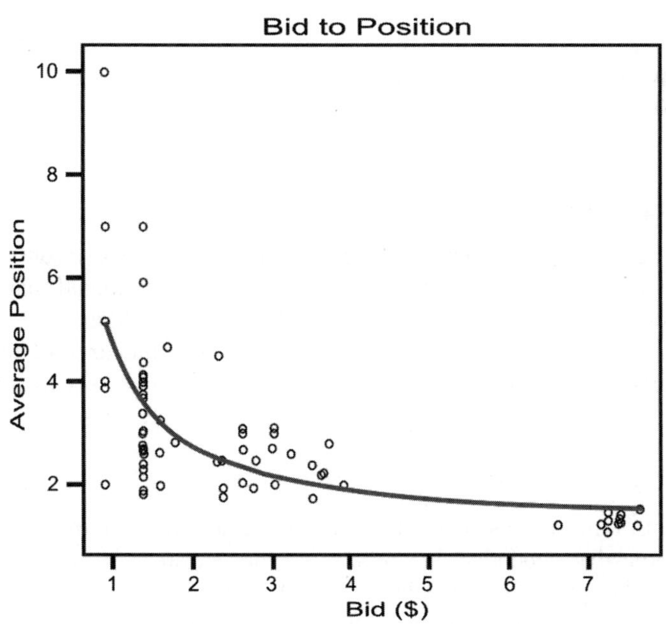

Figure 10.5 Position as a function of bid.

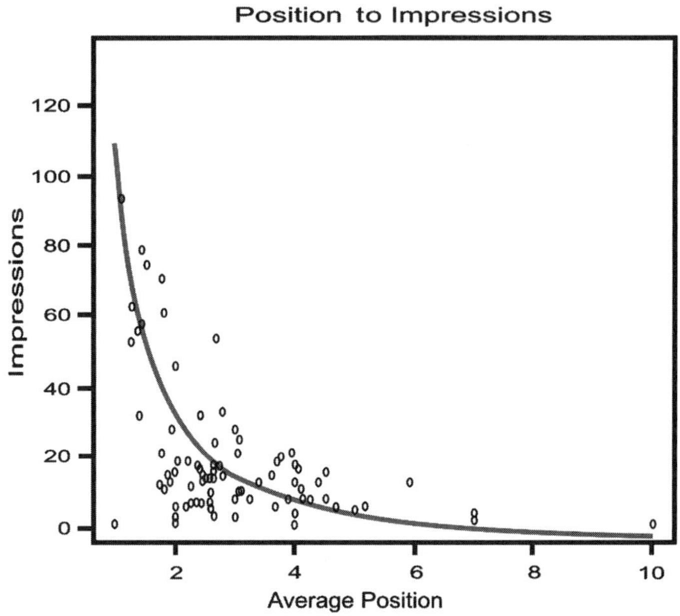

Figure 10.6 Impressions as a function of position.

In Figure 10.5, as the position tends toward 1, the curve flattens out. A minor change in position can be achieved only by a significant increase in bid. By contrast, Figure 10.6 indicates that such a minor change results in an exponential change in impressions. Such a scenario may result in upping one's bid to gain higher impressions, which will lead to unnecessary overspending.

10.2.6 Sparse Data

In general, the makeup of a campaign is such that over 80% of keywords have extremely sparse or no data at all. This long tail of keywords combined may sum up to large amounts of traffic and hence high cost. Under these circumstances statistical inference on the keyword level is of little use. In his Chapter 1 of this volume, Deepak Agarwal investigates the various solutions proposed for solving this problem.

10.2.7 Error Assessment

Once the profit function has been estimated, we would like our jump to the optimum bid to be somewhat dependant on the accuracy at that location. As we have seen, modeling the bid-profit relationship is often based on a number of intermediary dependent relationships. Estimation of the error distribution in our final relationship is therefore built up via a number of simultaneous error distributions using resampling techniques. In this way, we can obtain the estimated distribution of profit for each

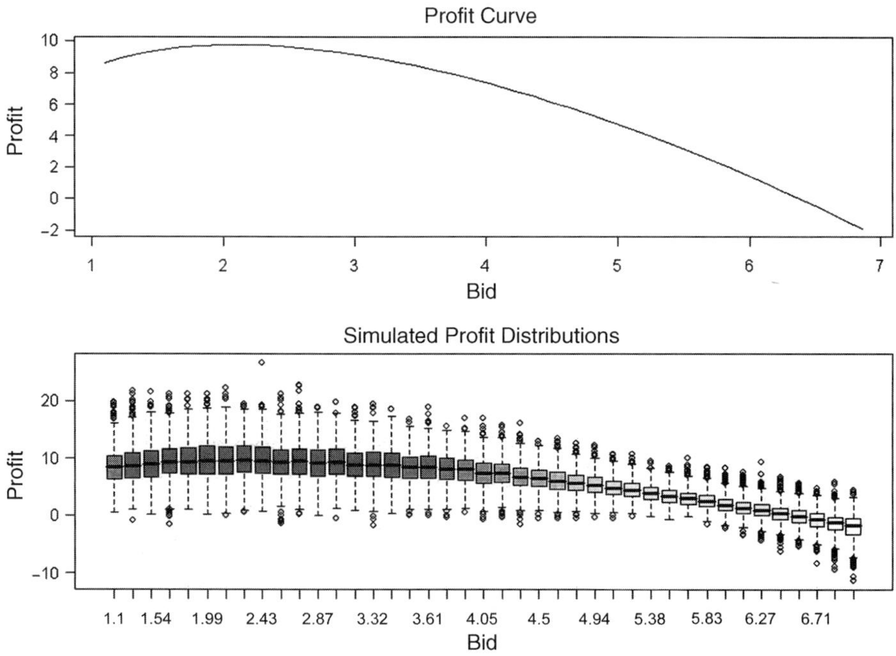

Figure 10.7 Estimating effective error through simulation.

possible bid and hence its inherent error. In Figure 10.7, though the optimum bid seems to be located around 2.10, the error at this location is actually relatively high, with the level of uncertainty increasing as bids decrease. This measure of uncertainty needs to be taken into account when deciding on a change of bids.

10.3 CASE STUDIES

10.3.1 Cost-Per-Click

Our aim here is to estimate, as a function of bidding, the actual average CPC. Since one cannot view competitors' bids and is basically required to bid blindfolded, the search engine provider guarantees to charge that CPC which is one cent above the bid of the lower-ranking competitor. In Figure 10.8, regression splines were considered in an attempt to locate areas of unnecessary overbidding.

In this example, the bid was the only predictor found to significantly influence the average CPC attained. In the first model, the intercept was found not to be significant, implying that there is not sufficient evidence for us to conclude that the intercept is not zero, an acceptable conclusion. Here the spline achieves the best fit, attaining the highest adjusted R^2 of 0.884 (Table 10.2).

However, as the bid exceeds that required to reach the top position, asymptotic behavior is expected, and this range needs to be modeled separately. Figure 10.9

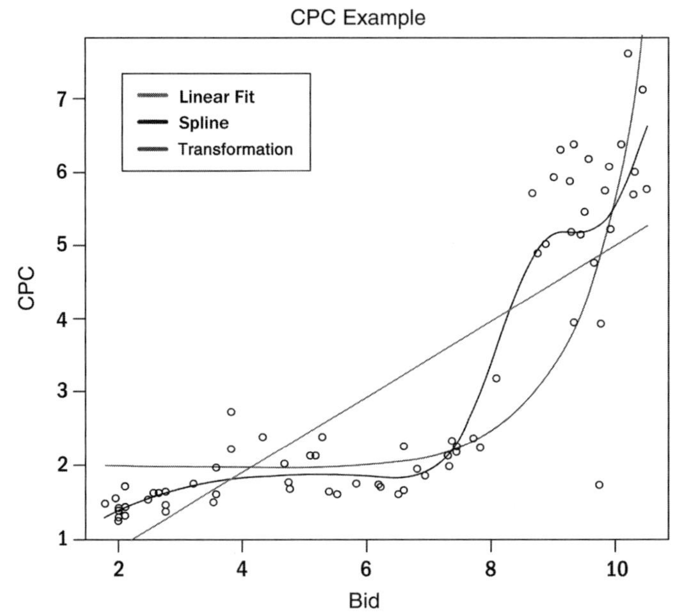

Figure 10.8 Best fit model selection.

shows various examples of linear fits to estimate the CPC based on the bid alone. Days where no clicks where received and therefore no costs were incurred are automatically removed from the training dataset.

10.3.2 Average Position

Our goal here is to estimate the position one's bid might attain. In Figure 10.10, the primary explanatory variable was again found to be the bid placed, transformed appropriately. Here the relationship between bid and average position was generally found to be exponentially decreasing in the sense that as the bid decreased, its position worsened exponentially. A position-bid relationship often changes with the competitor's behavior. Quantifying this rate of change provides a good measure of competitiveness. The amount of historic data to be included in the training dataset also needs to be carefully determined.

10.3.3 Impressions

The number of impressions received by each ad is a direct measure of Internet traffic, and as expected, routine testing revealed high dependence of impressions on day of the week, generally dropping over the weekend. Seasonal changes and time trends were also found to be significant through regression analysis. In this example, when aggregated on an adgroup or campaign level, weekday influence is quite apparent (Figure 10.11).

TABLE 10.2 Model Comparison

Linear Model	Linear Model (Transformation)	Spline
Residuals: Min 1Q Median 3Q Max −3.1440 −0.9292 0.3579 0.5544 2.5113	Residuals: Min 1Q Median 3Q Max −3.1908 −0.5008 −0.2555 0.2177 2.7457	Parametric coefficients: Estimate Std. Error t Value Pr(>|t|) (Intercept) 3.256 0.0796 40.90 <2e−16
Coefficients: Estimate Std. Error t Value Pr(>|t|) (Intercept) −0.17329 0.26783 −0.647 0.52 Bid 0.51492 0.03994 12.894 <2e−16	Coefficients: Estimate Std. Error t Value Pr(>|t|) (Intercept) 1.953e+00 1.346e−01 14.51 <2e−16 exp(Bid) 1.700e−04 1.244e−05 13.67 <2e−16	Approximate significance of smooth terms: edf Est.rank F p-Value s(Bid) 7.806 8 74.43 <2e−16
Residual standard error: 1.055 Multiple R^2: 0.6863 Adjusted R^2: 0.6821 F-statistic: 166.3 on 1 and 76 DF, p-value: < 2.2e−16	Residual standard error: 1.013 Multiple R^2: 0.7109 Adjusted R^2: 0.7071 F-statistic: 186.9 on 1 and 76 DF, p-value: < 2.2e−16	R^2 (adj) = 0.884 Deviance explained = 89.6% Scale est. = 0.40589 n = 78

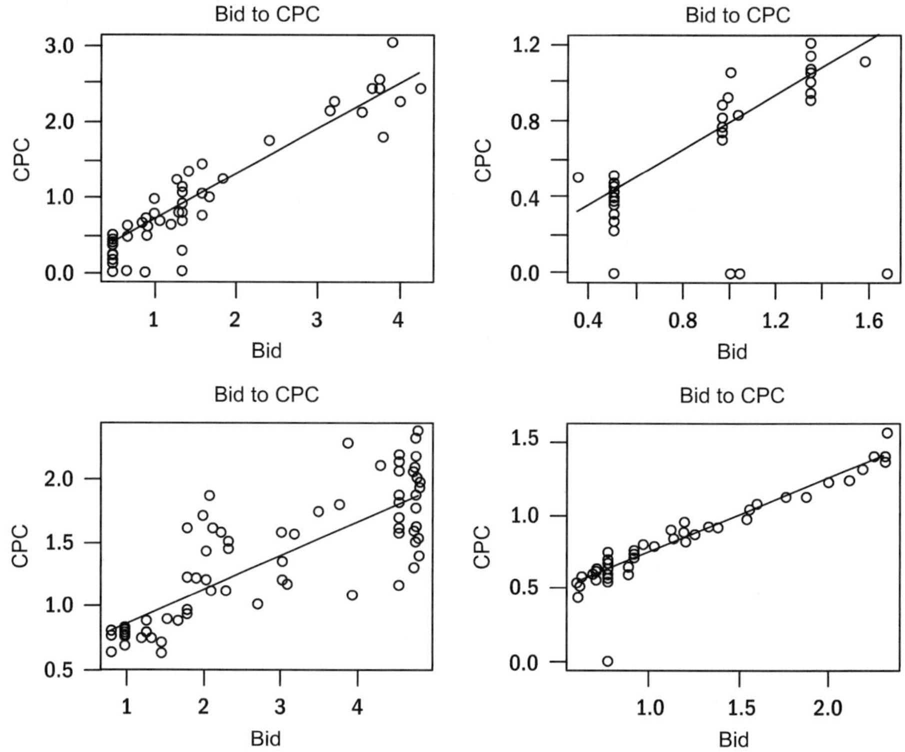

Figure 10.9 CPC often meets expectation.

Here we discover that about 64% of the variance in impressions is explained by a day-of-the-week effect (Table 10.3).

For some datasets, clustering techniques could be applied to group some of the days together in modeling this weekly variation—for example, workdays as opposed to weekends.

In the following example, in addition to various time variables, the number of impressions received was found to be highly correlated to the average position achieved, increasing exponentially as the position decreased. Figure 10.12 demonstrates the relationship of impressions to day of the week as well as average position; no interaction between the two was found. The observation located at the lower left was identified as an outlying cluster occurring in the past and was not included in the fit.

10.3.4 Click-Through Rate

In this case study, generalized linear models for binomial data were considered. The main explanatory variable found to be significant was again the average position attained. The nature of this relationship is generally product specific. Customers purchasing flowers, for example, generally do not shop around, so top positions

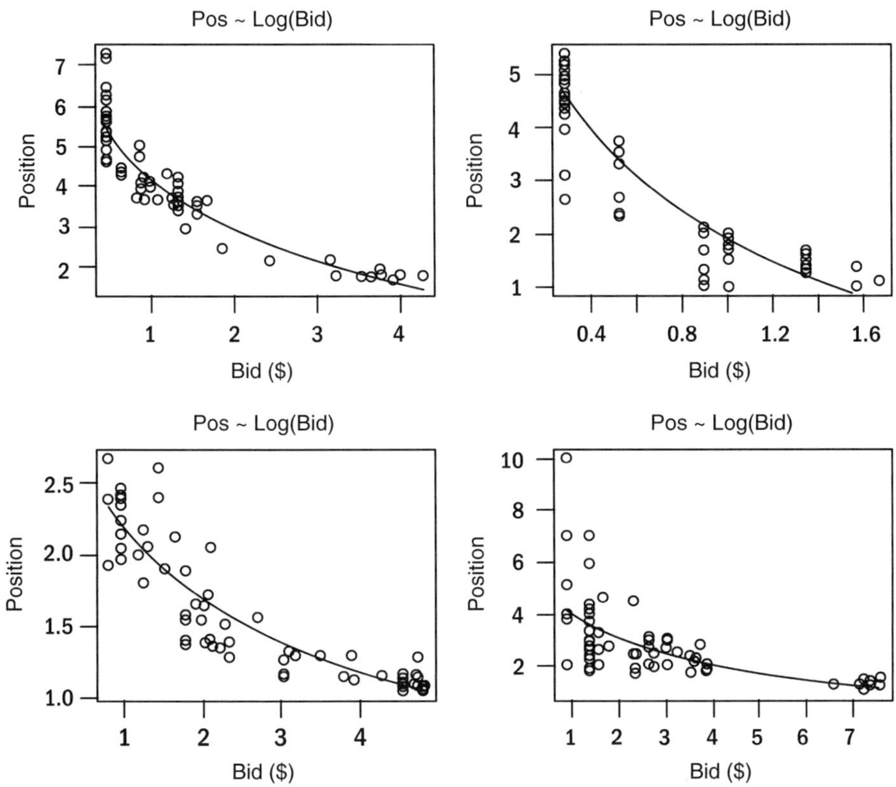

Figure 10.10 Examples of position as a function of bid.

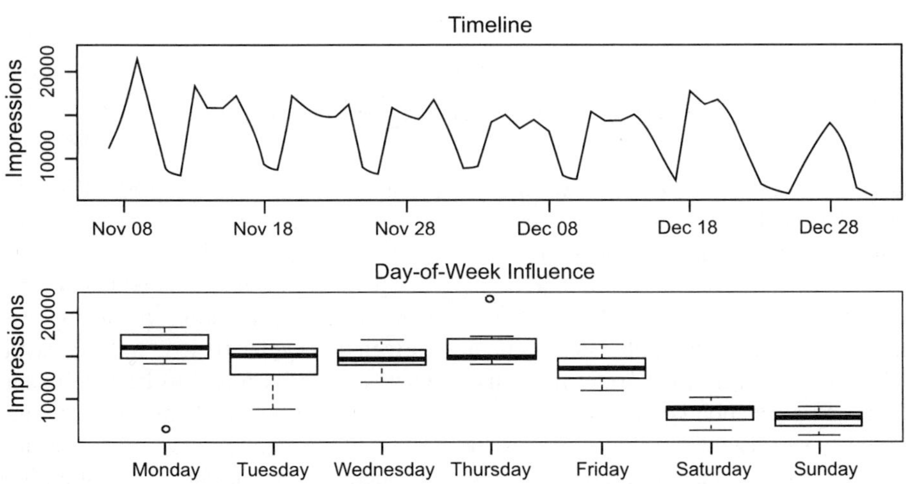

Figure 10.11 Weekday influence on impression volume.

TABLE 10.3 Weekday ANOVA Table

	Coefficients			
	Estimate	Std. Error	t Value	Pr(>|t|)
(Intercept)	13580.1	820.7	16.547	<2e−16***
Monday	1389.4	1201.4	1.157	0.2532
Saturday	−5191.9	1160.6	−4.473	4.72e−05***
Sunday	−5996.3	1160.6	−5.166	4.57e−06***
Thursday	2549.1	1160.6	2.196	0.0329*
Tuesday	517.8	1160.6	0.446	0.6575
Wednesday	1115.5	1160.6	0.961	0.3413

Multiple R^2: 0.6769; adjusted R^2: 0.6365

Codes: '***' <0.001, '**' <0.01, '*' <0.05; '·' <0.1.

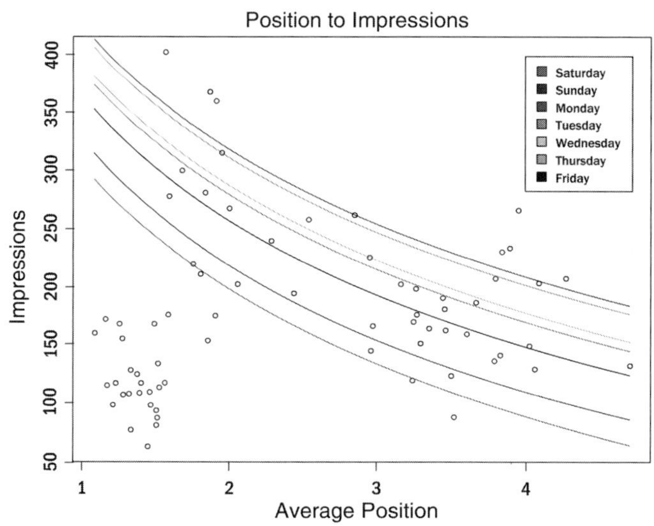

Figure 10.12 Identification of outlying clusters.

are crucial. On the other hand, when purchasing a car, customers may be willing to consider ads farther down the page; consequently, one may witness a steady click-through rate even at lower position.

In this example, an additional two overlapping keywords were added to the common adgroup splitting the shared impressions. In addition to losing its impressions, click-through rates were also affected (Figure 10.13). Indeed, with such dirty data, no monotonic relationship to position was found to be significant (Pearson correlation coefficient = 0.1467067). Using various methods for detecting outliers, we discovered that many of the poorly fitting data correspond to distant time points; by cleaning these data, we could identify the current relationship. To aid visualization of recent data, each observation in the scatterplot (Figure 10.14)

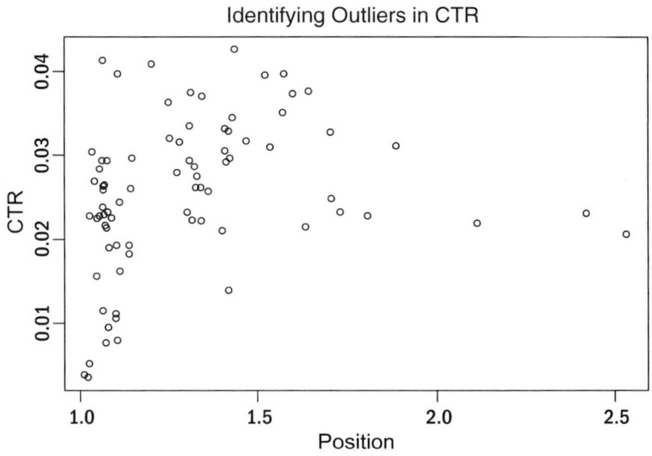

Figure 10.13 Identification of historic change in CTR.

has been resized in proportion to a time-weighting variable. In this scenario, a linear model proves the best fit with an exponential transformation on the dependent variable (Table 10.4). Here we see significance in our independent variable. All observations within the prediction interval now attain a correlation of -0.61 between click-through rate and position.

10.3.5 Conversion Rate

Conversion rate is another parameter that requires separate modeling and estimation. Often a day of the week influence may be present, as shown in Figure 10.15. For this specific keyword, a significant drop in impressions occurs over the weekend.

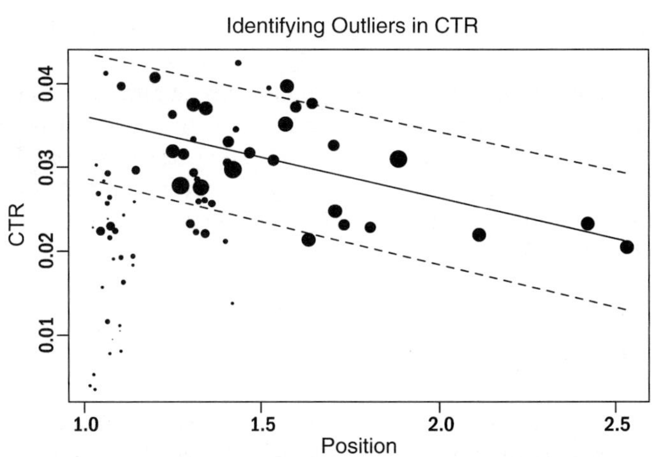

Figure 10.14 Observations re-scaled proportionate to time.

TABLE 10.4 Click-Through-Rate ANOVA Table

Coefficients				
	Estimate	Std. Error	t Value	Pr(>|t|)
(Intercept)	1.046728	0.004892	213.968	<2e−16***
AvgPosition	−0.009999	0.003056	−3.272	0.00311**

Residual standard error: 0.005613 on 25 degrees of freedom.
Multiple R^2: 0.2998, Adjusted R^2: 0.2718
F-statistic: 10.7 on 1 and 25 DF; p-value: 0.003115.

Codes: '***' <0.001, '**' <0.01, '*' <0.05; '.' <0.1.

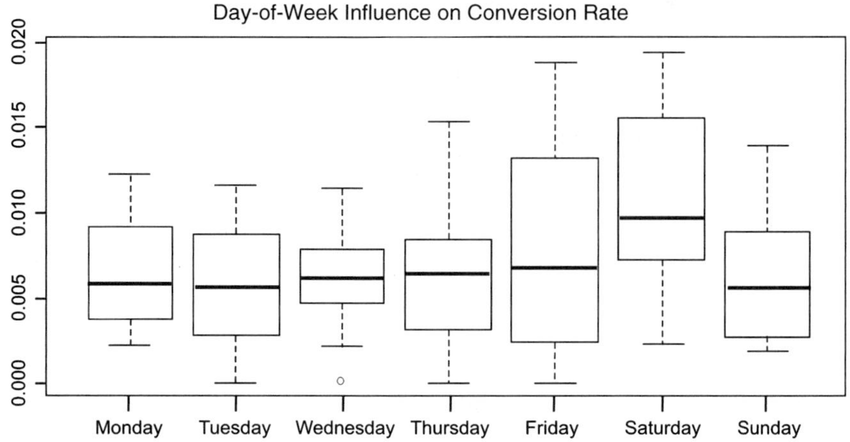

Figure 10.15 Day-of-week influence on conversion rate.

However conversion rates increase, leading to the conclusion that bids should be raised.

Depending on the product, the conversion rate may, at times, be dependent on the position of the ad. Higher positions may receive more clicks but may be less targeted, resulting in a drop in conversion rate. The converse may be true as well. Under certain circumstances, the search engine provider may also place top-performing ads in a highlighted color, resulting in a boost of the click-through rate, yet these clicks may, again, be less targeted.

10.4 CONCLUSION

In this chapter, we have attempted to show how statistical techniques can be applied to estimate the expected profit at each feasible bid, and that this function is often dependent on other explanatory variables. Once estimated, this function can be used to locate an optimum bid. We have described some of the challenges unique to pay-per-click data analysis and presented sample case studies. In summary, statistical inference may be used to aid the decision-making process of pay-per-click management, allowing for optimal returns on advertising costs.

ACKNOWLEDGMENT

We would like to express our sincere thanks to Professor David Steinberg for his continuous support of this project.

REFERENCES

Edelman, B., Ostrovsky, M., and Schwarz, M. (2006). Internet advertising and the generalized second price auction: Selling billions of dollars worth of keywords. Second Workshop on Sponsored Search Auctions, Ann Arbor, MI. June.

iProspect in conjunction with JupiterResearch (2007) Offline channel influence on online search behavior study.

Langer, N., Forman, C., Kekre, S., and Sun, B. (2007). Ushering buyers into electronic channels. Working Paper, Tepper School of Business, Carnegie Mellon University.

SECTION III

NEW METHODS FOR E-COMMERCE DATA

11

CLUSTERING DATA WITH MEASUREMENT ERRORS

Mahesh Kumar

Department of Decision and Information Technologies, R.H. Smith School of Business, University of Maryland, College Park, Maryland

Nitin R. Patel

Massachusetts Institute of Technology and Cytel Software, Cambridge, Massachusetts

11.1 INTRODUCTION

With the rapid growth of the Internet in the past decade, it has become very easy for e-commerce businesses to collect large amounts of data (often in the order of terabytes of data) related to customer demographics, their preferences, market trends, etc. This data must be analyzed accurately in order to make competitive business decisions in today's market environment. An important data analysis technique commonly used for many e-commerce applications is clustering, a process of organizing large volumes of data into groups (clusters) so that data points in the same group are more similar to each other than data points in different groups. Clustering has been successfully applied to a number of e-commerce applications such as customer segmentation, clustering data streams, clustering online auctions, and webpage classification (Zamir and Etzioni 1998; Guha et al. 2003; Jank and Shmueli 2005).

The clustering problems arising from e-commerce applications differ from traditional clustering problems in two ways. First, e-commerce generally deals with very large datasets; second, e-commerce data is often not available in the traditional Euclidean vector form (a requirement for most clustering methods). For example,

Statistical Methods in e-Commerce Research. Edited by W. Jank and G. Shmueli

TABLE 11.1 An Example of Web Navigation Data

User 1	session 1	1	2	2	2	5	4	2	6					
	session 2	4	1	3	1	1	4	1	5	6	2			
	session 3	2	3	4	7	2	3							
User 2	session 1	5	2	1	4	5	2	3	6					
	session 2	2	7	9	3	3	4	4	4	1	2	6	7	5
User 3	session 1	4	2	7	2	3	2	5	4	8	3	2		

consider clustering individuals based on web browsing behavior, as shown in Table 11.1.

The table contains sequences of webpage requests made by three users on a website, where each webpage is identified by a unique page number. We would like to cluster these users based on their web navigation patterns, but it is not clear how to obtain such clusters using standard clustering techniques.

Piccolo (1990), Cadez and Smyth (1999), Cadez et al. (2000), and Maharaj (2000) have suggested transforming these kinds of data into a Euclidean vector form via a preprocessing step and then applying any standard clustering method. By modeling the data for each individual as coming from a statistical model (e.g., a Markov chain model for each user in the above example), the preprocessing step generates a set of Euclidean vectors (one vector for each user) of model parameters. Then the individuals can be clustered based on similarity between the vectors of estimated parameters using standard clustering techniques, as shown in Figure 11.1. Since the estimated model parameters are typically much smaller in dimension than the original data, this method provides a natural way to reduce the size of the data to be clustered.

A commonly used method for estimating statistical model parameters is the maximum likelihood method, which also provides the covariance matrix (or error) associated with the estimate of the parameters. We show that incorporating this error information in the clustering process, as in Figure 11.2, can produce different and often better clusters than the ones produced by traditional clustering methods, such as k-means and Ward's hierarchical clustering (Anderberg 1973; Jain and Dubes 1988). We develop a theory and algorithms and present a range of illustrative applications for this new approach to clustering that we call *error-based clustering*.

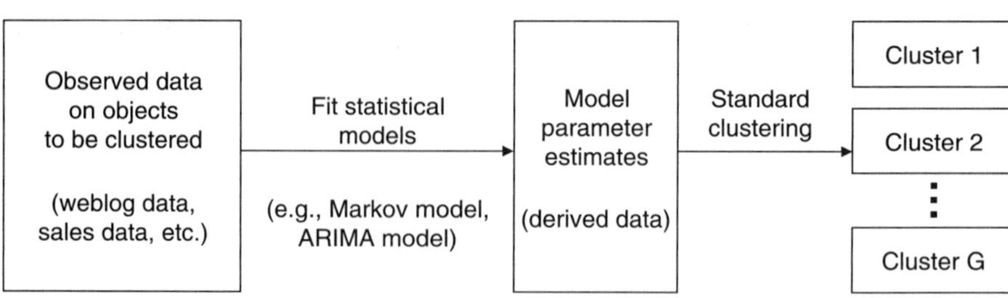

Figure 11.1 Clustering model parameters using standard clustering.

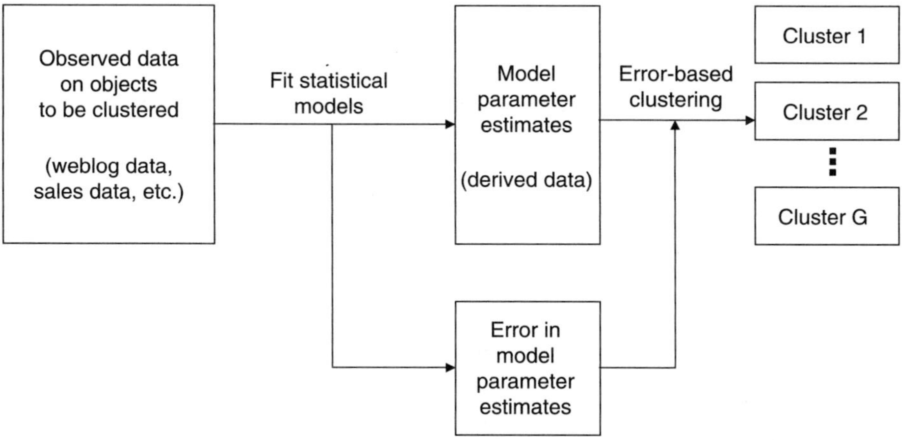

Figure 11.2 Clustering model parameters using error-based clustering.

Error-based clustering explicitly incorporates errors associated with data into cluster analysis.

The main contributions of this chapter can be divided into two parts. In the first part, we develop a general model and algorithms for incorporating error information into cluster analysis. We assume that the errors associated with data follow multivariate Gaussian distributions[1] and are independent between data points. Using this probability model for the data, we model the clustering problem as a likelihood maximization problem. The maximum likelihood principle provides us with a new criterion for clustering that incorporates information about the errors associated with data. The new criterion is used to develop two algorithms for error-based clustering: (1) *hError*, a hierarchical clustering algorithm that produces a nested sequence of clusters, and (2) *kError*, a partitioning algorithm that partitions the data into a specified number of clusters. These algorithms are generalizations of the popular hierarchical clustering algorithm of Ward 1963 and the k-means clustering algorithm (Jain and Dubes 1988), respectively. We also provide a heuristic method for selecting the number of clusters in the *hError* algorithm, which in itself is a challenging problem (e.g., Milligan and Cooper 1985).

In the second part, we describe settings where measurement errors associated with the data to be clustered are readily available and where error-based clustering is likely to be superior to clustering methods that ignore errors. We focus on clustering preprocessed data obtained by fitting statistical models to the observed data. We show that, for Gaussian distributed observed data, the optimal error-based clusters of preprocessed data are the same as the maximum likelihood clusters of the observed data. In other words, the clusters obtained in Figure 11.2 using a two-step decomposition procedure are the same as the clusters obtained in Figure 11.3 using single-step

[1]Error measurements obtained during preprocessing of data using the maximum likelihood method are approximately multivariate Gaussian distributed.

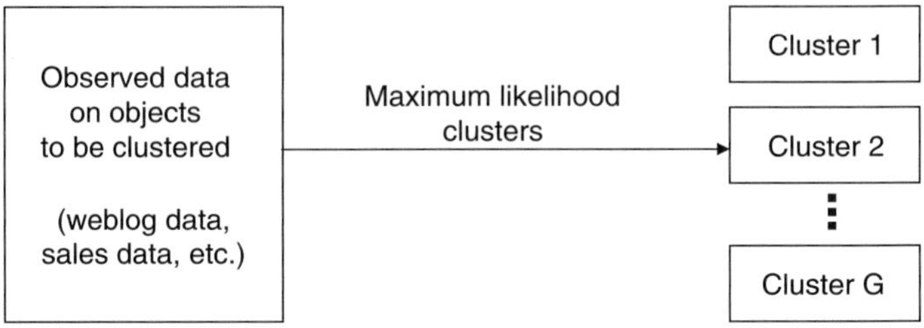

Figure 11.3 Maximum likelihood clusters of the observed data.

maximization. We also report briefly on two applications with real-world data and a series of simulation studies using four preprocessing methods based on (1) sample averaging, (2) multiple linear regression, (3) Auto-regressive integrated moving average (ARIMA) models for time series, and (4) Markov chains. These empirical studies suggest that error-based clustering performs significantly better than traditional clustering methods on these applications.

11.2 RELATED WORK

Probability models have been used for many years as a basis for cluster analysis (e.g., Scott and Symons 1971; McLachlan and Basford 1988; Banfield and Raftery 1993; Celeux and Govaert 1995; Gaffney and Smyth 1999; Cadez et al. 2000; Fraley and Raftery 2002). In these models, data are viewed as samples from mixtures of probability distributions, where each component in the mixture represents a cluster. The goal is to partition the data into clusters such that data points that come from the same probability distribution belong to the same cluster. Banfield and Raftery (1993), Cadez et al. (2000), and Fraley and Raftery (2002) have shown the effectiveness of such probability models in a number of practical applications, including clustering of medical data, gene expression data, weblog data, image data, and spatial data. While these authors provide a general probability model that allows any probability distribution for data, it is worth noting that they have found that a mixture of Gaussian distributions is applicable to many problems in practice.

The probability model used in error-based clustering is similar to the one used in model-based clustering (Banfield and Raftery 1993; Fraley and Raftery 2002). In model-based clustering, data points are modeled as arising from a mixture of multivariate Gaussian populations, where each population represents a cluster. The parameters of the mixture components are estimated by maximizing the likelihood of the observed data. We differ from standard model-based clustering because instead of modeling the *populations* as multivariate Gaussian, we model the *error associated with each data point* as multivariate Gaussian. In other words, in the special case when it is assumed that all data points in the same cluster have the same error distribution, error-based clustering is equivalent to model-based clustering. In that sense, error-based clustering is a generalization of model-based clustering. By allowing different amounts of error for each data point, error-based clustering explicitly models error information for incorporation into the clustering algorithms.

In our literature search, we have come across only two publications that explicitly consider error information in multivariate clustering. Chaudhuri and Bhowmik (1998) provide a heuristic solution for the case of uniformly distributed spherical errors associated with data. Kumar et al. (2002) provide a model for Gaussian errors, but their work does not consider correlation among the variables. We consider a general case of multivariate Gaussian errors and provide a formal statistical procedure to model them. Modeling errors using the Gaussian distribution has a long history (beginning with Gauss!) and is applicable to many problems in practice (e.g., Sarachik and Grimson 1993; Feng et al. 1996).

The work of Stanford and Raftery (2000) and Cadez et al. (2000) is similar to the work we have proposed in this chapter. These authors have proposed clustering of individuals that may not have the usual vector representation in a fixed dimension. The work in these papers is specific to their applications and is based on the expectation maximization (EM) algorithm. We have proposed an alternative approach that is easy to use computationally and is applicable to a broad class of problems.

While there is almost no prior published work on incorporating error information in multivariate cluster analysis, there has been significant work on this topic for one-dimensional data (e.g., Fisher 1958; Cox and Spjotvoll 1982; Cowpertwait and Cox 1992). Cowpertwait and Cox (1992) applied their technique to clustering univariate slope coefficients from a group of regressions and used it for predicting rainfall. We extend their work to clustering multivariate regression coefficients.

11.3 MODEL FOR ERROR-BASED CLUSTERING

The data to be clustered consist of n observations $x_1, \ldots, x_n$ (column vectors) in $\Re^p$ and n positive definite matrices $\Sigma_1, \ldots, \Sigma_n$ in $\Re^{p\times p}$, where x_i represents measurements on p characteristics and Σ_i represents the covariance matrix associated with the observed measurements of x_i. Suppose that the data points are independent and that each arises from a p-variate Gaussian distribution with one of K possible means $\theta_1, \ldots, \theta_K$, $K \leq n$, that is, $x_i \sim N_p(\mu_i, \Sigma_i)$, where $\mu_i \in \{\theta_1, \ldots, \theta_K\}$ for $i = 1, \ldots, n$. Our goal is to find the clusters $C_1, \ldots, C_K$ such that observations that have the same mean (μ_i) belong to the same cluster with $\mu_i = \theta_k$, where θ_k is the common value of θ_i for the observations in C_k, $k = 1, \ldots, K$.

Let $S_k = \{i|x_i \in C_k\}$; then $\mu_i = \theta_k$ for $\forall_i \in S_k$, $k = 1, \ldots, K$. Given data points $x_1, \ldots, x_n$ and the error matrices $\Sigma_1, \ldots, \Sigma_n$, the maximum likelihood principle leads us to choose $S = (S_1, \ldots, S_K)$ and $\theta = (\theta_1, \ldots, \theta_K)$ so as to maximize the likelihood

$$L(x|S, \theta) = \prod_{k=1}^{K} \prod_{i\in S_k} \frac{1}{(2\pi)^{\frac{p}{2}}} |\Sigma_i|^{-\frac{1}{2}} e^{-\frac{1}{2}(x_i-\theta_k)^t \Sigma_i^{-1}(x_i-\theta_k)}, \tag{11.1}$$

where $|\Sigma_i|$ is the determinant of Σ_i for $i = 1, \ldots, n$. It is easy to show that the likelihood in equation (11.1) is maximum for the partition $S_1, \ldots, S_K$ that solves

$$\min_{S_1,\ldots,S_K} \sum_{k=1}^{K} \sum_{i \in S_k} (x_i - \hat{\theta}_k)^t \, \Sigma_i^{-1} (x_i - \hat{\theta}_k), \tag{11.2}$$

where $\hat{\theta}_k$ is the maximum likelihood estimate (MLE) of θ_k given by

$$\hat{\theta}_k = \Big(\sum_{i \in S_k} \Sigma_i^{-1}\Big)^{-1} \Big(\sum_{i \in S_k} \Sigma_i^{-1} x_i\Big), \qquad k = 1, \ldots, K. \tag{11.3}$$

The minimization in equation (11.2) makes intuitive sense because each data point is weighted by the inverse of its error; that is, data points with smaller error get higher weight and vice versa. Notice that $\hat{\theta}_k$ is a weighted mean of the data points in C_k. We will refer to it as the *Mahalanobis mean* of cluster C_k (because it is very similar to the Mahalonobis distance function (Mahalanobis 1936)). Let Ψ_k denote the error (or covariance) matrix associated with $\hat{\theta}_k$. Using simple matrix algebra, we get

$$\Psi_k = Cov(\hat{\theta}_k) = Cov\Big[\Big(\sum_{i \in S_k} \Sigma_i^{-1}\Big)^{-1} \Big(\sum_{i \in S_k} \Sigma_i^{-1} x_i\Big)\Big] = \Big(\sum_{i \in S_k} \Sigma_i^{-1}\Big)^{-1}, \tag{11.4}$$

where $Cov(x)$ refers to the $p \times p$ covariance matrix associated with x.

It is useful to note two useful properties of the objective criterion of error-based clustering in equation (11.2). First, when all error matrices are equal and spherical, that is, $\Sigma_i = \sigma_2 I$ for all $i = 1, \ldots, n$, where I denotes an identity matrix, the criterion of error-based clustering is the same as the minimum squared Euclidean distance criterion used in k-means. In that sense, the error-based metric is a generalization of the the metric used in k-means. Second, the objective criterion of error-based clustering is invariant under an affine transformation of the data space because each term in the summation of equation (11.2) is similar to the Mahalanobis distance function, which is well known to be scale-invariant.

11.4 *hError* CLUSTERING ALGORITHM

The formulation in equation (11.2) is known to be Non-deterministic polynomial time (NP)-hard (Brucker 1977).[2] We present a greedy heuristic to optimize the objective criterion of error-based clustering, which we call the *hError* algorithm. The *hError* algorithm is similar to Ward's agglomerative hierarchical clustering algorithm (Ward 1963). We also provide a heuristic for finding the number of clusters in a dataset that is suitable for the *hError* algorithm.

[2]An optimal clustering can be obtained in polynomial time for one-dimensional data using dynamic programming (Fisher 1958).

11.4.1 A Hierarchical Greedy Heuristic for *hError*

The *hError* algorithm starts with n singleton clusters, each corresponding to a data point. At each stage of the algorithm, we merge a pair of clusters that leads to the minimum increase in the objective function of error-based clustering. The merging process can either continue until all data points are merged into a single cluster or stop when the desired number of clusters is obtained. Next, we show that the greedy heuristic at each stage of the *hError* algorithm is equivalent to combining the closest pair of clusters according to a distance function that is easy to compute.

Theorem 1. *At each step of the hError algorithm, we merge a pair of clusters C_u and C_v for which the distance*

$$d_{uv} = (\hat{\theta}_u - \hat{\theta}_v)^t(\Psi_u + \Psi_v)^{-1}(\hat{\theta}_u - \hat{\theta}_v) \tag{11.5}$$

is minimized, where $\hat{\theta}_u$ and $\hat{\theta}_v$ are the Mahalanobis means of clusters C_u and C_v, respectively, and Ψ_u and Ψ_v are the error matrices as defined in equation (11.4).

Proof of the theorem is provided in the Appendix. Note that the distance function presented in equation (11.5) is similar to the Mahalanobis distance function (Mahalanobis 1936). The *hError* algorithm is a generalization of Ward's method for hierarchical clustering. In the special case when $\Sigma_i = \sigma^2 I$ for all $i = 1, \ldots, n$, the *hError* algorithm specializes to Ward's algorithm. Like Ward's algorithm, *hError* has a time complexity of $O(n^2)$. In the next subsection, we present a heuristic for selecting the number of clusters in the *hError* algorithm.

11.4.2 Number of Clusters

In this section, we propose a new method for finding the number of clusters that is suitable for the *hError* algorithm. The method involves testing a series of hypotheses, one at each stage of the *hError* algorithm. Let the clusters at an intermediate stage of *hError* be $S_1, \ldots, S_K$. Then we test the hypothesis that the means (μ_i) are equal within each cluster, that is, we test the consistency with hypothesis of the form

$$H_K : \mu_i = \theta_k, \qquad \forall i \in S_k, \quad k = 1, \ldots, K. \tag{11.6}$$

From standard multivariate theory, we know that

$$Z_K^2 = \sum_{k=1}^{K} \sum_{i \in S_k} (x_i - \hat{\theta}_k)^t \, \Sigma_i^{-1} (x_i - \hat{\theta}_k) \tag{11.7}$$

follows a chi-square distribution with $(n - K)p$ degrees of freedom. Therefore, we reject H_K at a significance level of $(1 - \alpha)$ if $Z_K^2 > \chi^2_{\alpha,(n-K)p}$. The above hypothesis is clearly consistent at the first stage of the *hError* algorithm, when there are n clusters ($Z_K^2 = 0$ at the first stage). The algorithm stops merging clusters when it encounters the first hypothesis that is rejected, because further merging of the clusters will generally give a less consistent clustering (Cox and Spjotvoll 1982). In our implementation of *hError*, we have used $\alpha = 0.01$.

We note that the merging process of *hError* selectively puts data points that are near each other in the same cluster, whereas the Z_K^2 statistic assumes that the clusters are random sets of points.[3] Therefore, the value of the Z_K^2 statistic will generally be underestimated according to the $\chi^2_{\alpha,(n-K)p}$ measure. This makes the Z_K^2 statistic a "liberal" measure of the quality of clustering in the sense that the proposed method will accept a clustering even if it should be rejected with probability α. This implies that *hError* will tend to produce fewer clusters than the true number of clusters in the data. We have found this to be true in the simulation studies presented in Section 11.7.

11.5 *kError* CLUSTERING ALGORITHM

We present here another heuristic algorithm, *kError*, that is appropriate when the number of clusters, K, is given. *kError* is similar to the k-means algorithm. It is an iterative algorithm that starts with an initial partition of data and cycles through the following two steps:

Step 1: For a given set of K clusters, compute the cluster centers as the Mahalanobis means of the clusters.

Step 2: Reassign each data point to the closest cluster center using the distance formula in equation (11.8).

The distance of a data point, x_i, from a cluster center, $\hat{\theta}_k$, is given by

$$d_{ik} = (x_i - \hat{\theta}_k)^t \Sigma_i^{-1} (x_i - \hat{\theta}_k). \tag{11.8}$$

We note that the distance function in equation (11.8) is different from the one in equation (11.5). The distance function in equation (11.8) does not contain Ψ_k, the error term associated with $\hat{\theta}_k$. We have chosen this distance function because it guarantees a decrease in the objective function in each iteration of *kError*. The difference between these distance functions is analogous to the difference in the distance functions used in Ward's and k-means methods (Anderberg 1973).

The *kError* algorithm is a generalization of the k-means algorithm. In the special case when $\Sigma_i = \sigma^2 I$ for all $i = 1, \ldots, n$, the *kError* algorithm specializes to k-means. The time complexity of *kError* is linear in the number of data points and the number of iterations the algorithm makes. In our empirical studies, we have found that the algorithm generally converges after a few iterations (typically, less than 10 iterations);

[3]This situation is analogous to using the F test in stepwise multiple linear regression.

therefore, the *kError* algorithm is generally faster than the *hError* algorithm.[4] On the other hand, *kError* requires a priori specification of the number of clusters.

A drawback of the *kError* algorithm is that the final clusters may depend on the initial partition. It can also produce empty clusters if all points in a cluster are reassigned to other clusters, thereby reducing the number of clusters. The k-means algorithm also has these shortcomings (Pena et al. 1999). We propose the following solution, which is similar to the one that is often used in k-means. Run the *kError* algorithm a large number of times with different random initial partitions and pick the one that has the smallest value of the objective function. We ignore those solutions that contain one or more empty clusters. Pena et al. (1999) have shown that if k-means is run a large number of times, the resulting clusters will be close to optimal and insensitive to the initial partition. In our empirical studies, we have found that this is also true for *kError*.

11.6 CLUSTERING MODEL PARAMETERS USING ERROR-BASED CLUSTERING

In this section, we present a class of clustering problems where error information about data to be clustered is readily available and where error-based clustering results are typically superior to those achieved with standard clustering methods that ignore error information. We focus on clustering problems where the objects to be clustered (observed data) are modeled, or preprocessed, using statistical models. Each object in this case is identified by the parameters of a statistical model. A commonly used method for estimating the parameters of statistical models is the maximum likelihood method, which also provides the covariance matrix (or error information) associated with the estimates of the model parameters. If we wish to cluster these objects on the basis of the model parameter estimates, then error-based clustering becomes a natural clustering method. We must note that while covariance matrices are estimated in this application, error-based clustering assumes that covariance matrices are given. This kind of approximation is common in the statistics literature to simplify a model. A central result presented in this chapter is that, if the observed data are Gaussian distributed, then the optimal error-based clusters of the estimated model parameters are the same as the maximum likelihood clusters of the observed data.

Let X_i be the observed data for the ith object for $i = 1, \ldots, n$. Here X_i's do not have to be vectors of fixed dimension; for example, X_i is the page click data for the ith user in Table 11.1. We assume that X_i comes from a statistical model based on a set of parameters $\theta_i = (\theta_{i1}, \ldots, \theta_{ip})$. We further assume that the likelihood function of the observed data is Gaussian so that the log-likelihood, $\ell(X_i|\theta_i)$, is a quadratic function of θ_i, and the third and higher-order partial derivatives of $\ell(X_i|\theta_i)$ are identically equal to zero.

[4]It is easy to show that *kError* algorithm converges in a finite number of iterations. The proof is similar to the one for the finite convergence of k-means (Anderberg 1973).

Let $\hat{\theta}_i$ be the MLE of θ_i; then it satisfies

$$\left[\frac{\partial \ell(X_i|\theta)}{\partial \theta}\right]_{\hat{\theta}_i} = 0_p, \tag{11.9}$$

where 0_p is a $p \times 1$ vector of zeroes. An estimate of the covariance matrix associated with $\hat{\theta}_i$ is given by

$$\Sigma_i = \left[-\frac{\partial^2 \ell(X_i|\theta)}{\partial \theta \partial \theta'}\right]^{-1}_{\hat{\theta}_i}. \tag{11.10}$$

We cluster these n objects based on similarity in their model parameter estimates; that is, the input to the clustering algorithm consists of n sets of model parameter estimates, $\hat{\theta}_i$, and associated error matrices, Σ_i, for $i = 1, 2, \ldots, n$. Consider a cluster C_k that contains n_k objects with indices $S_k = \{i_1, i_2, \ldots, i_{n_k}\}$. Let us denote all of the observed data in cluster C_k by $X_{S_k} = (X_{i_1}, X_{i_2}, \ldots, X_{i_{n_k}})$. If we assume that all data in this cluster have a common model parameter, θ_{S_k}, then its MLE, $\hat{\theta}_{S_k}$, satisfies

$$\left[\frac{\partial \ell(X_{S_k}|\theta)}{\partial \theta}\right]_{\hat{\theta}_{S_k}} = 0_p, \tag{11.11}$$

where

$$\ell(X_{S_k}|\theta) = \ell(X_{i_1}, X_{i_2}, \ldots, X_{i_{n_k}}|\theta) = \sum_{i \in S_k} \ell(X_i|\theta), \tag{11.12}$$

and an estimate of the covariance matrix associated with $\hat{\theta}_{S_k}$ is given by

$$\Sigma_{S_k} = \left[-\frac{\partial^2 \ell(X_{S_k}|\theta)}{\partial \theta \, \partial \theta'}\right]^{-1}_{\hat{\theta}_{S_k}}. \tag{11.13}$$

Lemma 1. *For Gaussian distributed observed data, the MLE of the common model parameter in a cluster is equal to the Mahalanobis mean of the model parameter estimates of individual objects in the cluster, that is,*

$$\hat{\theta}_{S_k} = \left(\sum_{i \in S_k} \Sigma_i^{-1}\right)^{-1} \left(\sum_{i \in S_k} \Sigma_i^{-1} \hat{\theta}_i\right). \tag{11.14}$$

Proof. Since we know that third and higher-order partial derivatives of $\ell(X_i|\theta)$ are zero for Gaussian distributed data, we can write the Taylor series expansion of $\ell(X_i|\theta)$ as

$$\left[\frac{\partial \ell(X_i|\theta)}{\partial \theta}\right]_{\hat{\theta}_{S_k}} = \left[\frac{\partial \ell(X_i|\theta)}{\partial \theta}\right]_{\hat{\theta}_i} + \left[\frac{\partial}{\partial \theta}\left(\frac{\partial \ell(X_i|\theta)}{\partial \theta}\right)\right]_{\hat{\theta}_i} (\hat{\theta}_{S_k} - \hat{\theta}_i). \tag{11.15}$$

Using equations (11.9) and (11.10), the above equation reduces to

$$\left[\frac{\partial \ell(X_i|\theta)}{\partial \theta}\right]_{\hat{\theta}_{S_k}} = -\Sigma_i^{-1}(\hat{\theta}_{S_k} - \hat{\theta}_i). \tag{11.16}$$

From equations (11.11) and (11.12), we know that

$$0_p = \left[\frac{\partial \ell(X_{S_k}|\theta)}{\partial \theta}\right]_{\hat{\theta}_{S_k}} = \sum_{i \in S_k} \left[\frac{\partial \ell(X_i|\theta)}{\partial \theta}\right]_{\hat{\theta}_{S_k}} = -\sum_{i \in S_k} \Sigma_i^{-1}(\hat{\theta}_{S_k} - \hat{\theta}_i). \tag{11.17}$$

Rearranging terms in the above equation, we get the desired result. □

Lemma 2. *For Gaussian distributed observed data, the error matrix associated with $\hat{\theta}_{S_k}$ is the same as the error matrix associated with the Mahalanobis mean of the model parameter estimates of individual objects in the cluster, that is,*

$$\Sigma_{S_k} = \left(\sum_{i \in S_k} \Sigma_i^{-1}\right)^{-1}. \tag{11.18}$$

Proof. It follows from the Taylor series expansion of each term in the matrix $\left[\frac{\partial^2 \ell(X_i|\theta)}{\partial \theta\, \partial \theta'}\right]_{\hat{\theta}_{S_k}}$ around $\hat{\theta}_i$ and then using equations (11.12) and (11.13). □

Theorem 2. *For Gaussian distributed observed data, the optimal error-based clusters of the model parameter estimates are the maximum likelihood clusters of the observed data.*

Proof. The maximum likelihood clusters of the observed data are given by

$$\begin{aligned}\max_{S_1,\ldots,S_K;\theta_{S_1},\ldots,\theta_{S_K}} \sum_{k=1}^{K} \sum_{i \in S_k} \ell(X_i|\theta_{S_k}) &= \max_{S_1,\ldots,S_K} \sum_{k=1}^{K} \left[\max_{\theta_{S_k}} \sum_{i \in S_k} \ell(X_i|\theta_{S_k})\right] \\ &= \max_{S_1,\ldots,S_K} \sum_{k=1}^{K} \sum_{i \in S_k} \ell(X_i|\hat{\theta}_{S_k}). \end{aligned} \tag{11.19}$$

Since third and higher-order partial derivatives of $\ell(X_i|\theta)$ are zero for Gaussian distributed data, each term on the right side of equation (11.19) can be expanded using the Taylor series as

$$\begin{aligned}
\ell(X_i|\hat{\theta}_{S_k}) &= \ell(X_i|\hat{\theta}_i) + \left[\frac{\partial \ell(X_i|\theta)}{\partial \theta}\right]'_{\hat{\theta}_i} (\hat{\theta}_{S_k} - \hat{\theta}_i) \\
&\quad + \frac{1}{2}(\hat{\theta}_{S_k} - \hat{\theta}_i)' \left[\frac{\partial^2 \ell(X_i|\theta)}{\partial \theta\, \partial \theta'}\right]_{\hat{\theta}_i} (\hat{\theta}_{S_k} - \hat{\theta}_i) \\
&= \ell(X_i|\hat{\theta}_i) + 0 + \frac{1}{2}(\hat{\theta}_{S_k} - \hat{\theta}_i)'(-\Sigma_i^{-1})(\hat{\theta}_{S_k} - \hat{\theta}_i) \\
&= \ell(X_i|\hat{\theta}_i) - \frac{1}{2}(\hat{\theta}_{S_k} - \hat{\theta}_i)'\, \Sigma_i^{-1}(\hat{\theta}_{S_k} - \hat{\theta}_i). \qquad (11.20)
\end{aligned}$$

Substituting equation (11.20) in equation (11.19), we get

$$\begin{aligned}
\max_{S_1,\ldots,S_K;\theta_{S_1},\ldots,\theta_{S_K}} \sum_{k=1}^{K} \sum_{i \in S_k} \ell(X_i|\theta_{S_k}) &= \max_{S_1,\ldots,S_K} \sum_{k=1}^{K} \sum_{i \in S_k} [\ell(X_i|\hat{\theta}_i) \\
&\quad - \frac{1}{2}(\hat{\theta}_{S_k} - \hat{\theta}_i)'\Sigma_i^{-1}(\hat{\theta}_{S_k} - \hat{\theta}_i)] \\
&\equiv \min_{S_1,\ldots,S_K} \sum_{k=1}^{K} \sum_{i \in S_k} (\hat{\theta}_i - \hat{\theta}_{S_k})'\Sigma_i^{-1}(\hat{\theta}_i - \hat{\theta}_{S_k}), \qquad (11.21)
\end{aligned}$$

which is the optimal error-based clustering of the model parameter estimates. The equivalence above follows because $\sum_{k=1}^{K} \sum_{i \in S_k} \ell(X_i|\hat{\theta}_i)$ is a constant. □

In the remainder of this chapter, we present results from a series of empirical studies that suggest that error-based clustering is more appropriate than traditional clustering methods for clustering model parameters.

11.7 EMPIRICAL STUDY

In this section, we present results from simulation studies on four statistical models: (1) sample averaging, (2) multiple linear regression, (3) ARIMA time series, and (4) Markov chains. We also discuss two empirical studies on real-world datasets. We compared clustering results using *kError* and *hError* against those using k-means, Ward's method, and model-based clustering. Since k-means and Ward's method depend on the units of data measurement, we applied these methods after normalizing the data to unit variance on each variable. For model-based clustering, we used the MATLAB toolbox developed by Martinez and Martinez (2004). The number of clusters is assumed to be given for each clustering method. We evaluate a clustering method by its misclassification error, that is, the number of data points assigned to an incorrect cluster. In our study, we found that the misclassification error was significantly smaller for *kError* and *hError* than for the other methods. We also found that *kError* performed significantly better than *hError*. The reason is that we ran *kError* with 50 different random initial partitions and picked the one that achieved the best objective value. This helped *kError* achieve better objective value than *hError*.

In the simulation study presented here, we have considered four statistical models that are widely used in practice. The study can be easily extended to other models that use maximum likelihood to estimate model parameters and associated error matrices. We have used datasets with two or three clusters and models with two or three parameters. While the study can be easily extended to datasets with more than three clusters and models with more than three parameters, it was helpful to illustrate our idea with small examples that are easy to interpret.

11.7.1 Clustering Sample Means

There are many situations in which one has access only to aggregated data. This may happen because of the need to simplify data management—for example, in e-commerce data or census data—or for confidentiality reasons—for example, in data from clinical trials or surveys. The aggregated data are often represented by their sample mean and variance-covariance statistics (error information) associated with the sample mean. Our goal is to cluster these sample means.

11.7.1.1 Data Generation. Thirty samples of data were generated as follows. We start with three values of true mean $\mu = \mu_1, \mu_2, \mu_3$, where each value of μ corresponds to a cluster in our data. For each value of μ, we generate 10 samples of Gaussian distributed data with mean of μ and covariance of Σ_i for the ith sample. Here Σ_i is randomly generated as V_iV_i', where V_i is a 2×2 matrix of uniformly distributed random numbers between 0 and 2.5. Each sample contains 10 data points. We represent each sample by its sample mean and sample covariance based on these 10 data points. Given sample means and covariance matrix estimates for 30 samples (without the knowledge of their true means), our goal is to partition them into three clusters so that samples having same true mean belong to the same cluster. The values of μ's for this experiment were generated from a two-dimensional spherical Gaussian distribution with a variance of 1 on each dimension.

11.7.1.2 Clustering Results. The average misclassification errors using various clustering methods in 100 replications of the above experiment are reported in Table 11.2. The numbers in parentheses are the standard deviation of average

TABLE 11.2 Average Misclassification Error for Clustering of Sample Means

Clustering Method	Average Misclassification Error
kError	3.45 (0.32)
hError	4.52 (0.34)
Model-based	8.20 (0.48)
k-Means	8.02 (0.41)
Ward	12.25 (0.45)

TABLE 11.3 Misclassification Error with Diagonal Approximation for Sample Covariance

Clustering Method	Average Misclassification Error
kError with diagonal approximation	6.67 (0.41)
hError with diagonal approximation	6.99 (0.42)

misclassification errors. The misclassification errors are much smaller for *kError* and *hError* than for the other methods. In the above experiment, *hError* was able to find the correct number of clusters (three in this case) 81 times in 100 runs of the experiment. Of the remaining 19 runs, it found two clusters on 18 runs and one cluster on 1 run. This is consistent with the theory we presented in Section 11.4.2.

In this experiment, we assumed that one has access to the entire $p \times p$ covariance matrix associated with each sample mean. In many practical situations, one has access to only variance on each variable of the sample. In such cases, the $p \times p$ error matrix can be approximated to be a diagonal matrix. Table 11.3 shows the effect of this approximation in the above experiment. We note that *kError* and *hError* still produced better clusters than k-means, Ward's method, and model-based clustering, but (unsurprisingly) the results were somewhat worse than those achieved when we used the entire covariance matrix for each sample mean.

11.7.2 Clustering Multiple Linear Regression Models

For application of error-based clustering on multiple linear regression models, we consider a clustering problem that is common in the finance industry. Suppose we want to cluster a set of stocks based on similarity in their performance against overall market performance. A commonly used model to measure the performance of a stock against the market performance is the capital asset pricing model (CAPM), described below (Jones 1991).

$$R_{it} - R_f = \alpha_i + \beta_i(R_{mt} - R_f) + \epsilon_{it} \quad i = 1, \ldots, N, \quad t = 1, \ldots, T, \tag{11.22}$$

where R_{it} is the rate of return on stock i at time t, R_f is the rate of return on a risk-free investment, and R_{mt} is the rate of return on a market benchmark (e.g., the S&P 500) at time t. Here β_i for stock i is a measure of the risk profile of the stock (higher β's represent riskier stocks), and α_i is a measure of how much better (or worse) the stock did than the CAPM predicted.

Given R_f and the observed values of R_{mt} and R_{it} for a sequence of time periods for stock i, we can estimate its α and β using ordinary linear regression. In many applications, it is useful to cluster stocks that have similar values of α and β.[5] Through a study on simulated datasets, we demonstrate that error-based clustering produces better clusters than standard clustering methods in this application.

[5]Stock ratings are often developed based on similarity in α and β (Lui et al. 2007).

11.7.2.1 Data Generation. We generate data for 30 stocks as follows. We start with three values of (α, β) that are generated from a two-dimensional spherical Gaussian distribution with a variance of 1 on each dimension. The three values of (α, β) correspond to three clusters of stock. For each value of (α, β), we generate data for 10 stocks as follows. For each stock, we generate its return in 10 time periods using equation (11.22), where R_{mt} is randomly generated from a uniform distribution between 3% and 8% ε_{it} is generated from a normal distribution $N(0, 1)$, and R_f is taken to be zero. Given stock return and market return data for the 30 stocks, our goal is to partition them into three clusters so that stocks that have common values of (α, β) belong to the same cluster.

Given data for 10 time periods for a stock, its (α, β) is estimated using the least square method. Let the estimate be ($\hat{\alpha}_i$, $\hat{\beta}_i$) and the associated covariance matrix be Σ_i, for the ith stock, $i = 1, \ldots, 30$. This constitutes the data to be clustered.

11.7.2.2 Clustering Results. Table 11.4 presents the average misclassification error for various clustering methods in 100 replications of the above experiment. In this example, the observed data (stock returns) are vectors of fixed (10) dimensions. Therefore, we also present the results when Ward's and k-means methods were applied directly to the observed data.

We found that *kError* and *hError* performed significantly better than the other methods. We also found that clustering on model parameters performed significantly better than clustering on observed data. Using the method described in Section 11.4.2, *hError* was able to find the correct number of clusters (three in this case) 90 times in 100 runs of the above experiment. On the remaining 10 runs it found two clusters.

11.7.3 Clustering Time Series Models

For clustering of time series models, we illustrate the proposed methodology using autoregressive (AR) models of order p,

$$y_{it} = \phi_{i1} y_{it-1} + \cdots + \phi_{ip} y_{it-p} + \varepsilon_{it}, \qquad i = 1, \ldots, n, \tag{11.23}$$

TABLE 11.4 Average Misclassification Error for Clustering Regression Models

Clustering Method	Average Misclassification Error
kError	1.50 (0.28)
hError	1.64 (0.31)
Model-based	7.02 (0.61)
k-Means	7.63 (0.36)
Ward	11.70 (0.39)
k-Means on observed data	13.51 (0.21)
Ward on observed data	13.96 (0.20)

where y_{it} is the value of the ith time series at time t; $\phi_{i1}, \ldots, \phi_{ip}$ are the model parameters for this time series; and ε_{it} are independent and Gaussian distributed random errors. First, we estimate the parameters $\hat{\phi}_i = (\hat{\phi}_{i1}, \ldots, \hat{\phi}_{ip})$ and the associated covariance matrices Σ_i for $i = 1, \ldots, n$, and then we cluster these p-dimensional vectors, $\phi_1, \ldots, \phi_n$ using error-based clustering.

11.7.3.1 Data Generation. For this experiment we generate data for 30 time series from AR(2) models as follows. We start with three values for (ϕ_{i1}, ϕ_{i2}) that are generated from a uniform distribution between 0 and 1 (we ignore values where $\phi_{i1} + \phi_{i2} \geq 1$). The three values of (ϕ_{i1},ϕ_{i2}) correspond to three clusters of time series. For each value of (ϕ_{i1},ϕ_{i2}), we generate 10 time series using equation (11.23), giving a total of 30 time series. For each time series we generate data for 50 time points. ε_{it} is chosen to be from a Gaussian distribution $N(0, 0.01)$. Given 50 data points for a time series, the MLE of its parameters $\hat{\phi}_i = (\hat{\phi}_{i1}, \hat{\phi}_{i2})$ and the associated 2×2 covariance matrix Σ_i are obtained using the System Identification Toolbox in MATLAB. Clusters of time series are obtained based on $\hat{\phi}_i$ and Σ_i, $i = 1, \ldots, 30$.

11.7.3.2 Clustering Results. The average misclassification error using various clustering methods in 100 replications of the above experiment is reported in Table 11.5. Here again the observed data (time series values) are vectors of fixed (50) dimensions. Therefore, we also present results when Ward's and k-means methods were applied directly to the observed data. We see that the misclassification error is smaller for *kError* and *hError* than for the other methods. Further, clustering on model parameters performed significantly better than clustering on observed data. On this application, the *hError* algorithm was able to find the correct number of clusters (three in this case) 87 times in the 100 runs of the experiment. Of the remaining 13 runs, it found two clusters on 11 runs and one cluster on 2 run.

11.7.4 Clustering Markov Chain Models

For this application, we illustrate the proposed methodology using a four-state Markov chain model for an online shopping website, as shown in Figure 11.4.

TABLE 11.5 Average Misclassification Error for Clustering of Time Series Models

Clustering Method	Average Misclassification Error
kError	6.32 (0.36)
hError	6.89 (0.38)
Model-based	8.49 (0.40)
k-Means	7.78 (0.39)
Ward	10.91 (0.43)
k-Means on observed data	15.58 (0.17)
Ward on observed data	18.21 (0.10)

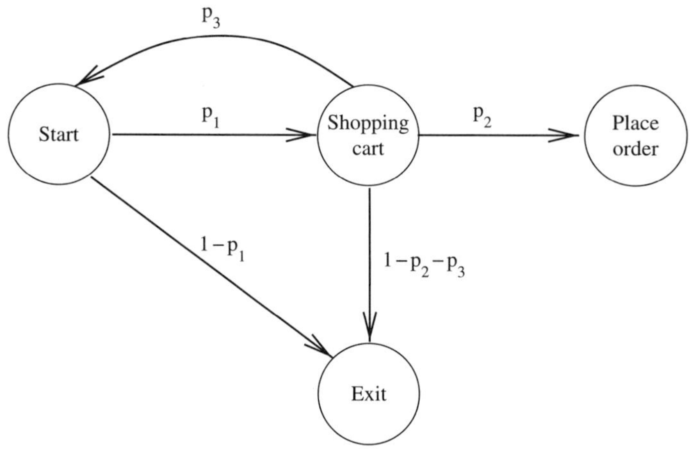

Figure 11.4 Markov chain model for online users.

Here each user is assumed to have different transition probabilities between states of the Markov chain. The online behavior of a user is completely determined by its probability vector (p_1, p_2, p_3).

11.7.4.1 Data Generation. Data for 60 users are generated as follows. We start with two values of transition probabilities (p_1, p_2, p_3) that are generated from a uniform distribution between 0 and 1 (we ignore values where $p_2 + p_3 \geq 1$). The two values of (p_1, p_2, p_3) correspond to two clusters of users. Thirty users are generated for each value of (p_1, p_2, p_3), giving us a total of 60 users. We generate 20 sessions for each user using the multinomial distribution based on the transition probabilities. We estimate $(\hat{p}_1, \hat{p}_2, \hat{p}_3)$ and the associated covariance matrix for each user based on page transition data on then twenty sessions and then cluster the users based on these estimates.

11.7.4.2 Clustering Result. The average misclassification error using various clustering methods in 100 replications of the above experiment is reported in Table 11.6. Here again we find that the misclassification error is much smaller for *kError* and *hError* than for the other methods. On this application, the *hError* algorithm was able to find the correct number of clusters (two in this case) 83 times in the 100 runs of the experiment. On the remaining 17 runs, it found one cluster.

11.7.5 Real Data Studies

We conducted two empirical studies on real-world datasets, where error-based clustering produced significantly better clusters than traditional clustering methods that do not account for error measurements. In the first study, our objective was to forecast sales of items in a retail store based on seasonal patterns (or seasonality) of these

TABLE 11.6 Average Misclassification Error for Clustering of Markov Chain Models

Clustering Method	Average Misclassification Error
kError	7.44 (0.34)
hError	9.76 (0.38)
Model-based	14.03 (0.63)
k-Means	13.11 (0.55)
Ward	19.36 (0.60)

items. Current methods for estimating seasonality produce a large number of seasonality estimates, one for each class of merchandise. If only a small amount of data is available to estimate seasonality, the estimates of seasonality will have large errors which lead to poor sales forecasts. We found that the seasonality estimates, and consequently the sales forecasts, can be improved by clustering seasonality estimates from different classes of merchandise. In this application, we found that error-based clustering performed significantly better than traditional clustering methods in improving the sales forecasts. On a sample of real datasets from a large national retail store, we found that error-based clustering improved the sales forecasts by about 43%, whereas k-means and Ward's methods improved the forecasts by only about 24%. The details of this study are available in Kumar et al. (2002).

In the second study, we measured the effectiveness of error-based clustering on a personal income dataset, which is a collection of 25 time series representing per capita personal income during 1929–1999 in 25 U.S. states.[6] These data were first studied by Kalpanis et al. (2001), who used the Euclidean distance between ARIMA time series as a basis for clustering. The authors believe that the dataset has two true clusters: One consists of the East Coast states, California, and Illinois, where there was a high growth rate in personal income, and the other consists of the Midwestern states, where there was a low growth rate.[7] Our goal is to identify the two clusters based on income data for these time series.

The per capita income time series are nonstationary in mean as well as variance. To remove this nonstationarity, we applied the preprocessing steps used in Kalpanis et al. (2001). We smoothed the original series by taking a window average over a window of size 2 and then took logarithms of the smoothed time series. We fitted ARIMA(1,1,0) models to the resulting time series. ARIMA(1,1,0) has only one parameter; thus, clusters of time series were formed on the basis of the values of the estimated parameter of ARIMA(1,1,0) and the associated variance terms. The misclassification error is reported in Table 11.7. We also compared our findings with those of the (CEP) method proposed in Kalpanis et al. (2001). We found

[6]The dataset can be obtained from http://www.bea.gov/bea/regional/spi. The states included in this data are Connecticut, Washington, D.C., Delaware, Florida, Massachusetts, Maine, Maryland, North Carolina, New Jersey, New York, Pennsylvania, Rhode Island, Virginia, Vermont, West Virginia, California, Idaho, Iowa, Indiana, Kansas, North Dakota, Nebraska, Oklahoma, and South Dakota.

[7]The first 17 states form the first cluster and the last 8 states form the second cluster.

TABLE 11.7 Misclassification Error and SSE Imrovement on Personal Income Data

Clustering Method	Misclassification Error	Reduction in SSE (%)
kError	0	6.1
hError	0	6.1
Model-based	5	0.8
k-Means	3	2.0
Ward	5	0.8
CEP	3	2.0
No clustering	-	0.0

that *kError* and *hError* were able to discover the true clusters, whereas the other methods did not.

We also conducted an out-of-sample study as follows. The first 55 years of data for each state were used to estimate the time series parameters, and then income for the last 15 years was predicted using the estimated common parameter for each cluster obtained using various clustering methods. The accuracy of the predicted income was measured by the sum of squared errors (SSE). Table 11.7 presents the percentage reduction in SSE for each clustering method from the base case when no clustering was used. We found that standard clustering methods reduced SSE (or improved the forecast over no clustering) to a small extent, while error-based clustering reduced it more substantially.

11.8 SUMMARY AND DISCUSSION

In this chapter, we have developed a new clustering technique that recognizes errors associated with data. We developed a probability model for incorporating error information in the clustering process and provided algorithms for error-based clustering that are generalizations of the popular Ward's and k-means algorithms.

While error-based clustering can be applied directly to observed data, we believe a major area for applications involves clustering model parameters (preprocessed data) obtained by fitting statistical models to observed data. Currently, clustering problems in such applications are solved using the EM algorithm. The key difference between the EM approach and the error-based clustering approach for clustering model parameters is that the EM approach simultaneously clusters the individuals and estimates model parameters for each individual, whereas error-based clustering decomposes it into two steps: (1) estimate model parameters and associated covariance matrices for each object and (2) cluster estimated parameters using error-based clustering. An advantage of the decomposition process is that once we have estimated model parameters, we can choose from a variety of clustering algorithms—for example, the *hError* and *kError* algorithms proposed in this chapter—while the EM approach is restricted to a k-means-like algorithm. Another advantage of the two-step process of error-based clustering is that the clustering method is applied to estimated parameters, where the data are much smaller than the original observed data. Thus, the two-step process provides a natural way to reduce the dimensions of the observed data. Moreover, we have shown in this chapter that when the observed data are normally distributed, the two-step process of error-based clustering produces exactly the same clusters as the single step maximum likelihood clusters of the EM algorithm.

Finally, we demonstrated the effectiveness of error-based clustering in a series of empirical studies on clustering model parameters for four statistical models: sample averaging, multiple linear regression, ARIMA time series, and Markov chains. While we studied these four models in this chapter, the concept of using error-based clustering for clustering model parameters is very general and can be applied to the large class of statistical models where MLE is used to estimate parameters.

Here are a few future directions for this research that we are exploring. We would like to extend the theory we developed for Gaussian distributed data to the generalized linear models. We believe that the theory can also be extended to give approximation results for general probability distributions. We wish to do a detailed study with real-world data on clustering other statistical models.

REFERENCES

Anderberg, M.R. (1973). *Cluster Analysis for Applications.* Academic Press, New York, NY.

Banfield, J.D. and Raftery, A.E. (1993). Model-based Gaussian and non-Gaussian clustering. *Biometrics*, 49: 803–821.

Brucker, P. (1977). On the complexity of clustering problems. In *Optimization and Operations Research* (R. Henn, B. Korte, and W. Oletti, eds.). Springer-Verlag, Berlin.

Cadez, I., Gaffney, S., and Smyth, P. (2000). A general probabilistic framework for clustering individuals. *Proceedings of the Sixth ACM SIGKDD International Conference on Knowledge Discovery and Data Mining.*

Cadez, I. and Smyth, P. (1999). Probabilistic clustering using hierarchical models. ICS Technical Report, University of California at Irvine.

Celeux, G. and Govaert, G. (1992). A classification EM algorithm for clustering and two stochastic versions. *Computational Statistics and Data Analysis*, 14: 315–332.

Celeux, G. and Govaert, G. (1995). Gaussian parsimonious clustering models. *Pattern Recognition*, 28: 781–793.

Chaudhuri, B.B. and Bhowmik, P.R. (1998). An approach of clustering data with noisy or imprecise feature measurement. *Pattern Recognition Letters*, 19: 1307–1317.

Cowpertwait, P.S.P. and Cox, T.F. (1992). Clustering population means under heterogeneity of variance with an application to a rainfall time series problem. *The Statistician*, 41: 113–121.

Cox, D.R. and Spjotvoll, E. (1982). On partitioning means into groups. *Scandinavian Journal of Statistics*, 9: 147–152.

Feng, Z., McLerran, D., and Grizzle, J. (1996). A comparison of statistical methods for clustered data analysis with Gaussian error. *Statistics in Medicine*, 15(16): 1793–806.

Fisher, W.D. (1958). On grouping for maximum homogeneity. *Journal of the American Statistical Association*, 53: 789–798.

Fraley, C. and Raftery, A.E. (2002). Model-based clustering, discriminant analysis, and density estimation. *Journal of the American Statistical Association*, 97: 611–631.

Gaffney, S. and Smyth, P. (1999). Trajectory clustering using mixtures of regression models. *Proceedings of the Fifth ACM SIGKDD International Conference on Knowledge Discovery and Data Mining.*

Guha, S., Meyerson, A., Mishra, N., Motwani, R., and O'Callaghan, L. (2003). Clustering data streams: theory and practice. *Transactions on Knowledge and Data Engineering*, 15(3): 515–528.

Jain, A.K. and Dubes, R.C. (1988). *Algorithms for Clustering Data.* Englewood Cliffs, NJ: Prentice Hall.

Jank, W. and Shmueli, G. (2005). Profiling price dynamics in online auctions using curve clustering. Working Paper, RHS-06-004, Robert H Smith School, University of Maryland.

Jones, P. (1991). *Investment Analysis and Management.* New York: Wiley.

Kalpakis, K., Gada, D., and Puttagunta, V. (2001). Distance measures for effective clustering of ARIMA time-series. *Proceedings of the IEEE International Conference on Data Mining.*

Kumar, M., Patel, N.R., and Woo, J. (2002). Clustering seasonality patterns in the presence of errors. *Proceedings of the Eighth ACM SIGKDD International Conference on Knowledge Discovery and Data Mining.*

Lui, D., Markov, S., and Tamayo, A. (2007). What makes a stock risky? Evidence from sell-side analysts' risk ratings. *Journal of Accounting Research*, 45(3): 629–665.

Mahalanobis, P.C. (1936). On the generalized distance in statistics. *Proceedings of the National Institute of Science of India.*

Maharaj, E.A. (2000). Clusters of time series. *Journal of Classification*, 17: 297–314.

Martinez, A.R. and Martinez, W.L. (2004). Model-based clustering toolbox for MAT-LAB. Available at http://www.stat.washington.edu/fraley/mclust.

McLachlan, G.J. and Basford, K.E. (1988). *Mixture Models: Inference and Applications to Clustering.* New York: Marcel Dekker.

Milligan, G.W. and Cooper, M.C. (1985). An examination of procedures for determining the number of clusters in a data set. *Psychometrika*, 50: 159–179.

Pena, J., Lozano, J., and Larranaga, P. (1999). An empirical comparison of four initialization methods for the k-means algorithm. *Pattern Recognition Letters*, 50: 1027–1040.

Piccolo, D. (1990). A distance measure for classifying Arima models. *Journal of Time Series Analysis*, 11(2): 153–164.

Sarachik, K.B. and Grimson, W.E.L. (1993). Gaussian error models for object recognition. *IEEE Conference on Computer Vision and Pattern Recognition.*

Scott, A.J. and Symons, M.J. (1971). Clustering methods based on likelihood ratio criteria. *Biometrics*, 27: 387–397.

Stanford, D.C. and Raftery, A.E. (2000). Finding curvilinear features in spatial point patterns: Principal curve clustering with noise. *IEEE Transactions on Pattern Analysis and Machine Intelligence*, 22(6): 601–609.

Ward, J.H. (1963). Hierarchical grouping to optimize an objective function. *Journal of the American Statistical Association*, 58: 236–244.

Zamir, O. and Etzioni, O. (1998). Web document clustering: A feasibility demonstration. *Proceedings of SIGIR.*

APPENDIX: PROOF OF THEOREM 1

Let E_k be the contribution of cluster C_k to the objective function, that is,

$$\begin{aligned} E_k &= \sum_{i \in S_k} (x_i - \hat{\theta}_k)^t \Sigma_i^{-1} (x_i - \hat{\theta}_k) \\ &= \sum_{i \in S_k} x_i^t \Sigma_i^{-1} x_i + \hat{\theta}_k^t \Big(\sum_{i \in S_k} \Sigma_i^{-1} \Big) \hat{\theta}_k - 2\hat{\theta}_k^t \Big(\sum_{i \in S_k} \Sigma_i^{-1} x_i \Big) \\ &= \sum_{i \in S_k} x_i^t \Sigma_i^{-1} x_i + \hat{\theta}_k^t \Psi_k^{-1} \hat{\theta}_k - 2\hat{\theta}_k^t \Psi_k^{-1} \hat{\theta}_k \\ &= \sum_{i \in S_k} x_i^t \Sigma_i^{-1} x_i - \hat{\theta}_k^t \Psi_k^{-1} \hat{\theta}_k, \qquad k = 1, \ldots, K. \end{aligned} \tag{11.A.1}$$

Suppose we choose to merge clusters C_u and C_v during an iteration of *hError* and let the resulting cluster be C_w. The net increase in the objective function is given by

$$\Delta E_{uv} = E_w - E_u - E_v = -\hat{\theta}_w^t \Psi_w^{-1} \hat{\theta}_w + \hat{\theta}_u^t \Psi_u^{-1} \hat{\theta}_u + \hat{\theta}_v^t \Psi_v^{-1} \hat{\theta}_v. \tag{11.A.2}$$

From equations (11.3) and (11.4) and the fact that $S_w = S_u \cup S_v$, it follows that

$$\Psi_w = (\Psi_u^{-1} + \Psi_v^{-1})^{-1}, \tag{11.A.3}$$

$$\hat{\theta}_w = (\Psi_u^{-1} + \Psi_v^{-1})^{-1} (\Psi_u^{-1} \hat{\theta}_u + \Psi_v^{-1} \hat{\theta}_v). \tag{11.A.4}$$

Multiplying by $(\Psi_u^{-1} + \Psi_v^{-1})^{\frac{1}{2}}$ and taking the dot product with itself on both sides of equation (11.A.4) gives

$$\begin{aligned} \hat{\theta}_w^t (\Psi_u^{-1} + \Psi_v^{-1}) \hat{\theta}_w &= (\Psi_u^{-1} \hat{\theta}_u + \Psi_v^{-1} \hat{\theta}_v)^t (\Psi_u^{-1} + \Psi_v^{-1})^{-1} (\Psi_u^{-1} \hat{\theta}_u + \Psi_v^{-1} \hat{\theta}_v) \\ &= \hat{\theta}_u^t \Psi_u^{-1} (\Psi_u^{-1} + \Psi_v^{-1})^{-1} \Psi_u^{-1} \hat{\theta}_u + \hat{\theta}_v^t \Psi_v^{-1} (\Psi_u^{-1} + \Psi_v^{-1})^{-1} \Psi_v^{-1} \hat{\theta}_v \\ &\quad + 2\hat{\theta}_u^t \Psi_u^{-1} (\Psi_u^{-1} + \Psi_v^{-1})^{-1} \Psi_v^{-1} \hat{\theta}_v. \end{aligned} \tag{11.A.5}$$

The last term of equation (11.A.5) can be rewritten as

$$
\begin{aligned}
2\hat{\theta}_u^t \Psi_u^{-1}(\Psi_u^{-1} + \Psi_v^{-1})^{-1}\Psi_v^{-1}\hat{\theta}_v = {} & \hat{\theta}_u^t \Psi_u^{-1}(\Psi_u^{-1} + \Psi_v^{-1})^{-1}\Psi_v^{-1}\hat{\theta}_u \\
& + \hat{\theta}_v^t \Psi_u^{-1}(\Psi_u^{-1} + \Psi_v^{-1})^{-1}\Psi_v^{-1}\hat{\theta}_v \\
& - (\hat{\theta}_u - \hat{\theta}_v)^t \Psi_u^{-1}(\Psi_u^{-1} + \Psi_v^{-1})^{-1} \\
& \times \Psi_v^{-1}(\hat{\theta}_u - \hat{\theta}_v). \qquad (11.A.6)
\end{aligned}
$$

Substituting equation (11.A.6) into equation (11.A.5) gives

$$
\begin{aligned}
\hat{\theta}_w^t(\Psi_u^{-1} + \Psi_v^{-1})\hat{\theta}_w = {} & \hat{\theta}_u^t \Psi_u^{-1}(\Psi_u^{-1} + \Psi_v^{-1})^{-1}(\Psi_u^{-1} + \Psi_v^{-1})\hat{\theta}_u \\
& + \hat{\theta}_v^t(\Psi_u^{-1} + \Psi_v^{-1})(\Psi_u^{-1} + \Psi_v^{-1})^{-1}\Psi_v^{-1}\hat{\theta}_v \\
& - (\hat{\theta}_u - \hat{\theta}_v)^t \Psi_u^{-1}(\Psi_u^{-1} + \Psi_v^{-1})^{-1}\Psi_v^{-1}(\hat{\theta}_u - \hat{\theta}_v) \\
= {} & \hat{\theta}_u^t \Psi_u^{-1}\hat{\theta}_u + \hat{\theta}_v^t \Psi_v^{-1}\hat{\theta}_v - (\hat{\theta}_u - \hat{\theta}_v)^t \\
& \times \Psi_u^{-1}(\Psi_u^{-1} + \Psi_v^{-1})^{-1}\Psi_v^{-1}(\hat{\theta}_u - \hat{\theta}_v) \\
= {} & \hat{\theta}_u^t \Psi_u^{-1}\hat{\theta}_u + \hat{\theta}_v^t \Psi_v^{-1}\hat{\theta}_v - (\hat{\theta}_u - \hat{\theta}_v)^t \\
& \times (\Psi_u + \Psi_v)^{-1}(\hat{\theta}_u - \hat{\theta}_v) \qquad (11.A.7)
\end{aligned}
$$

The last equality follows from the fact that $A^{-1}(A^{-1} + B^{-1})^{-1}B^{-1} = (A + B)^{-1}$ for any two matrices A and B.

Substituting equations (11.A.3) and (11.A.7) into equation (11.A.2) gives

$$
\Delta E_{uv} = (\hat{\theta}_u - \hat{\theta}_v)^t(\Psi_u + \Psi_v)^{-1}(\hat{\theta}_u - \hat{\theta}_v). \qquad (11.A.8)
$$

Minimizing ΔE_{uv} is therefore the same as minimizing the distance $d_{uv} = (\hat{\theta}_u - \hat{\theta}_v)^t(\Psi_u + \Psi_v)^{-1}(\hat{\theta}_u - \hat{\theta}_v)$ among all possible pairs of clusters C_u and C_v.

12

FUNCTIONAL DATA ANALYSIS FOR SPARSE AUCTION DATA

BITAO LIU AND HANS-GEORG MÜLLER

Department of Statistics, University of California, Davis, California

12.1 INTRODUCTION

eBay.com is today's biggest global online auction marketplace. It provides a convenient environment for millions of sellers and buyers to carry out real-time transactions on the Internet. The fact that eBay makes complete auction information available to the public provides opportunities for researchers to analyze online bidding behavior. This may lead to improved transaction strategies, benefiting both sellers and bidders. Although there are millions of different auctioned products available on eBay at any moment, we focus here on a small subset, on online transactions of a similar or same type of product, auctioned under the same modalities (same duration setting and same billing currency (U.S. dollars)) and during the same time period.

We refer to Jank and Shmueli (2005a) and Reddy and Dass (2006) for some prior functional data analysis approaches for online auction data; this previous work is largely based on approaches described in Ramsay and Silverman (2002, 2005). The time spacing of the bids from each auction is usually sparse for long periods during the auction and often becomes very dense as the auction is nearing its end. This phenomenon, caused by bidders who place their bids at the last moment, when an auction is near its close, is known as *sniping*. Since the auctioned products that we are interested in are very much alike, the corresponding realizations of the price process for these products at different auctions will be considered to be i.i.d., although at a more detailed level, some dependencies between auctions that are

Statistical Methods in e-Commerce Research. Edited by W. Jank and G. Shmueli

close in calendar time might be present. We ignore such possible dependencies in the following discussion and make the assumption that the observed bids from each auction can be viewed as measurements of the realization of an underlying random smooth stochastic process, the *price process.*

Functional data analysis, and especially functional principal component modeling, provides useful tools for analyzing such data (see also Jank and Shmueli, 2006). Traditional functional principal component analysis (FPCA) requires dense and regular design data. Due to the high degree of data sparsity and time irregularity present in the auction data, it is therefore necessary to adjust this methodology. For this purpose, we adopt the principal analysis through conditional expectation (PACE) method developed in Yao et al. (2005). This will allow us to recover price trajectories even if only one or two bids are available for a given auction, provided that the pooled timings of bids from all auctions included in the data form a dense grid. An alternative is to fit a functional random effects model (James et al. 2000; Rice & Wu 2001), which handles sparse functional data through a prespecified function basis such as the B-spline basis.

The chapter is organized as follows. In Section 12.2, we provide an overview of the auction system on eBay.com. In Section 12.3, we review the methodology used to recover individual trajectories by the PACE method, which will be applied to the estimation of both the log price process and the log price ratio process. Using a modified approach, we adapt PACE to summarize the bid history at varying current times during an ongoing auction, providing time-varying principal component scores. These can be used to predict the closing price while an auction is still ongoing, providing instant updates when a new bid is registered during an ongoing auction. In Section 12.4, we illustrate our methods with data for 158 Palm M515 personal digital assistant auctions. We also present the prediction of the closing price for each auction by functional linear regressions, using the time-varying functional principal component scores as predictors. Other issues, such as monotonicity of the fitted price curves, will be briefly discussed.

12.2 ONLINE AUCTIONS AT eBAY.COM

The most common auctions on eBay are single-item auctions, which are organized as second-price auctions in which bidders submit confidential bids and the bidder who first offers the highest bid wins the auction. However, the winning bidder is only obliged to pay the second-highest bid. Dutch auctions are used by eBay in the multiple item case, where the price starts out high, as set by the seller, and drops until someone wins the item. In this chapter, we consider only the single-item auction case.

The major sources of eBay's revenue are auction listings and selling fees. All sellers are charged a nonrefundable insertion fee to enter the auction listing, with the amount depending on the starting price or reserve price (fully refundable if an item is sold) of an item. If an item is sold, the seller is also charged a final value fee, the amount depending on a certain percentage of the closing price. Moreover, eBay offers a list of optional features, such as *Buy It Now*, *eBay Picture Services*

Fees, etc., for sellers to enhance their listings for a fee. A seller can choose the listing duration to be either one, three, five, seven, or ten days. Here, we consider seven-day auctions only, the most popular auction for single-item listings.

eBay uses an automatic bidding system for the single-item auctions that we consider here. Bidders enter the maximum amount they are willing to pay for the item, the so-called willing-to-pay (WTP) values or proxy bids. The current price, corresponding to the real-time price displayed in eBay's webpage, is referred to as the *live bid* by Jank and Shmueli (2005b). We adopt this convention, using *current price* and *live bid* synonymously. This price plays a key role in decision making for any bidder participating in an auction. The bidding system requires that new WTP values be at least equal to the current price plus the preset bid increment, which is determined by the current price. During an auction, eBay places new bids on behalf of a bidder according to the preset bid increment table, so that the bidder can bid against other bidders' maximum bid until his or her WTP value is reached.

The WTP values form the *bid history*, available on eBay's online auction listing during an ongoing auction. However, the real-time highest WTP values, which are viewable online, do not permit one to extract the listed bid history. The actual current highest WTP value can be higher than the current price, but its level is not disclosed until another higher WTP value from a different bidder is received. A consequence is that the winner's WTP value is never revealed. eBay stores the completed listings for all auctions that ended during any preceding 15 days period and makes them available to any registered eBay user. These listings provide a good data source for statistical analysis. The completed listing for each auction contains a detailed bid history, which includes a description of the item, item number, closing price, quantity of items auctioned, number of proxy bids received during the auction, and the WTP values (except for the winner's proxy bid), displayed in descending order, along with the corresponding bidder information, date, and time.

It is important to observe that because the actual highest WTP value at any given time is not available during an ongoing auction, it is not uncommon that some of the earlier WTP values are higher than some later WTP values. For instance, assume that the live bid for an auction is currently \$30. Suppose Bidder A is currently the highest WTP value holder, say at \$50. Because the amount of WTP \$50 is invisible to other bidders, Bidder B joins the auction and places \$35 as his or her WTP value, which causes the current price to jump to \$35 (the second highest WTP) + 1 (the bid increment corresponding to \$35) = \$36. The system will now show the current price as \$36, and Bidder B has been outbid. Therefore, although the live bids themselves are monotone increasing over the course of an auction, the corresponding WTP values often are not monotone.

From the data collection point of view, WTP values are much easier to obtain than real-time live bids, since the complete bid history of any eBay auction that ended in any prior 15-day period can be searched and retrieved from the publicly available eBay webpages. During an ongoing auction, the live bid (or current price) is displayed dynamically as a single number in eBays's auction webpage. This number increases whenever there is an update on the second highest WTP value for the auction. Information from eBay is needed to obtain the entire live bid history for any completed auction. However, even without access to this information, one can recover almost all live bids from the WTP information available in the bid history by making use of the bid increment table, which is defined at http://pages.ebay.com/help/buy/bid-increments.html.

The one auction characteristic about which information cannot be recovered, to the best of our knowledge, is the presence of a secret reserve price for a given auction; it appears that the completed listings from eBay do not contain information about the existence of this feature of an auction, and it is only viewable by the bidders in an on-going auction listing. If an auction features a secret reserve price, the converted live bids will differ from the actual live bids by at most one record, namely, a live bid where the secret reserve price has come into effect. As an example, assume that an auction contains a secret reserve price of \$100. Suppose the current live bid is \$80, with the highest WTP value being \$90 bid by Bidder A. Now Bidder B comes in with a WTP of \$120. eBay's automatic bidding system will institute a special price jump due to the secret reserve price, that is, the live bid will jump to \$100, while without the presence of a secret reserve price, it would have been \$90 (the second highest WTP) + \$1 (the bid increment corresponding to \$90) = \$91. However, the next converted bid will no longer be influenced by the secret reserve price, so only this one-time increment of the live bid will differ from the value it would have had in the absence of a secret reserve price. Continuing this example, an incoming Bidder C will be prompted by the eBay system to enter a WTP value that is \$102.5 or more, where \$2.5 is the preset bid increment based on \$100. Suppose Bidder C enters \$105; then the converted bid will be \$105 (the second highest WTP) + \$2.50 (the bid increment corresponding to \$105) = \$107.50, which is the correct converted live bid even if, during the conversion of the data from WTP values to live bids, it is not known that a secret reserve price was present.

Since the secret reserve price feature is a costly option for sellers, and since the unknown size of the special price jump when the reserve is met is potentially discouraging for bidders participating in such auctions, this feature is used more often in auctions of relatively expensive products. The auction data in this chapter were collected from WTP values, and in the absence of knowledge about the presence of a secret reserve, the converted live bids can be assumed to reflect the price process reasonably well. We will discuss further details of the conversion of auction bids from WTP to live bids in the case study section. From now on, *bids* is used synonymously with *live bids*.

12.3 FUNCTIONAL METHODS FOR SPARSE AUCTION DATA

12.3.1 Recovering Longitudinal Trajectories Through FPCA

The observed sequence of live bids is assumed to be generated by realizations of an underlying smooth random function, denoted as the price process $X(\cdot)$, where X is a square integrable function, defined on a domain $\mathcal{T} = [0,T]$ for a $T > 0$. The price process X is assumed to have an unknown smooth mean function $EX(t) = \mu(t) \in$

L^2 and an unknown smooth covariance function $\mathrm{Cov}(X(s), X(t)) = G(s, t) \in L^2$, $s, t \in \mathcal{T}$. The covariance function $G(s,t)$ is assumed to have an orthogonal expansion with nonincreasing eigenvalues λ_k and eigenfunctions ϕ_k of the autocovariance operator A_G, defined by

$$(A_G f)(t) = \int_{\mathcal{T}} G(s, t) f(s) ds \tag{12.1}$$

for any f in L^2. Since G is symmetric and nonnegative definite, this linear operator in the Hilbert space $L^2(\mathcal{T})$ is a Hilbert-Schmidt operator (see Courant and Hilbert 1953). The orthogonal expansion of G is

$$G(s, t) = \sum_{k=1}^{\infty} \lambda_k \phi_k(s) \phi_k(t), \tag{12.2}$$

and the Karhunen-Loève representation (Ash and Gardner 1975; Rice and Silverman 1991) of a random trajectory X_i is then given by

$$X_i(t) = \mu(t) + \sum_{k=1}^{\infty} \xi_{ik} \phi_k(t). \tag{12.3}$$

Here the $\xi_{ik} = \int_{\mathcal{T}} (X_i(t) - \mu(t)) \phi_k(t) dt$ are the functional principal component (FPC) scores of X_i, for $k = 1, 2, \ldots$. These are random variables with $\mathrm{E}(\xi_{ik}) = 0$, $\mathrm{E}(\xi_{ik}\xi_{ik'}) = 0$ for $k \neq k'$ and $\mathrm{E}\xi_{ik}^2 = \lambda_k$, where $\sum_k \lambda_k < \infty$ and $\lambda_1 \geq \lambda_2 \geq \cdots \geq 0$. In other words, the FPCs form a sequence of uncorrelated random variables with decreasing variances which are given by the eigenvalues.

In practice, the sequence of observed bids differs from the values of the smooth underlying trajectory, and the differences correspond to a *measurement error*. In the case of auction data, these differences are best interpreted as random aberrations of prices around the smooth underlying price trajectory rather than as physical measurement errors. Let T_{ij} be time points at which bids are placed for price trajectory X_i, where $i = 1, \ldots, n$, $j = 1, \ldots, n_i$, and $T_{ij} \in \mathcal{T}$. The bid counts $n_1, n_2, \ldots, n_n$ for the n trajectories are assumed to form an i.i.d. sequence of random variables, independent of all other random variables. With an additive measurement error assumption (see also Rice and Wu 2001), equation (12.3) can be connected with actual bids Y_{ij} observed at times T_{ij} through

$$\begin{aligned} Y_{ij} = Y_i(T_{ij}) &= X_i(T_{ij}) + \varepsilon_{ij} \\ &= \mu(T_{ij}) + \sum_{k=1}^{\infty} \xi_{ik} \phi_k(T_{ij}) + \varepsilon_{ij}, \end{aligned} \tag{12.4}$$

where the ε_{ij} denote the aberrations of the bids from the underlying smooth trajectories, assumed to be i.i.d. with mean 0 and constant variance σ^2, and such that ε_{ij} is independent of ξ_{ik} for all i,k.

Identifying the components of (12.4) starts with the nonparametric estimation of the mean and covariance functions, following the proposal in Yao et al. (2005). We use local linear scatterplot smoothers for the mean function and local linear surface smoothers for the covariance function (see Fan and Gijbels 1996). By pooling all observed bids together across all auctions, we obtain the estimated mean function $\hat{\mu}(t)$ by minimizing

$$\sum_{i=1}^{n}\sum_{j=1}^{n_i}\kappa_1\left(\frac{T_{ij}-t}{h_\mu}\right)\{Y_{ij}-\beta_0-\beta_1(t-T_{ij})\}^2 \tag{12.5}$$

with respect to β_0 and β_1, where $\hat{\mu}(t)=\hat{\beta}_0(t)$, $t\in\mathcal{T}$. Here $\kappa_1(\cdot)$ is a univariate kernel function, usually chosen as a symmetric (around 0) density function. The bandwidth h_μ can be selected through leave-one-curve-out cross-validation (CV) (Rice and Silverman 1991) or variants, such as generalized cross-validation (GCV). Often, a subjective choice through visualization is more adequate.

The assumptions about the bid aberrations ε_{ij} present in model (12.4) imply that only the diagonal elements of the covariance are affected, as

$$\mathrm{Cov}(Y_{ij},\,Y_{il}|T_{ij},\,T_{il})=\mathrm{Cov}(X(T_{ij}),\,X(T_{il}))+\sigma^2\delta_{jl},$$

where δ_{jl} is the Kronecker delta. For "raw covariances" $G_i(T_{ij},T_{il})=(Y_{ij}-\hat{\mu}(T_{ij}))(Y_{ij}-\hat{\mu}(T_{il}))$, $\hat{\mu}(\cdot)$ being the estimated mean function, one finds $\mathrm{E}[G_i(T_{ij},T_{il})\mid T_{ij},T_{il}]\approx\mathrm{Cov}(Y_{ij},Y_{il}\mid T_{ij},T_{il})$ and this motivates smoothing the raw covariances while omitting the diagonal terms. Let $\hat{G}(s,t)$ be a smooth estimate of the covariance surface, for example obtained by minimizing

$$\sum_{i=1}^{n}\sum_{1\le j\ne l\le n_i}\kappa_2\left(\frac{T_{ij}-s}{h_{G_1}},\frac{T_{il}-t}{h_{G_2}}\right)\{G_i(T_{ij},T_{il})-(\beta_0+\beta_{11}(s-T_{ij})+\beta_{12}(t-T_{il}))\}^2 \tag{12.6}$$

with respect to β_0, β_{11}, and β_{12}, where the surface estimate is $\hat{G}(s,t)=\hat{\beta}_0(s,t)$, $s,t\in\mathcal{T}$, according to the local least squares method. Here $\kappa_2(\cdot,\cdot)$ is a bivariate density, and usually one chooses the smoothing parameters such that $h_{G_1}=h_{G_2}=h_G$. Similarly to the situation for the mean, bandwidths h_G can be selected through CV, GCV, or visually.

Estimating eigenfunctions and eigenvalues in model (12.4) corresponds to finding the solutions $\hat{\phi}_k$ and $\hat{\lambda}_k$ of the eigen-equations,

$$\int_{\mathcal{T}}\hat{G}(s,t)\hat{\phi}_k(s)ds=\hat{\lambda}_k\hat{\phi}_k(t), \tag{12.7}$$

where the $\hat{\phi}_k$ are subject to $\int_{\mathcal{T}}\hat{\phi}_k(t)^2dt=1$ and $\int_{\mathcal{T}}\hat{\phi}_k(t)\hat{\phi}_m(t)dt=0$ for $m\ne k$. The estimated eigenvalues and eigenfunctions are then obtained by spectral decomposition of the discretized smoothed covariance (Rice and Silverman 1991; Capra and Müller 1997).

Traditional FPCA uses numerical integration to estimate the FPC scores $\xi_{ik}=\int_{\mathcal{T}}(X_i(t)-\mu(t))\phi_k(t)dt$. For price trajectories X_i, the estimated FPC scores via the integration method would be $\hat{\xi}^I_{ik}=\sum_{j=1}^{n_i}(Y_{ij}-\hat{\mu}(T_{ij}))\,\hat{\phi}_k(T_{ij})(T_{ij}-T_{i,j-1})$, with $T_{i,j-1}=0$, assuming the T_{ij} are ordered by size. This method requires the observed bids to be dense and regularly spaced, and recorded without aberrations from the price trajectories. In this situation, $Y_{ij} = X_{ij}$. However, this estimator suffers seriously if the observed bids are noisy, sparse, or irregularly spaced.

As an alternative to numerical integration, Yao et al. (2003) proposed a shrinkage estimator to estimate the ξ_{ik} for dense and noisy data with missing values. The PACE method developed by Yao et al. (2005) aims at the estimation of ξ_{ik} for sparse, irregularly observed, and noisy data and thus provides a natural approach for the auction data. Here we give a brief overview of some pertinent details. Let $\tilde{\mathbf{X}}_i = (X_i(T_{i1}), \ldots, X_i(T_{in_i}))^T$, $\tilde{\mathbf{Y}}_i = (Y_{i1}, \ldots, Y_{in_i})^T$, $\mu_i = (\mu(T_{i1}), \ldots, \mu(T_{in_i}))^T$, $\phi_i = (\phi(T_{i1}), \ldots, \phi(T_{in_i}))^T$, and assume that ξ_{ik} and ε_{ij} in (12.4) are jointly Gaussian. The best prediction of the kth FPC score ξ_{ik} of X_i, given the observed bids for this price trajectory, is the conditional expectation, which is (Yao et al. 2005)

$$\xi_{ik}^* = \mathrm{E}[\xi_{ik} | \tilde{\mathbf{Y}}_i] = \lambda_k \phi_{ik}^T \Sigma_{\mathbf{Y}_i}^{-1} (\tilde{\mathbf{Y}}_i - \mu_i), \tag{12.8}$$

where $\Sigma_{\mathbf{Y}_i} = \mathrm{Cov}(\tilde{\mathbf{Y}}_i, \tilde{\mathbf{Y}}_i) = \mathrm{Cov}(\tilde{\mathbf{X}}_i, \tilde{\mathbf{X}}_i) + \sigma^2 \mathbf{I}_{n_i}$. Substituting estimates for $\lambda_k, \phi_{ik}, \Sigma_{\mathbf{Y}_i}$, and μ_i, the estimates for the predicted ξ_{ik} then become

$$\hat{\xi}_{ik} = \hat{\mathrm{E}}[\xi_{ik} | \tilde{\mathbf{Y}}_i] = \hat{\lambda}_k \hat{\phi}_{ik}^T \Sigma_{\mathbf{Y}_i}^{-1} (\tilde{\mathbf{Y}}_i - \hat{\mu}_i), \tag{12.9}$$

where the (j,l)th element of $\hat{\Sigma}_{\mathbf{Y}_i}$ is $(\hat{\Sigma}_{\mathbf{Y}_i})_{j,l} = \hat{G}(T_{ij}, T_{il}) + \hat{\sigma}^2 \delta_{jl}$. Here $\hat{G}(T_{ij}, T_{il})$ can be estimated from (12.6) and $\hat{\sigma}^2$ is estimated as described in equation (2) of Yao et al. (2005).

The idea for the estimate of $\hat{\sigma}^2$ is simply to compare a smoothed version of just the diagonal elements of the covariance matrix with the diagonal of the resulting surface estimate along the direction perpendicular to the diagonal of the covariance matrix. Under the Gaussian assumption, the estimator (12.9) targets the best prediction, though not the actual values of the FPC scores. Assuming the major modes of variation of the infinite-dimensional price processes X_i correspond to the first K eigenfunctions, the estimate for the predicted price trajectory X_i is

$$\hat{X}_i^K(t) = \hat{\mu}(t) + \sum_{k=1}^{K} \hat{\xi}_{ik} \hat{\phi}_k(t). \tag{12.10}$$

The estimators appearing in (12.10) are shown to be asymptotically consistent under mild conditions in Yao et al. (2005), where asymptotic pointwise and simultaneous confidence bands for the predicted individual trajectories are also derived. To select a reasonable number of K eigenfunctions to approximate the infinite-dimensional process, one can compute the one-curve leave-out score (Rice and Silverman 1991), aiming at minimizing the leave-one-curve-out prediction error

$$CV(K) = \sum_{i=1}^{n} \sum_{j=1}^{n_i} \left\{ Y_{ij} - \hat{Y}_i^{(-i)}(T_{ij}) \right\}^2, \tag{12.11}$$

where $\hat{Y}_i^{(-i)}(T_{ij}) = \hat{\mu}^{(-i)}(T_{ij}) + \sum_{k=1}^{K} \hat{\xi}_{ik}^{(-i)} \hat{\phi}_k^{(-i)}(T_{ij}), j = 1, \ldots, n_i$ are the predicted bids for price trajectory X_i. These values are estimated after removing the data for the trajectory X_i itself, where the estimated mean function and eigenfunctions are evaluated at the observed bid times for this trajectory.

The construction of the CV score for a given number of K components involves fitting model (12.4) n times, which can be computationally expensive when n is large. Moreover, in practice, the final choice of K by the CV method tends to be large and often unstable, as CV does not sufficiently restrict the degrees of freedom. As an alternative to the CV method, Yao et al. (2005) proposed Akaike Information Criterion (AIC) (Shibata 1981) and Bayesian Information Criterion (BIC) (Schwarz 1978) types of criteria based on a pseudo-Gaussian log-likelihood, computed conditionally on the observed $\hat{\xi}_{ik}$,

$$\hat{L} = \sum_{i=1}^{n} \left\{ -\frac{n_i}{2}\log(2\pi) - \frac{n_i}{2}\log\hat{\sigma}^2 - \frac{1}{2\hat{\sigma}^2}\left(\tilde{\mathbf{Y}}_i - \widehat{\mu}_i - \sum_{k=1}^{K} \hat{\xi}_{ik}\widehat{\phi}_{ik}\right)^T \times \left(\tilde{\mathbf{Y}}_i - \widehat{\mu}_i - \sum_{k=1}^{K} \hat{\xi}_{ik}\widehat{\phi}_{ik}\right)\right\}.$$

Criteria for the selection of K are minimizing either $AIC(K) = -2\hat{L} + 2K$ or $BIC(K) = -2\hat{L} + K\log(N)$ where $N = \sum_{i=1}^{n} n_i$. Note that for each given K one may compute $AIC(K)$ or $BIC(K)$ by fitting the model (12.4) once only, which is computationally more efficient than the CV method.

In practice, we found the choices obtained from these criteria to be reasonable and usually better than those achieved with CV. In addition to the above criteria, the choice of K by means of a scree plot is relatively simple and often quite adequate. To construct a scree plot, one needs to fit a model with a relatively large number of FPCs, say M components, so that λ_M is close to zero. Next, one plots λ_k against k for $k = 1, \ldots, M$. The idea is to choose the number K such that eigenvalues beyond λ_K can be ignored in terms of the fraction of additional variance they would explain.

The proportion of variation left unexplained by the truncated expansion in (12.10) can be defined as

$$V(K) = 1 - \frac{\sum_{k=1}^{K} \lambda_k}{\sum_{k=1}^{M} \lambda_k} \tag{12.12}$$

and is estimated by plugging in eigenvalue estimates. The selected K will be a point where, ideally, the function $V(K)$ starts to flatten after an initial decline.

12.3.2 Analyzing Evolving Bid Trajectories

For nearly identical products that are auctioned repeatedly over a given period of time in similar auction settings (same duration of an auction, same currency in use, etc.), one can model the bid history for each auction as a price process using the FDA approach, as described in the previous section, under the assumption that these auctions are independent of each other.

Using the PACE method, one may recover the price trajectory for each online auction. Sometimes the trajectories underlying a bid history observed up to a current time t of an ongoing auction are of particular interest, especially for online applications where an auction has been incompletely observed but one aims at characterizing the ongoing auction at an intermediate time. The PACE method can be readily adapted to this setting.

Let t, where $0 < t < T$, be the current length of time an auction has been running. Then the price process for an auction observed up to time t is denoted by $X(s,t)$, $0 \leq s \leq t$ (see also Müller and Zhang 2005). The corresponding mean and covariance functions are (i) $\mathrm{E}X(s,t) = \mu(s,t)$, with $\mu(s,t) = \mu(s)$ for $s \leq t$, and (ii) $\mathrm{Cov}(X(s_1,t), X(s_2,t)) = G_t(s_1,s_2)$, with $G_t(s_1,s_2) = G(s_1,s_2)$ for $0 \leq s_1, s_2 \leq t$, respectively. The orthogonal expansion of the covariance function is $G_t(s_1, s_2) = \sum_{k=1}^{\infty} \lambda_{kt}\phi_{kt}(s_1)\phi_{kt}(s_2)$, where $\lambda_{1t} \geq \lambda_{2t} \geq \cdots \geq 0$ are eigenvalues and $\phi_{1t}(\cdot), \phi_{2t}(\cdot), \ldots,$ are orthonormal eigenfunctions of the autocovariance operator A_{G_t}. The eigenvalues and eigenfunctions correspond to the solution of the eigen-equations $\int_0^t G_t(s,u)\phi_{kt}(u)du = \lambda_{kt}\phi_{kt}(s)$. Then the Karhunen-Loève representation of price trajectory of X_i up to time t corresponds to

$$X_i(s, t) = \mu(s, t) + \sum_{k=1}^{\infty} \xi_{ikt}\phi_{kt}(s), \quad 0 \leq s \leq t, t \geq 0, \tag{12.13}$$

where $\xi_{ikt} = \int_0^t \{X_i(s, t) - \mu(s, t)\}\phi_{kt}(s)ds$, the FPC for the trajectory X_i in $[0, t]$. As usual, we have $\mathrm{E}(\xi_{ikt}) = 0$ and $\mathrm{E}(\xi_{ikt}\xi_{ik't}) = 0$ for $k \neq k'$ and $\mathrm{E}(\xi_{ikt}^2) = \lambda_{kt}$ with $\sum_k \lambda_{kt} < \infty$.

The bids again may be viewed as contaminated values of the price trajectory as above, and estimation follows the same principles described in Section 12.3.1, with the modification that the input data (or observed bids) Y_{ijt} are restricted to $T_{ij} \in [0, \mathrm{t}]$. As t varies, one thus obtains a time-varying version of the PACE method, which provides continuous updates for the characteristics of the price process as it evolves over time.

12.3.3 Predicting the Closing Price Through Bid Histories

In addition to modeling the evolution of price trajectories for each auction, we are interested in predicting the final price using currently available bid information at each time t, $0 < t < T$. Specifically, instead of using only a current bid at or near time t as a predictor, we aim at relating the closing price of an auction, i.e., the value of the price process at time T, to the entire price process from 0 to current time t. The underlying rationale for this approach is that the price history contains much richer information than the current price alone.

If price processes are well approximated by projecting on the function space spanned by the first K eigenfunctions, it is sufficient to use the first K time-varying functional principal component scores ξ_{kt} to represent trajectories (price processes)

X_i up to current time t. We may assume various regression models with the final price Y^* as response and ξ_{kt} as predictors. For example, Hastie and Tibshirani (1993) proposed a series of varying coefficient generalized linear models for cross-sectional data, while Hoover et al. (1998) and Fan and Zhang (1999, 2000) considered modeling varying-coefficient functions for longitudinal data.

When $g_t(\cdot)$, the time-varying link function, is chosen as the identity function, the varying coefficient linear model becomes

$$E(Y^*|\xi_{1t},\ldots,\xi_{Kt}) = \beta_{0t} + \sum_{k=1}^{K} \xi_{kt}\beta_{kt}, \tag{12.14}$$

where Y^* is assumed to be an independent Gaussian random variable, the varying intercept function β_{0t} is the mean of the price function at time t , the ξ_{kt} are the uncorrelated individual random components with zero mean, and β_{kt} are the corresponding varying coefficients for ξ_{kt}. The least squares estimates of the varying coefficients of $\beta_t = (\beta_{0t}, \ldots, \beta_{Kt})^T$ will be usually obtained for each t, $\hat{\beta}_t = \text{argmin}_{\beta_t} \sum_{i=1}^{n}\{Y_i^* - (\beta_{0t} + \sum_{k=1}^{K} \hat{\xi}_{ikt}\beta_{kt})\}^2$. The predicted final price for an auction with price process $X_i(s), s \leq t$ observed up to current time t is then

$$\hat{Y}_{it}^* = \hat{\beta}_{0t} + \sum_{k=1}^{K} \hat{\xi}_{ikt}\hat{\beta}_{kt}. \tag{12.15}$$

An alternative that we considered but that did not lead to good results in predicting the final price was to use an additive model instead of the linear model (12.15).

To assess the change in prediction error as time progresses, we use the leave-one-out mean square prediction error (MSPE) function,

$$\text{MSPE}(t) = \frac{1}{n}\sum_{i=1}^{n}\left(Y_i^* - \hat{Y}_{it}^{*(-i)}\right)^2, \tag{12.16}$$

where $\hat{Y}_{it}^{*(-i)} = \hat{\beta}_{0t}^{(-i)} + \sum_{k=1}^{K} \hat{\xi}_{ikt}\hat{\beta}_{kt}^{(-i)}$ for model (12.14). As t moves closer to the auction end time T, the $\hat{\xi}_{ikt}$ contain more and more information about the auction; thus, MSPE is expected to be monotone falling as t increases, apart from random fluctuations. We will illustrate this feature for the Palm M515 personal digital assistant auction example in the following case study section.

12.4 CASE STUDY

12.4.1 Preprocessing of Data on Personal Digital Assistant Auctions

The data used in this chapter were collected at http://www.smith.umd.edu/ceme/statistics/data.html. The site contains 158 auctions of Palm M515 personal digital assistants (PDAs) that took place between March and May 2003. The bid values correspond to WTP values. The WTP values were converted into live bids according to the following conversion rules: (i) the first bid (opening bid) in each auction record, as set by the seller at the start of each auction, is considered the first live bid; (ii) the WTP value placed by the first bidder is considered the opening bid; (iii) any other current live bid is equal to the current second highest WTP value plus the bid increment corresponding to this price, as discussed above, with the constraint that the sum does not exceed the current highest WTP value; (iv) any bids that do not lead to an increase in the live bid are ignored; (v) the closing price is the same as the winner's bid (the converted live bid corresponding to the second to last WTP value). This conversion rule relies on the assumption that no secret reserve price is set in any of the auctions. We discuss this assumption below. One auction that did not contain any information about the bidder's identify was excluded, since rule (iv) cannot be correctly applied in this case. We further removed another auction that contained identical bids of the same value ($199) placed by the same and only bidder for the auction. Thus, the analysis reported here is based on the 156 remaining auctions.

The start of all auctions was recalibrated to time 0, so that the first day of an auction ends after 24 hours, and all auctions end at the end of day 7, corresponding to 168 hours. Although the recorded bid times are accurate up to seconds, we converted the time unit into hours, that is, the time domain in our analysis is between 0 and 168 hours for all auctions. Since the difference between the minimum bid ($0.01) and the maximum bid ($283.50) is large, we decided to transform the bids to the log-scale, with live bid values ranging from −4.61 and 5.65 and making the price data more normal. The predictions of individual trajectories in the PACE method rely on the Gaussian assumption. The analysis reported here was therefore implemented for the log price process.

12.4.2 Analysis of Log-Bids

We applied the PACE method to model the log price process for each auction on the time domain [0,168) (time units in hours). Auctions are indexed by $i = 1, \ldots, 156$ and bids are indexed by $j = 1, \ldots, n_i$, with the jth bid time in the ith auction denoted by t_{ij}. Though the closing prices for each auction were recorded, we used only the live bids that were registered before the end of the 168th hour (the end of day 7) as input data.

The number of live bids per auction ranges from 9 to 52. Figure 12.1 shows the box plot of the daily number of aggregated bids from all 156 auctions, based on the seven-day scale. The median number of bids placed on the first day of an auction is three, which drops to one for the second and third days. On day 4, the median number of bids per auction falls to essentially zero, followed by a small increase on day 5, when the median number of bids per auction rises to one. On day 6, the median climbs to two and it shoots up rapidly to nine on day 7, the last day of an auction. The highly irregular spacing of the times when bids were recorded reflects the high time variability of bidding behavior and the sniping phenomenon described earlier.

When modeling individual price trajectories, the classical FPCA approach does not work well for the estimation of the principal component scores due to the difficulties caused by the irregular bid times for the numerical integration step. The PACE method, however, allows recovery of individual trajectories as long as

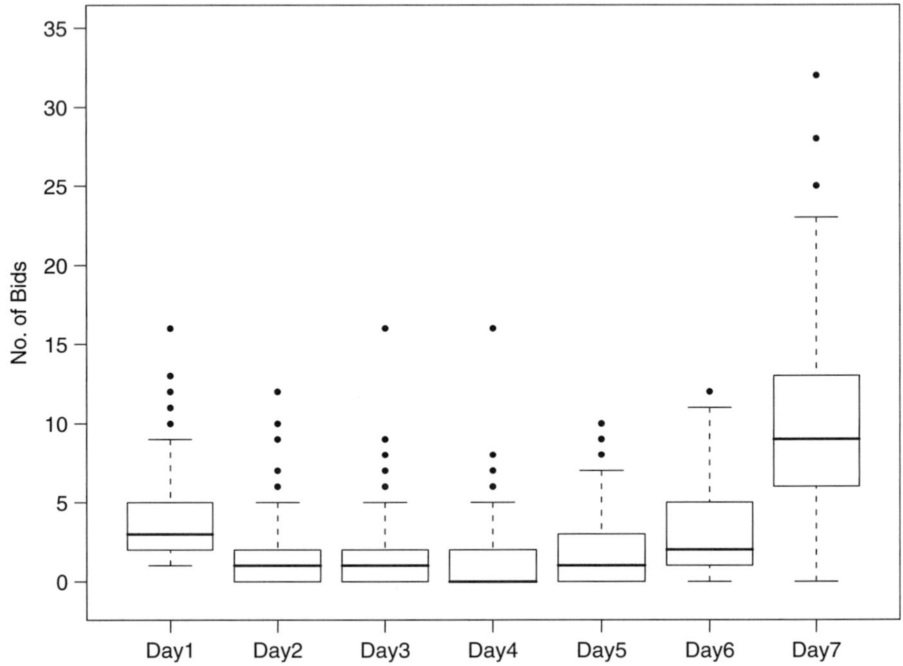

Figure 12.1 Box plots for numbers of daily bids for 156 seven-day auctions. The solid black dots are potential outliers, and the bold horizontal lines represent the median daily numbers of bids.

the pooled bid times over all auctions are dense in the domain, meaning that the maximal gap between adjacent bid times (formally, $\sup_{i,j}\inf_{i',j',T_{ij}\neq T_{i'j'}}|T_{ij}-T_{i'j'}| = O_p(\frac{1}{n})$) is relatively small and all pairs of bid times are dense on the domain squared, meaning that the distance between nearest neighbors of all pairs is relatively small, analogous to the one-dimensional case. We note that designs for which bid times are dense in the domain do not necessarily have the property that pairs of design points for the same subject are dense in the domain squared: Simply consider the case with two observations per auction which are randomly spaced but at most one day apart from each other.

Both design assumptions are empirically satisfied for the auction data. To model log price processes, we assume the observed bids are contaminated by aberrations added to the underlying smooth log price trajectories at each design point. As mentioned before, no secret reserve price is included in the modeling for any of the auctions. Should this assumption be violated for some of the auctions, the resulting bid conversion error will simply correspond to one of the bid aberrations that our model allows for.

As a first step in the analysis of the auction data, we pooled all bids together and used (12.5) to estimate the mean function for the log price trajectories (using the Epanechnikov kernel as weight function κ_1). The bandwidth choices resulting from both one-curve-leave-out CV or GCV appeared to be undersmoothing, and we therefore augmented them through visual assessment. We found $h_\mu = 12$ h to provide a reasonable fit for the mean function, which can be seen in the top right corner of Figure 12.2. The bids aggregated from all auctions are displayed as gray dots, and the estimated mean function is shown overlaid as a solid black curve. On average, log prices are seen to increase rapidly around the first day of an auction, and then the increases taper off until the final phase of an auction.

Next, we use (12.6) to obtain the estimated smooth covariance function for the log price trajectories. Here the input data are the raw covariances obtained by pooling all auction data together and removing the diagonal elements contaminated by bid aberrations. The bivariate kernel in (12.6) is chosen as the product of two one-dimensional Epanechnikov kernels. Bandwidth choices obtained from CV appeared to be undersmoothing, while the GCV bandwidths $(h_G, h_G) = (42\text{ h}, 42\text{ h})$ were found to be adequate for use in (12.6). We then apply spectral decomposition to the smooth covariance function at a pre-selected time grid to obtain estimated eigenvalues and eigenfunctions, the latter defined on this time grid. The estimated predictions for the FPC scores were then computed via (12.9).

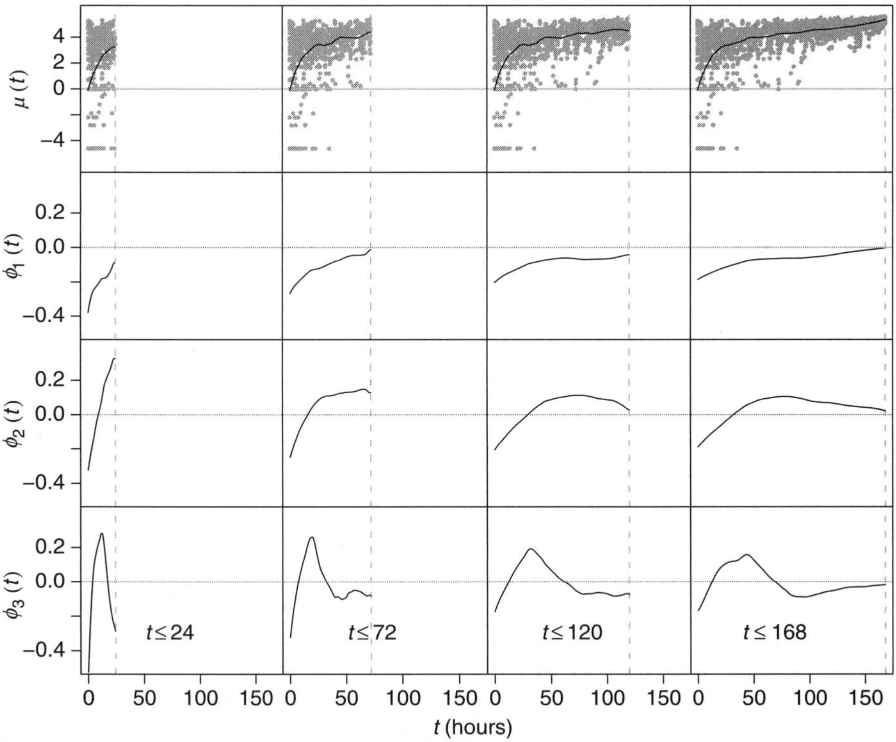

Figure 12.2 Top panels: Time-varying estimated mean function estimates $\hat{\mu}$ and the aggregated bids from all auctions for four different current times t, where $t = 24$, 72, 120, and 168 hours. Lower panels: Time evolution of the first three eigenfunctions, $\hat{\phi}_1$ (second row panels), $\hat{\phi}_2$ (third row panels), and $\hat{\phi}_3$ (bottom panels) for bid histories observed up to time t.

In the spectral decomposition step, it is crucial to determine a reasonable number K of included components, representing the relevant modes of variation of the log price trajectories. Choices from CV, AIC, or BIC were found to be too large. As an alternative selection tool, we applied the scree plot approach based on the fraction of variance explained (12.12) to determine K. It turned out that the choice $K = 3$ (out of $M = 20$) was satisfactory, as the first three components accounted for 97.65% of the total variation. The resulting estimated eigenfunctions can be seen in the last column of Figure 12.2. The first eigenfunction (72.69% of the total variation) represents an overall trend and has a shape similar to that of the mean function. The second eigenfunction (22.44% of the total variation) increases sharply in the first three days and then declines slowly afterward. This component may represent the dynamics of early price increments up to midauction. The third eigenfunction (2.52% of the total variation) increases in the first two days, decreases in the next two days, and then increases slowly until the end of the auction.

Figure 12.3 demonstrates the obtained fits for 16 estimated log price trajectories corresponding to 16 randomly selected auctions. The trajectories were fitted through (12.10) with $K = 3$. Overall, the price trajectories fit the data reasonably well. As the observed bids are monotone increasing within an auction, we might assume that the same holds for the log price process. The PACE method does not enforce such a constraint in the estimation procedure, and the fitted log price trajectories are not guaranteed to be monotone increasing. For example, the four graphs in the bottom panel display cases of nonmonotone fitted log price curves. The estimated trajectories from PACE are displayed as broken lines. A simple device to monotonize the log price trajectories is to apply the pooled-adjacent-violators algorithm (PAVA), introduced in Barlow et al. (1972), to the fitted curves obtained from the PACE method. Any nonincreasing estimated bids will be successively pooled together with their adjacent values and replaced by the average of the pooled values until no more nonmonotone estimated bids are encountered. The *R* function **isoreg()** is one of the implementations of the PAVA algorithm and is used here to obtain the monotonized curves for the four auctions shown in the bottom panel of Figure 12.3. The monotonized trajectories are displayed as solid black lines in the graphs. The flat line segments in each of these graphs reflect the PAVA averaged values of the estimated bids for nonmonotone segments. For further discussion of PAVA in the context of nonparametric smoothing methods, see Friedman and Tibshirani (1984).

12.4.3 Time-Varying Approach and Prediction of the Closing Price

To further study the auction dynamics, we also applied the time-varying PACE method (12.13), choosing various current times t. For demonstration purposes, we illustrate the evolution of the mean function and the first three eigenfunctions for the log price process based on bid histories observed up to current time t, where $t = 24$ h, 72 h, and 120 h, as illustrated in Figure 12.2. The estimated mean functions are shown in the top panels, obtained with bandwidths $h_\mu = 11$ h, 9 h, 9 h, respectively, selected via visual choice. From these graphs, we find that the estimated mean function for log price increases sharply within the first day ($t \leq 24$ h) of an auction, then flattens as time progresses, as seen in the graphs for $t \leq 72$ h (day 3)

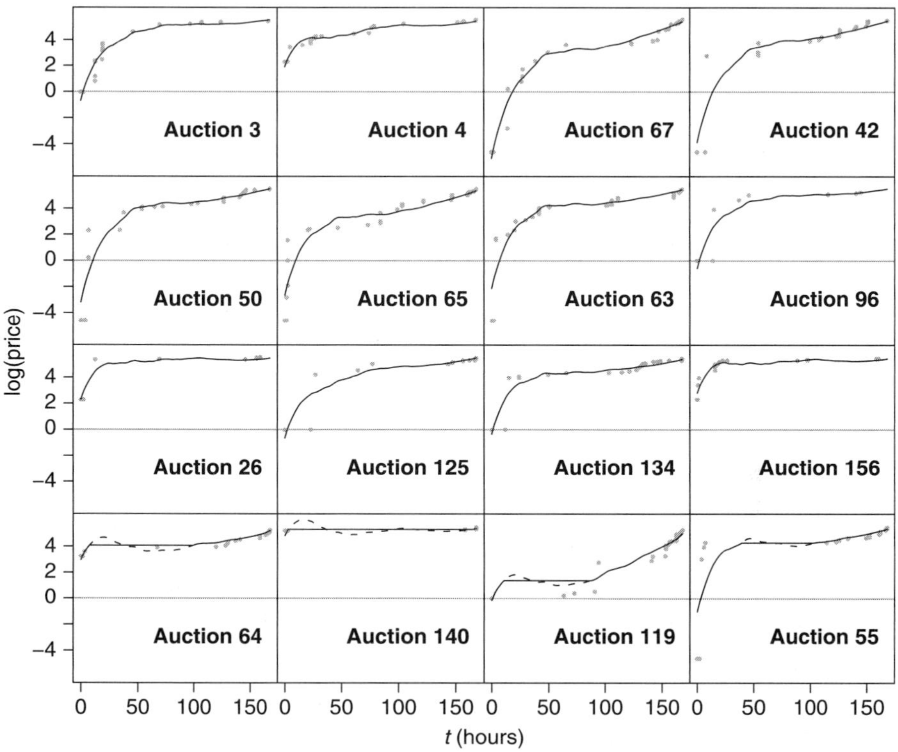

Figure 12.3 Estimated log(price) trajectories for 16 randomly selected Palm M515 PDA auctions. The observed log(bids) are displayed as gray dots, and the black solid lines represent the fitted trajectories obtained by the PACE method. The four plots in the bottom panel illustrate the results of monotonization. The broken lines represent the estimated log price functions obtained by the PACE method, and the black solid lines correspond to monotonized functions resulting from the PAVA algorithm.

and $t \leq 120$ h (day 5). The first three estimated eigenfunctions are displayed in the next three rows of Figure 12.2. They were obtained from the spectral decomposition of the three estimated covariance surfaces, choosing bandwidths $h_G = 11$ h (visually), 18 h (GCV), and 30 h (GCV), respectively. For each time t, the first three eigenfunctions explain about 96% to 97% of the total variation, with the first eigenfunction explaining about 70%, the second about 20%, and the third about 3% to 5% (except for $t \leq 24$ h, where it explains about 8%). The evolution of all eigenfunctions is seen to be quite smooth and gives rise to interpretations similar to those for the fits over the entire time domain of an auction.

One of the advantages of the time-varying PACE method is that one can summarize the bid history up to a current time t via the time-varying FPC scores and then use these scores as predictors of the closing price. We implemented such a scheme and obtained time-varying FPCs for the bid histories observed up to $24, \ldots, 168$ hours. All bandwidths were selected through GCV and the number of included components (FPC scores) through AIC, which led to an average of about 10 included components. To predict the closing price from bid histories to current time t, we compared two regression models with the closing price as the response and the time-varying FPC scores as predictors. One model was a linear regression using **lm()** in R for (12.15). As a nonlinear alternative, we also fitted a generalized additive model (GAM) (Hastie and Tibshirani 1990) (using the mgcv package in R). In this model, the unknown smooth additive functions f_j are estimated through cubic smoothing splines, with smoothing parameters chosen by GCV (Wood 2000). To check the goodness-of-fit of the time-varying predictions, we use (12.16) to calculate the mean squared prediction error (MSPE) for different current times t. The time-varying GAM modelling overfitted the data and resulted in unstable Mean squared prediction error (MSPE) values. Therefore, the linear model was found to be preferable. The result of the time-varying linear regression model is shown in Figure 12.4. As expected, prediction errors are seen to decline as time progresses. These declines seem to occur in jumps rather than continuously, with a steep drop near the end of the bidding process, which provides a quantitative explanation for observed bid sniping.

12.4.4 Analysis of Log Price Increments

As an alternative to modeling the log price process, we also considered modeling the log price increment process. Here the observed log bid ratios are used as input data

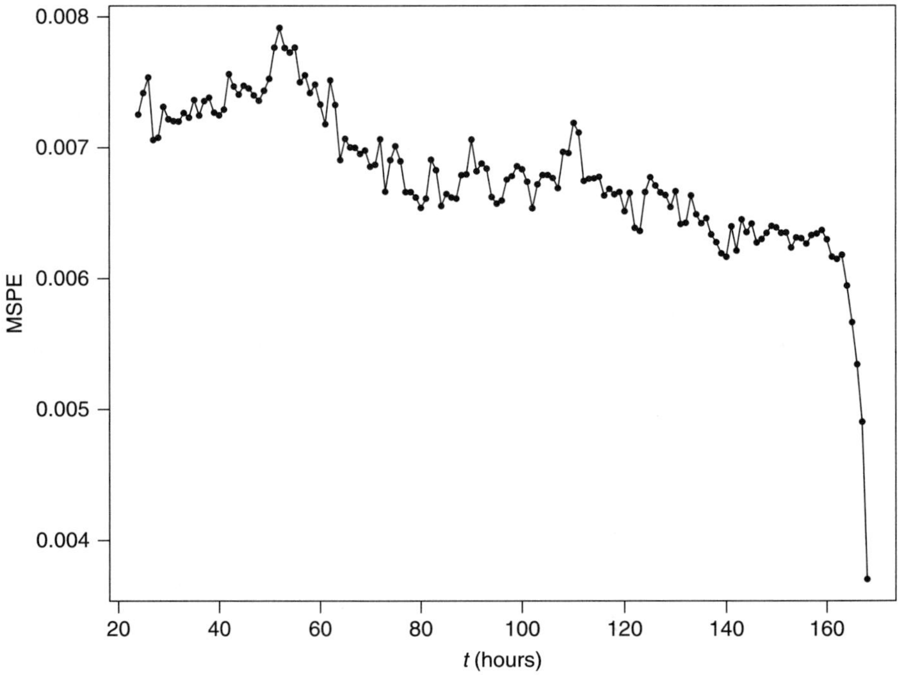

Figure 12.4 Mean squared prediction error (MSPE) versus current time t for time-varying predictions, using functional linear regression with time-varying FPC scores obtained from PACE for bid histories up to current time t as predictors and log(closing price) as response.

rather than log bid values. We removed any observed bids that were \$1 or smaller or were not strictly larger than previous bids. All second live bids fall into this category, since they are the same as the opening bids. If Y_{ij} is a bid observed at time $T_{ij} \in [0,168)$ (time in hours) and $Y_{ij} > 1$, we define the log bid ratios as

$$q_{ij} = \log \frac{Y_{ij}}{Y_{i,j-1}} \quad \text{with} \quad Y_{i0} \equiv 1, \quad i = 1, \ldots, 156, \; j = 1, \ldots, n_i.$$

Then $q_{i1} = \log\, Y_{i1}$ so that the first ratio is relative to \$1.

The analysis and estimation procedures follow exactly those described for the log price process in Section 12.4.2, substituting responses q_{ij} for Y_{ij}. The resulting estimated mean function for the log price ratio processes is shown in the top left corner of Figure 12.5, with $h_\mu = 12$ h (chosen visually). Not surprisingly, the estimated mean function for the log price increment is positive throughout, since the observed bids are monotone increasing. The log bid increments are largest in the beginning and drop off sharply after the first day, after which they continue to decrease mildly. Eigenfunctions are again obtained through spectral decomposition of the smooth covariance surface with $h_G = 12$ h (chosen visually). The scree plot approach indicates that choice of $K =$ two or three components is adequate.

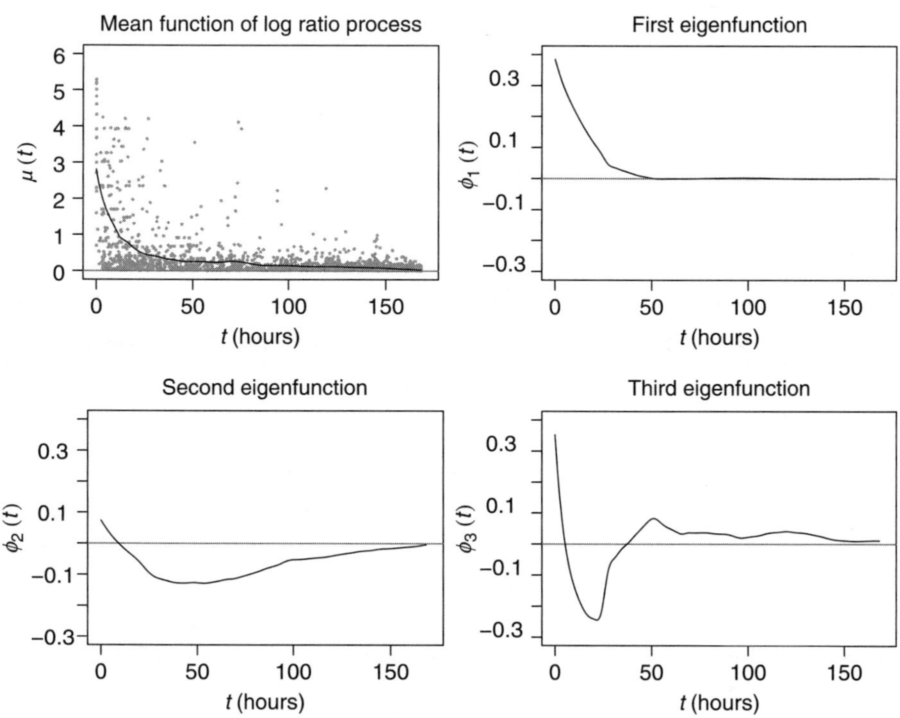

Figure 12.5 Estimated mean function $\hat{\mu}$ for the log price ratio process and observed log (bid ratios), aggregated over all auctions (left upper panel), and estimated first three eigenfunctions (remaining panels).

The resulting estimated eigenfunctions are displayed in Figure 12.5. The fractions of total variation that are explained by the first three eigenfunctions are 89.33%, 5.77%, and 2.89%, respectively. The shape of the first eigenfunction is very similar to that of the mean function. The second eigenfunction first decreases in the first two days and then slowly increases. The third eigenfunction declines rapidly in the first day, followed by a large increase during the second day, and then flattens. The estimated log price ratio trajectories for the selected auctions shown in Figure 12.3 are illustrated in Figure 12.6, based on three FPCs. Note that the estimated trajectories fit the data reasonably well, even with the relatively small number of repeated measurements (observed log bid ratios) per auction. The log bid increments appear to be largest up to the first or second day and then become almost zero. It appears that the effect of different opening prices is attenuated after the second day of the auction. We conclude from this analysis that the log price process itself contains more information about the overall online auction dynamics than does the increment process, which mainly reflects the dynamics at the beginning of an auction.

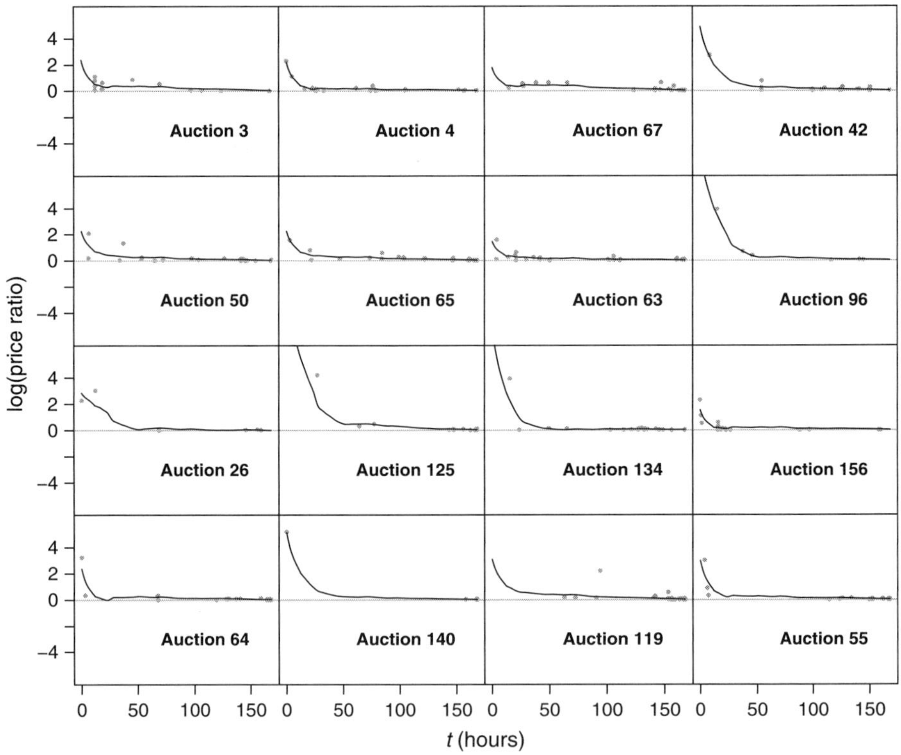

Figure 12.6 Estimated log(price ratio) trajectories for the selected Palm M515 PDA auctions displayed in Figure 12.3. The observed log(bid ratios) are defined as $\log \frac{Y_{ij}}{Y_{i,j-1}}$ with $Y_{i0} \equiv 1$, where Y_{ij} denotes the bid observed at time $T_{ij} \in [0,168)$ (time units in hours). The aggregated log(bid ratios) are displayed as gray dots, and the solid black lines represent fitted trajectories obtained by the PACE method.

12.5 CONCLUSIONS AND DISCUSSION

We show how a recent FPCA approach designed for the recovery of trajectories from sparse, irregular, and noisy longitudinal repeated measurements can be easily adapted to auction data. The implementation of this approach through the PACE method is found to be useful to fit log price processes and log bid ratio processes. In particular, the fits for the log price process and the associated eigenfunctions provide interesting insights into the time dynamics of online auctions. Adding a PAVA step easily leads to monotonized fitted trajectories. Alternative monotonized versions with more smoothness could be obtained by coupling PACE with other recent monotonization methods (e.g., Hall and Huang 2001), and connecting such methods with PACE provides a topic for future research.

In addition to studying online auction dynamics, prediction of the closing price is clearly an important aspect. In Section 12.3.3, we consider the concept of time-varying FPC scores, which summarize the bid history from the beginning to current time t of an auction. This is a particularly attractive feature for online auctions, because one usually has to make a decision at a current time t, based on available information about the auction history to time t. As we show, prediction based on these scores works well and can be computed for an arbitrary current time t. Our results from the case study indicate that the time-varying multiple linear regression provides good predictions and outperforms the time-varying GAM model.

There are many possible extensions of these approaches. For example, one can study the inclusion of other variables involved in an online auction. These can be time-dependent, such as the intensity of bids over time and feedback scores over time, or cross-sectional, such as opening bid, seller's rating, indicator for auctions that end during a weekend, etc. Future research may explore the use and extension of existing techniques for functional regression with functional response or generalized functional linear modeling with a scalar response when the model contains one or more predictor functions in addition to other scalar components.

ACKNOWLEDGMENTS

We are indebted to Wolfgang Jank for allowing us to use his auction data, for introducing the auction topic to us, and for many valuable discussions. This research was supported by NSF Grants DMS03-54448 and DMS05-05537.

REFERENCES

Ash, R.B. and Gardner, M.F. (1975). *Topics in Stochastic Processes*. Probability and Mathematical Statistics, Vol. 27. New York: Academic Press.

Barlow, R.E., Bartholomew, D.J., Bremner, J.M., and Brunk, H.D. (1972). *Statistical Inference Under Order Restrictions. The Theory and Application of Isotonic Regression*. Wiley Series in Probability and Mathematical Statistics. New York: Wiley.

Capra, W.B. and Müller, H.-G. (1997). An accelerated-time model for response curves. *Journal of the American Statistical Association*, 92: 72–83.

Courant, R. and Hilbert, D. (1953). *Methods of Mathematical Physics, Vol. I.* New York: Wiley.

Fan, J. and Gijbels, I. (1996). *Local Polynomial Modelling and Its Applications*. Monographs on Statistics and Applied Probability, Vol. 66. London: Chapman and Hall.

Fan, J. and Zhang, W. (1999). Statistical estimation in varying coefficient models. *Annals of Statistics*, 27: 1491–1518.

Fan, J. and Zhang, J.-T. (2000). Two-step estimation of functional linear models with applications to longitudinal data. *Journal of the Royal Statistical Society: Series B (Statistical Methodology)*, 62: 303–322.

Friedman, J. and Tibshirani, R. (1984). The monotone smoothing of scatterplots. *Technometrics*, 26: 243–250.

Hall, P. and Huang, L.-S. (2001). Nonparametric kernel regression subject to monotonicity constraints. *Annals of Statistics*, 29: 624–647.

Hastie, T. and Tibshirani, R. (1990). *Generalized Additive Models*. Monographs on Statistics and Applied Probability, Vol. 43. London: Chapman and Hall.

Hastie, T. and Tibshirani, R. (1993). Varying-coefficient models. *Journal of the Royal Statistical Society, Series B (Statistical Methodology)*, 55: 757–796. With discussion and a reply by the authors.

Hoover, D.R., Rice, J.A., Wu, C.O., and Yang, L.-P. (1998). Nonparametric smoothing estimates of time-varying coefficient models with longitudinal data. *Biometrika*, 85: 809–822.

James, G.M., Hastie, T.J., and Sugar, C.A. (2000). Principal component models for sparse functional data. *Biometrika*, 87: 587–602.

Jank, W. and Shmueli, G. (2005a). Profiling price dynamics in online auctions using curve clustering. *SSRN eLibrary*. Available at http://ssrn.com/paper=902893.

Jank, W. and Shmueli, G. (2005b). Visualizing online auctions. *Journal of Computational and Graphical Statistics*, 14: 299–319.

Jank, W. and Shmueli, G. (2006). Functional data analysis in electronic commerce research. *Statistical Science*, 21: 155–166.

Müller, H.-G. and Zhang, Y. (2005). Time-varying functional regression for predicting remaining lifetime distributions from longitudinal trajectories. *Biometrics*, 61: 1064–1075.

R Development Core Team (2006). *R: A Language and Environment for Statistical Computing*. Vienna: R Foundation for Statistical Computing.

Ramsay, J.O. and Silverman, B.W. (2002). *Applied Functional Data Analysis*. Springer-Verlag Series in Statistics. New York: Springer-Verlag.

Ramsay, J.O. and Silverman, B.W. (2005). *Functional Data Analysis* (2nd ed.). Springer-Verlag Series in Statistics. New York: Springer-Verlag.

Reddy, S.K. and Dass, M. (2006). Modeling on-line art auction dynamics using functional data analysis. *Statistical Science*, 21: 179–193.

Rice, J.A. and Silverman, B.W. (1991). Estimating the mean and covariance structure nonparametrically when the data are curves. *Journal of the Royal Statistical Society, Series B (Statistical Methodology)*, 53: 233–243.

Rice, J.A. and Wu, C.O. (2001). Nonparametric mixed effects models for unequally sampled noisy curves. *Biometrics*, 57: 253–259.

Schwarz, G. (1978). Estimating the dimension of a model. *Annals of Statistics*, 6: 461–464.

Shibata, R. (1981). An optimal selection of regression variables. *Biometrika*, 68: 45–54.

Wood, S.N. (2000). Modelling and smoothing parameter estimation with multiple quadratic penalties. *Journal of the Royal Statistical Society, Series B (Statistical Methodology)*, 62: 413–428.

Yao, F., Müller, H.-G., Clifford, A.J., Dueker, S.R., Follett, J., Lin, Y., Buchholz, B.A., and Vogel, J.S. (2003). Shrinkage estimation for functional principal component scores with application to the population kinetics of plasma folate. *Biometrics*, 59: 676–685.

Yao, F., Müller, H.-G., and Wang, J.-L. (2005). Functional data analysis for sparse longitudinal data. *Journal of the American Statistical Association*, 100: 577–590.

13

A FAMILY OF GROWTH MODELS FOR REPRESENTING THE PRICE PROCESS IN ONLINE AUCTIONS

VALERIE HYDE

Applied Mathematics and Scientific Computation Program, University of Maryland, College Park, Maryland

GALIT SHMUELI AND WOLFGANG JANK

Department of Decision and Information Technologies, R.H. Smith School of Business, University of Maryland, College Park, Maryland

13.1 INTRODUCTION AND MOTIVATION

With the advent of the Internet came online auction marketplaces, such as the popular eBay.com, which allow consumers and businesses to sell, buy, and bid on a variety of different goods. eBay is the largest consumer-to-consumer (C2C) marketplace and touts net revenues that topped \$1 billion for the first time for the first quarter of 2005 and were close to \$1.4 billion (36% higher) for the first quarter of 2006. On any given day, several million items, dispersed across thousands of categories, are available for sale on eBay. Indeed, eBay's slogan is "Whatever it is, you can find it on eBay." Buyers and sellers may be located on different continents and still conduct business since the online auction marketplace is always open and available. The wide and growing popularity of online auctions creates enormous amounts of publicly available auction bid data, providing an interesting and important topic for research. These data also pose special statistical challenges because of their special structure. In this chapter, we focus on the price process during an auction.

Statistical Methods in e-Commerce Research. Edited by W. Jank and G. Shmueli

Bids during an online auction arrive at unevenly spaced discrete time points chosen by bidders. We are interested in recovering the underlying continuous price evolution, i.e., the price at any time during the auction. From a visualization point of view, a series of discrete, unevenly spaced bids (e.g., presented as a scatterplot of bid vs. time) loses the conceptual as well as continuous nature of the price process. Furthermore, such plots do not scale to multiple auctions. A better representation is a continuous function with only a few parameters. Such a representation is conceptually more appealing, is more parsimonious, and can be further used for analyses such as clustering or regression models. Finally, the underlying continuous process may depend on derivatives of the function. A smooth price curve allows the calculation of derivatives (i.e., the first derivative is price velocity and the second derivative is price acceleration).

Functional data analysis (FDA) has become popular due to the seminal works of Ramsay and Silverman (2002, 2005). At its core, FDA deals with continuous objects such as curves or shapes as the observations of interest. For example, the temperature at a weather station over a period of one day may be considered a functional observation (curve) since it arises from a continuous process (the temperature at any moment in time over that day) even if it is measured at discrete time points (i.e., hourly). The curves representing daily temperature at several different weather stations can be considered a set of functional observations (Ramsay and Silverman 2002, 2005). Similarly, the price during an online auction is recorded at discrete time points but is inherently continuous. We therefore treat a set of auctions as a set of functional observations.

In FDA, nonparametric smoothing techniques are used to recover a functional object from discrete measurements. Examples are kernel smoothers, polynomial splines, and monotone splines (Ramsay 1998; Ramsay and Silverman 2005). Smoothing should be done in such a way that the resulting underlying object adequately represents the continuous process. In the auction setting, we know that the price is always positive and monotonically nondecreasing, so our smoothed price curve must reflect this.

To date, all of the smoothing methods applied to online auctions have proven to involve the specification of many parameters, such as the number and position of knots, a roughness penalty parameter, and the polynomial order. Resulting curves capture the price curves reasonably well; however, the fit often requires manual visual inspection (for parameter selection). Smoothing splines, which suffer from edge fitting, often result in a deteriorated fit at the start and end of the price curve and produce curves that are not necessarily monotone. Monotone smoothing splines produce monotone curves; however, they are computationally intensive and require lengthy run times for even a moderate number of auctions. Lastly, nonparametric smoothing does not provide an explanatory model for the price process. This motivated us to explore meaningful parametric representations of the price process. A parametric approach would be more elegant in the sense that it would provide a theoretical explanation of the process, it would potentially be computationally fast, and it would provide a more parsimonious representation.

In Hyde et al. (2006), for example, monotone smoothing is used for recovering the price curves in a set of Palm M515 auctions. They find three distinct shapes of price curves: j, stretched-out s, and straight line curves. The most popular shape in their dataset is a concave-up j-shaped price curve which represents auctions with gradual price increases until mid-to-late auction and then a price jump toward the end; the second most popular shape is the straight line, where the angle depends on the ratio of the opening and closing prices; and the third typical shape is a stretched-out s-shaped curve which reflects auctions where the price increases slowly, jumps up during midauction, and slowly increases to the close. These price curve shapes are the result of several bidding styles documented by Bapna et al. (2004b). Determining the number of each type of curve requires visual inspection of every single price curve. Clearly, a better method for grouping curves is needed beyond physical examination.

We can learn a lot about the auction price process by being able to group similar price curves as distinct parametric functions. For example, we can compare the price process distribution for distinct products or for market versus intrinsic values goods. We can also see if the distribution changes over time, which may suggest that bidders are evolving (since the price process is driven by bidders). We can also search for patterns in datasets using this additional information or use the particular growth function as a modeling variable.

The goal of this chapter is to introduce, within an FDA framework, a new *parametric* family of growth functions that describe the price processes in online auctions. The chapter is organized as follows: Section 13.2 discusses the characteristics of online auction data available on eBay.com, the set of data that we use throughout this work, and the representation of bid data as continuous curves. In Section 13.3, we discuss two nonparametric smoothing methods that have been used to describe auction price evolution and their limitations. In Section 13.4, we describe two popular growth models (exponential and logistic) and two additional useful growth modes (logarithmic and reflected logistic) and show how our parametric family of growth models is used for representing auction price growth. Section 13.5 introduces an automated selection procedure to choose the best growth model. Section 13.6 compares the quality of curves obtained through nonparametric methods with those from our parametric family of curves. In Section 13.7, we suggest several further uses of growth models. We conclude in Section 13.8 with final remarks and future research.

13.2 DATA FROM ONLINE AUCTIONS

The number of online auction sites is growing steadily. Despite different formats and rules, there is a common data structure that can be found across most sites. This structure comprises a time series that describes the bids placed over time (the bid history) and an associated set of features that describe the auction setting, such as the seller rating, the auction duration, and the item category. We refer to these features as the *auction attributes*. Figure 13.1 presents a snapshot of a closed auction from eBay.com showing the auction attributes (top) and the bid history (bottom). We see that this is a seven-day auction for a Vintage Rolex Submariner Black Dial Men's Wristwatch. The seller *tutleandwabbit* has a feedback rating of 100, with

eBay.com Bid History for
Vintage Rolex Submariner Black Dial Men's Wristwatch (Item # 320068139586)
Listed in category: Jewelry & Watches > Watches > Wristwatches

Winning bid: US $2,600.00
Ended: Jan-10-07 19:43:14 PST
Starting time: Jan-03-07 19:43:14 PST
History: 13 bids
Starting bid: US $0.99

Winning bidder: keefer10 (120)

Seller: tutleandwabbitt (100)
Feedback: 100% Positive
Member: since Jan-06-04 in United States
Item location: Bayside, New York, United States

Bidder	Bid Amount	Date of bid
Bidder 8	US $2,600.00	Jan-10-07 18:22:24 PST
Bidder 9	US $2,600.00	Jan-10-07 19:35:08 PST
Bidder 7	US $2,500.00	Jan-10-07 13:07:05 PST
Bidder 6	US $2,300.00	Jan-10-07 05:46:18 PST
Bidder 5	US $2,250.00	Jan-09-07 01:23:57 PST
Bidder 3	US $2,150.00	Jan-04-07 11:55:31 PST
Bidder 4	US $2,000.00	Jan-04-07 13:43:07 PST
Bidder 3	US $1,500.00	Jan-04-07 11:55:24 PST
Bidder 1	US $1,000.00	Jan-04-07 06:40:26 PST
Bidder 3	US $1,000.00	Jan-04-07 11:55:14 PST
Bidder 3	US $500.00	Jan-04-07 11:55:06 PST
Bidder 3	US $100.00	Jan-04-07 11:54:54 PST
Bidder 2	US $55.56	Jan-04-07 09:38:09 PST

Figure 13.1 Time series and attributes for a men's Rolex wristwatch auction. Note that bids are arranged in descending order by bid amount. This order, however, does not reflect the arrival of the bids. Rather, it reflects the current auction high bid.

100% positive feedback. The closing price is $2600, and there are a total of 13 bids from nine bidders. In this case, eBay has decided to alias the bidders' user names in order to protect them from fake offers. Aliasing is performed quite often for auctions of high-end merchandise.

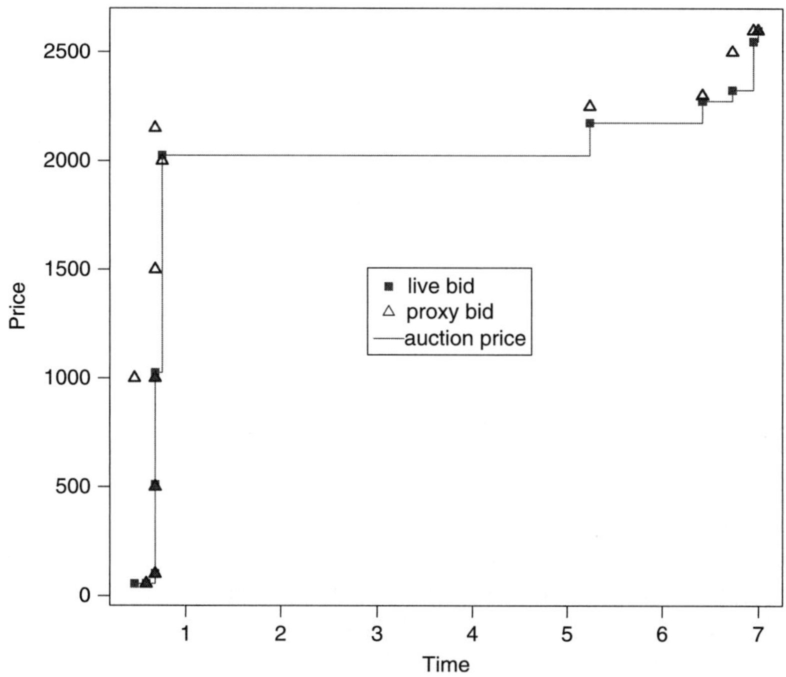

Figure 13.2 Proxy bids (triangles) and live bids (squares) for a men's Rolex wristwatch auction. The line connecting the live bids represents the current auction price.

To understand the structure of bid history data, it is necessary to understand the auction rules and bidding mechanism. On eBay, the majority of auctions are *second-price auctions*, which means that the winner is the bidder who placed the highest bid, but she or he pays the second highest price plus an increment. In our auction, Bidder 8 placed the highest bid but paid only Bidder 9's bid. (eBay does not disclose the highest bid (here, by Bidder 8)). Furthermore, eBay uses a *proxy bidding* system where bidders place the highest value that they are willing to pay, and then eBay bids on their behalf by increasing the current price by only an increment.[1] During the auction, the *current high bid* displayed is actually the second highest bid at the time plus an increment. Figure 13.2 shows this for the auction in Figure 13.1. We call the current price the *live bid*. We see that Bidder 9 placed a proxy bid of $2600 at 19:35:08; however, Bidder 8 placed an unknown higher bid at 18:22:24, and so is credited with the highest bid and wins the auction.

13.2.1 Luxury Wristwatch Data

Our data contain information on 472 completed seven-day luxury wristwatch (375 Rolex and 97 Cartier) auctions on eBay.com that transacted between September

[1]For further details see http://pages.ebay.com/help/buy/proxy-bidding.html.

TABLE 13.1 Descriptive Statistics for 472 Completed Seven-Day eBay Wristwatch Auctions

Variable	Mean (Std)	Median	Minimum	Maximum
Closing price	\$2019.00 (\$2561.89)	\$1300	\$70.00	\$24,000
Opening price	\$509.70 (\$896.96)	\$100	\$0.01	\$6500
Number of bids	14.65 (9.61)	13	2.00	57
Number of unique bidders	7.38 (4.05)	7	2.00	21
Unique bidders rating	64.62 (179.14)	11	−4.00	2648
Seller rating	571.99 (1,505.94)	107	−2.00	9055

15, 2001, and October 27, 2001. Our sample includes a variety of items in terms of make and model, new and used, and closing price. The average selling price for all the auctions is \$2019, with a median of \$1300 and a standard deviation of \$2561.89. We know from the literature (Bapna et al. 2004b; Wang et al. 2007) that auction attributes such as opening price, seller experience, number of bids, etc. affect not only the closing price of an auction but also the entire price process. Indeed, our sample is varied in all of those attributes. The range of opening prices is \$0.01 to \$6,500, the number of bids ranges from 2 to 57, and the average seller experience is 571.99, with a standard deviation of 1505.94. Descriptive statistics are provided in Table 13.1.

The left panel of Figure 13.3 shows the distribution of the number of bids for each day in the auction. Typically, there is some bidding activity at the auction start (first day), followed by a period of very little activity, cumulating in a surge of bidding at the very end of the auction (Shmueli et al. 2004). This last-moment bidding is often referred to as *sniping* (Bajari and Hortascu 2003; Roth and Ockenfels 2002; Bajari and Mortascu 2003). Figure 13.3 shows a similar bid distribution for two additional datasets: Palm Pilot (middle) and Xbox (right), which will be discussed later. For all three products, most of the bidding occurs on the final day of the auction, with the next most active bidding occurring on the first day. Clearly, when modeling the price process, the start and end of the auction are of particular importance.

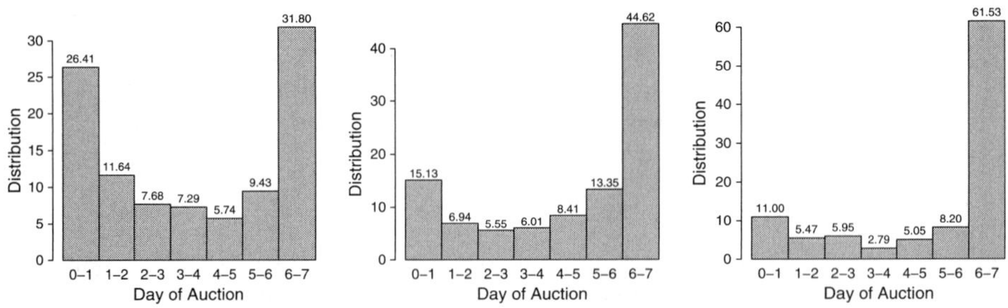

Figure 13.3 Distribution of daily bids over the course of seven-day auctions for 472 luxury wristwatch auctions (left), 134 Palm Pilot auctions (middle), and 85 Xbox auctions (right).

13.2.2 Representing Price Evolution as a Continuous Curve

Bids in online auctions are placed at varying time points. The resulting bid histories are therefore time series that are unevenly spaced, sometimes with very sparse and at other times very dense areas. Instead of considering the discrete set of bids in an auction as a vector, we use them to estimate the complete continuous price evolution that takes place during the auction. While we could simply "connect the dots" to obtain the price of the auction at any given time, this would overfit the data (i.e., model the noise), thereby providing a poor representation of the underlying continuous price process. An alternative is to represent the price as a continuous smooth curve. This type of curve representation is prevalent in FDA (Ramsay and Silverman 2005). The first step is therefore to represent/estimate the continuous price function from the discrete bid data.

Let y_j be the recording of an observation at time t_j, $j = 1, \ldots, n$ in an auction with n bids. We convert the raw data into a continuous function, $f(t)$, that allows for the evaluation of the price at any point t during the auction. As with any measurement, there is error, so we have

$$y_j = f(t_j) + e(t_j), \tag{13.1}$$

where $e(t_j)$ is considered white noise (i.e., $e(t_j) \sim (0, \sigma^2)$). Different smoothing methods exist to recover the price function $f(t)$ and will be discussed in Section 13.3. In Section 13.4, we propose a parametric alternative which also produces continuous price curves.

The advantage of the curve representation is that it treats price evolution as a single continuous entity. It captures the complete price evolution in a more compact and easier-to-visualize way than raw bid data. The price process can then be described by a few coefficients. Furthermore, an appealing feature of smooth curves is that we can gauge their derivatives (the first derivative is the price velocity and the second derivative is the price acceleration) in order to learn how price dynamics behave during the auction.

13.3 REPRESENTING PRICE EVOLUTION NONPARAMETRICALLY

There have been predominantly two approaches for representing the price process in online auctions. Jank et al. (2007) use penalized polynomial smoothing splines (p-splines), and Hyde et al. (2006) use penalized monotone splines. A comparison of the two for auction data is given in Alford and Urimi (2004). We now describe each of the two methods and their properties.

13.3.1 Smoothing Splines

Each auction has bids placed at different times. Rather than applying smoothers to the raw data directly, we apply them to a derived dataset that is sampled on the same set of time points for all auctions as follows. Consider the observed price during an auction, represented by the step function in Figure 13.2, where there is a new step every time a bid is placed. We use this step function to obtain our sampled data by selecting a set of knots at times $\tau_1, \tau_2, \ldots, \tau_L$ and determining the corresponding auction prices at those knots, $y_1^*, y_2^*, \ldots, y_L^*$. The noisy bid data are then discarded and replaced with the observations (τ_i, y_i^*). This transformation then allows us to use the same method to recover the price process in different auctions using the same smoother method.

The polynomial spline (Green and Silverman 1994) of order p is given by

$$f(t) = \beta_0 + \beta_1 t + \beta_2 t^2 + \cdots + \beta_p t^p + \sum_{l=1}^{L} \beta_{pl}(t - \tau_l)_+^p, \tag{13.2}$$

where $u+ = uI(u \geq 0)$ is the positive part of the function u. Many smoothers of this type tend to fit the data too closely (and thus model the noise); therefore, a *roughness penalty* approach is commonly employed. This method takes into account the trade-off between data fit (i.e., minimizing $f(t) = \sum_j (y_j^* - f(t_j))^2$) and function smoothness. A popular measure of roughness, which measures the degree of departure from a straight line, is of the form

$$PEN_m(t) = \int [D^m f(t)]^2 dt, \tag{13.3}$$

where $D^m f$, $m = 1, 2, 3, \ldots$ denotes the mth derivative of the function f. A highly variable function will yield a high value of $PEN_m(t)$. If the highest derivative of interest is m, then using $m + 2$ as the polynomial order will assure m continuous derivatives. For online auctions, where the first and second derivatives have been shown to be of interest, we use polynomials of order $m = 4$. The penalized smoothing spline f minimizes the penalized squared error

$$PENSSE_{\lambda,m} = \int (y(t) - f(t))^2 + \lambda PEN_m(t). \tag{13.4}$$

When the roughness parameter is set to $\lambda = 0$, the penalized squared error drops out, and the function fits the data. Larger values of λ penalize the function for being curvy, and as $\lambda \to \infty$, the fitted curves approach a linear regression. Ramsay and Dalzell (1991) and Ramsay (1998) suggest that the smoothing parameter λ can often be chosen by inspection of the smoothes or through optimizing metrics such as generalized cross-validation (GCV).

We have encountered a number of challenges using penalized smoothing splines in the online auction context. First, and most detrimental, is that the created functions are not always monotone nondecreasing, as auction price necessarily must be. Second, the functions are often very wiggly, especially at the ends. This is particularly egregious in the online auction context, where the start and end of the auction are of major importance. Third, there tends to be a large number of coefficients to estimate (due to adequate choices of knots and the polynomial order). Finally, there are multiple decisions about smoothing parameters that must be made in advance—the number and position of knots, the polynomial order, and the roughness penalty parameter λ—in order to provide a reasonable fit to the entire set of auctions. An alternative, where each auction is fit separately using a different set of parameters, will lead to confounding in the analysis results (see Jank et al. 2007).

13.3.2 Monotone Splines

Since the bidding process by nature is nondecreasing, Hyde et al. (2006) use monotone smoothing splines to represent the price process. The idea behind monotone smoothing (Ramsay 1998) is that monotone increasing functions have a positive first derivative. The exponential function has this property and can be described by the differential equation $f'(t) = w(t)f(t)$. This means that the rate of change of the function is proportional to its size. Consider the linear differential equation

$$D^2f(t) = w(t)Df(t). \tag{13.5}$$

Here $w(t) = \frac{D^2f(t)}{Df(t)}$, which is the ratio of acceleration and velocity. It is also the derivative of the logarithm of velocity which always exists (because we define velocity to be positive by the equation $Df(t) = e^{w(t)}$). The differential equation has the following solution:

$$f(t) = \beta_0 + \beta_1 \int_{t_0}^{t} \exp\left(\int_{t_0}^{v} w(v)dv\right)du, \tag{13.6}$$

where t_0 is the lower boundary over which we are smoothing. After some substitutions (see Ramsay and Silverman 2005), we can write

$$f(t) = \beta_0 + \beta_1 e^{w(t)} \tag{13.7}$$

and estimate β_0, β_1, and $w(t)$ from the data. Since $w(t)$ has no constraints, as $f(t)$ does in the form of the differential equation, it may be defined as a linear combination of K known basis functions (i.e., $w(t) = \sum_k c_k\phi_k(t)$). Examples of a basis functions are $\phi_k(t) = t$, which represents a linear model, or $\phi_k(t) = \log(t)$, which is a nonlinear transformation of the inputs. The penalized least squares criterion is thus

$$PENSSE_\lambda = \sum_i [y_i - f(t)]^2 + \lambda \int_0^T [w^2(t)]^2 dt. \tag{13.8}$$

While monotone smoothing solves the wiggly problem of the penalized smoothing splines, some of the same challenges remain and new ones arise. First, monotone smoothing is computationally intensive. The more bids there are, the longer the fitting process. Second, we still cannot fit the original raw data but rather the derived data (τ_i, y_i^*) (obtained from the step function). Finally, as with smoothing splines, the researcher must determine the number and location of knots and the roughness parameter λ that provide a reasonable fit to the entire set of auctions.

13.4 REPRESENTING PRICE EVOLUTION PARAMETRICALLY

We now introduce a parametric family of four growth models that are able to capture price evolution in many types of auctions. These are exponential growth, logarithmic growth, logistic growth, and reflected-logistic growth functions. Exponential and logistic growth functions have long been used to model population growth, dissemination of information, spread of disease, and more.

Our parametric approach is elegant, computationally fast, and parsimonious. It allows automated fitting, and there is no need to specify any parameters in advance. We choose models that are theoretically relevant in terms of monotonicity (to accurately reflect the price process) and that provide insight into the price process in online auctions.

Our approach, although motivated by online auctions, actually provides an alternative to the nonparametric smoothing methods that are customary in the field of FDA. This research opens the door for parametric curves to be the basis of FDA.

In the following, we describe each of the four models. Their functional form, derivative form, and parameters are summarized in Table 13.2.

13.4.1 Exponential Growth

13.4.1.1 Exponential Model. Exponential growth has been used to describe a variety of natural phenomena including the dissemination of information, the spread of disease, and the multiplication of cells in a petrie dish. In finance, the exponential equation is used to calculate the value of interest-bearing accounts compounded continuously. The idea behind exponential growth is that the rate of growth is proportional to the function's current size; that is, growth follows the

TABLE 13.2 Price Evolution, Velocity, and Acceleration Functions for Growth Models

Growth Model	Price Evolution	Price Velocity	Price Acceleration	Parameters
Exponential	$Y(t) = Ae^{rt}$	$Y'(t) = Are^{rt}$	$Y''(t) = Ar^2e^{rt}$	A, r
Logarithmic	$Y(t) = \frac{\ln\left(\frac{1}{A}\right)}{r}$	$Y'(t) = \frac{1}{rt}$	$Y''(t) = \frac{-1}{rt^2}$	A, r
Logistic	$Y(t) = \frac{L}{1 + Ce^{rt}}$	$Y'(t) = \frac{-LCre^{rt}}{(1 + Ce^{rt})^2}$	$Y''(t) = \frac{-LCr^2e^{rt}(1-Ce^{rt})}{(1 + Ce^{rt})^3}$	C, r
Reflected-logistic	$Y(t) = \frac{\ln(\frac{L}{t} - 1) - \ln(C)}{r}$	$Y'(t) = \frac{-L}{rt^2(\frac{L}{t} - 1)}$	$Y''(t) = \frac{L(L\text{-}2t)}{rt^4(\frac{L}{t} - 1)}$	C, r

differential equation

$$Y'(t) = rY(t) \tag{13.9}$$

or the equivalent equation

$$Y(t) = Ae^{rt}, \tag{13.10}$$

where t is time and $r > 0$ is the growth constant. Equivalently, exponential decay, when $r < 0$, can model phenomena such as the half-life of an organic event.

From a theoretical standpoint, exponential growth can describe a price process for auctions where there are gradual price increases until mid-to-late auction and a price jump toward the end. This is reminiscent of the j-shaped price curves that Hyde et al. 2006 find. An example of an auction that is captured well with exponential growth is shown in Figure 13.4 (top). The corresponding velocity curve (middle) and acceleration curve (bottom) are proportional to the price curve, as the differential equation implies. The price velocity and acceleration are zero or small during most of the auction before spiking at the end.

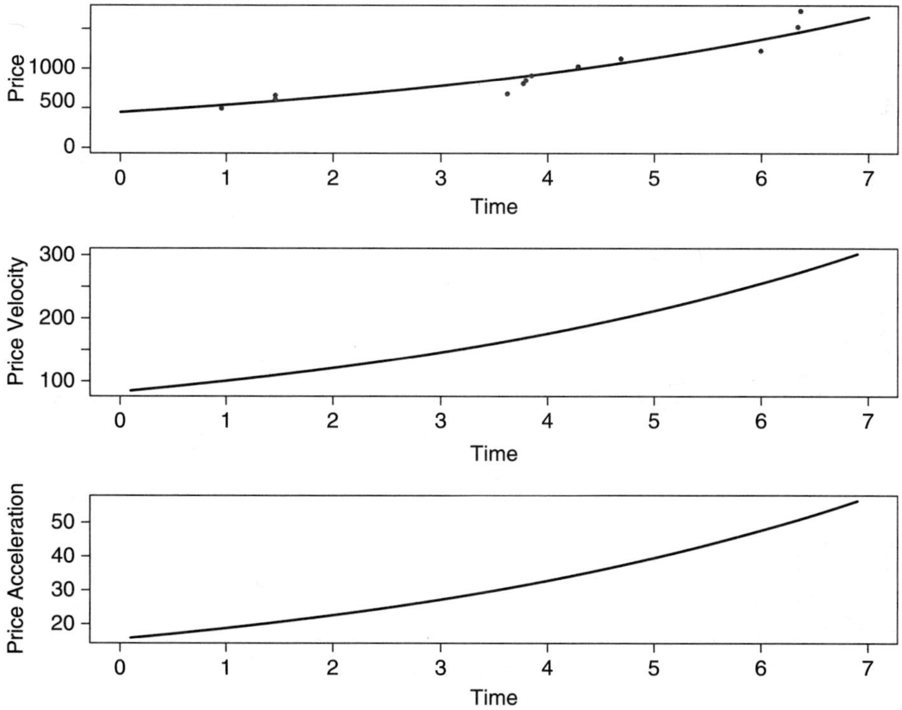

Figure 13.4 Price (top), velocity (middle), and acceleration (bottom) for an auction fit with exponential growth.

13.4.1.2 Logarithmic Model. We also find that the inverse of the exponential function,

$$Y(t) = \frac{\ln\left(\frac{t}{A}\right)}{r}, \tag{13.11}$$

which is called *logarithmic growth*, approximates price processes well. We choose a form of the logarithmic model that maps the original exponential model over the line $y = x$. This type of growth occurs when early bidding increases the price early in the auction, but because of the existence of a market value, the price flattens out for the remainder of the auction (as shown in Figure 13.5). This type of price behavior tends to be rare, as most bidders do not wish to reveal their bids early in the auction. However, inexperienced bidders who do not understand the proxy bidding mechanism on eBay have been shown to bid high early (Bapna et al. 2004a). In this model, the velocity starts at its maximum and then decays to little or zero as the auction progresses. The acceleration is always negative, since the price increases more slowly throughout the auction, and approaches zero at the end (where very little change in price is occurring).

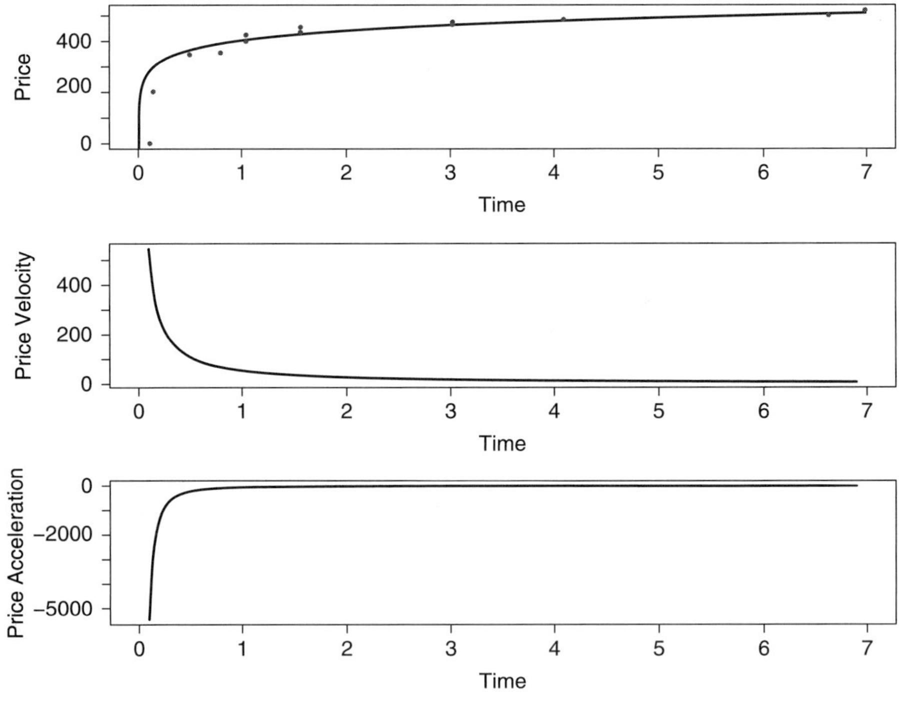

Figure 13.5 Price (top), velocity (middle), and acceleration (bottom) for an auction fit with logarithmic growth.

13.4.2 Logistic Growth

13.4.2.1 Logistic Model. While exponential growth often makes sense over a fixed period of time, in many cases growth cannot continue indefinitely. For example, there are only a finite number of people to spread information or disease; the petrie dish can only hold a certain number of cells. A typical application of the logistic equation is in population growth. In the beginning, there are seemingly unlimited resources and population grows increasingly fast. At some point, competition for food, water, land, and other resources causes growth to slow down; however, population is still growing. Finally, overcrowding, lack of food, and susceptibility to disease limit the population to some maximal carrying capacity.

In auctions, it is possible for growth to start exponentially with a big price increase in the middle of the auction, perhaps due to an inexperienced bidder. In the presence of a market value (or *carrying capacity*), the increased price slows competition, as smart bidders will make sure not to overpay for the item, and it is necessary for the price to flatten out near the end of the auction. The closing price that we witness is analogous to the carrying capacity L in the logistic growth function.

The logistic model is given by

$$Y(t) = \frac{L}{1 + Ce^{rt}}, \tag{13.12}$$

and the differential equation is

$$Y'(t) = rY(t)\left(\frac{Y(t)}{L} - 1\right), \tag{13.13}$$

where L is the carrying capacity, t is time, r is the growth rate, and C is a constant. Logistic growth forms a stretched-out s-shaped curve, discussed by Hyde et al. (2006), where the price increases slowly, then jumps up during midauction, and finally levels off toward the end of the auction. This price process and dynamics can been seen in Figure 13.6. The velocity is small or zero at the beginning and end of the auction, where there is little change in price, with a peak in midauction corresponding to the price increase. The acceleration is zero for most of the beginning and end of the auction. It peaks during the first part of the growth spike, where the price is increasing at an increasing rate, followed by a valley during the second part of the growth spike, where the price is increasing at a decreasing rate.

13.4.2.2 Reflected-Logistic Model. Another common price process in online auctions is the inverse of logistic growth, or reflected-logistic growth, given by the function

$$Y(t) = \frac{\ln\left(\frac{L}{t} - 1\right) - \ln(C)}{r}. \tag{13.14}$$

This type of growth occurs when there is some bidding early in the auction that results in a price increase, followed by little to no bidding in the middle of the auction and

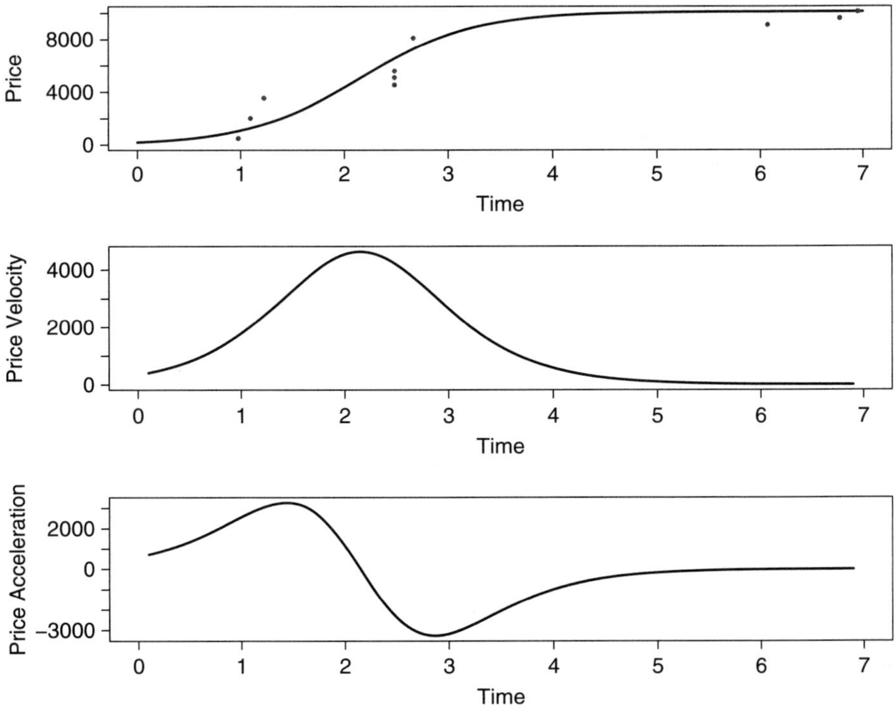

Figure 13.6 Price (top), velocity (middle), and acceleration (bottom) for an auction fit with logistic growth.

then another price increase as the auction progresses toward its close. The early price increases is indicative of early bidding by inexperienced bidders (Bapna et al. 2004a), and the price spike at the end may be caused by sniping (Roth and Ockenfels 2002; Bajari and Hortascu 2003). An example of reflected-logistic growth is shown in Figure 13.7. The velocity has peaks at the start and end of the auction, where there are jumps in price and little or no velocity during the middle of the auction, where the price does not change significantly. The acceleration curve is similar in shape to the price curve; however, it is negative in the beginning (where the price increases at a decreasing rate) and positive at the end (where the price increases at an increasing rate).

13.4.3 Fitting Growth Models

Unlike the nonparametric smoothers, we fit the growth models directly to the live bids (t_j, y_j). A simple and computationally efficient method for fitting each of the growth models is by linearizing the function and then performing least squares analysis. Since we are especially interested in obtaining an accurate fit at the beginning and end of the auction, it is usually necessary to add two additional points representing the price at the start and close of the auction: $(t = 0,\ y = \min(y_j))$ and $(t = l,$

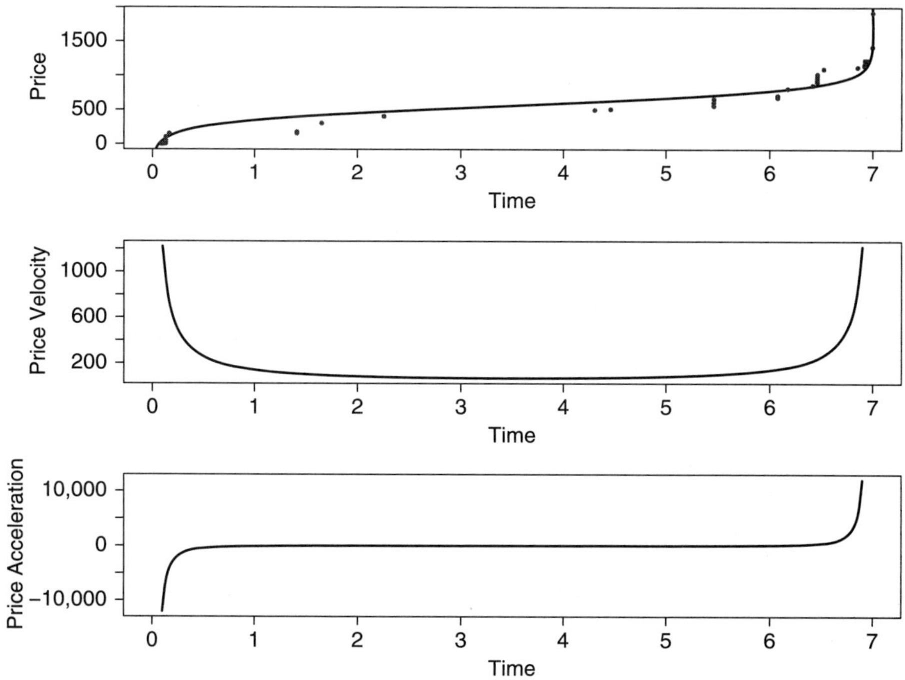

Figure 13.7 Price (top), velocity (middle), and acceleration (bottom) for an auction fit with reflected-logistic growth.

$y = \max(y_j)$), where l[2] is the duration of the auction and y_j is the value of bid j. Note that $\min(y_j)$ = opening price and $\max(y_j)$ = closing price. It is not necessary to add these extra points for logistic growth, where the maximum price is already incorporated into the function by defining $L = \max(y_j)$. Further, empirical evidence shows that auctions whose underlying price process is logistic tend to start close to zero, where logistic growth must start.

13.4.3.1 Fitting Exponential Growth. The exponential growth model from equation (13.10) can be linearized as

$$\ln Y = \ln A + rt. \tag{13.15}$$

We fit this model directly to the live bids with the two additional points in order to constrain the price at the start and end of the auction. In this case, two parameters,

[2]If all the auctions have the same duration, say seven days, then l would be replaced by 7. If the auctions have different durations, we can either vary l accordingly for each auction or standardize the time for all auctions to be between 0 and 1.

A and r, are estimated. Although we can fix A as the opening price, since $Y(t = 0) = Ae^{0r} = A$, we choose to estimate both parameters for two reasons: First, empirical evidence shows that the two-parameter estimation allows a better fit at the end of the auction; second, the other three growth models require estimation of two parameters. Thus, it is easier to compare model fit.

The straight-line curves discussed in Hyde et al. (2006) are a special case of exponential growth. When $r = 0$, we get a horizontal straight line, and when r is close to 0, we get a price curve that resembles a straight line with a positive slope.

13.4.3.2 Fitting Logarithmic Growth. The logarithmic growth model is given in equation (13.11). As with exponential growth, the two extra points are added to the live bids to ensure a good fit at the start and end of the auction. Notice that we cannot reduce this function to one parameter because when $t = 0$, ln(0) does not exist. Also, we cannot linearize this function. Therefore, rather than using optimization methods for parameter estimation where guesses of the initial value are necessary, we fit $T(y) = Ae^{ry}$ and linearize as in the exponential growth case. This time, least squares minimizes over time instead of price.

13.4.3.3 Fitting Logistic Growth. The logistic growth model from equation (13.12), where L is the distribution's asymptote, can be linearized as

$$\ln\left(\frac{L}{y} - 1\right) = \ln(C) + rt. \tag{13.16}$$

We know that $\lim_{t\to 1} = L$ (since for logistic growth $r < 0$ and $L > 0$). Define $L = \max(price) + \delta$, where $\delta = 0.01$ is needed so that the left-hand side (LHS) is defined over all bids y. In this case, there is no need to add the starting and closing points to the live bids because defining the asymptote takes care of the fit at the end. Auctions whose underlying price process can be described by logistic growth tend to start out low, so there is also no need to set the start value.

13.4.3.4 Fitting Reflected-Logistic Growth. The reflected-logistic function is given in equation (13.14). As with logarithmic growth, we cannot linearize this function. Instead, we fit $T(y) = \frac{L}{1+Ce^{ry}}$, where $L = 1 + \varepsilon$ ($\varepsilon = 0.00001$), to obtain the coefficients for C and r. We need $\varepsilon << \delta$ since the time range is much smaller than the price range. As with logarithmic growth, least squares minimizes over time instead of price. Note that here, the extra points are ($t = 0.000001$, $y = \min(y_j)$) and ($t = l$, $y = \max(y_j)$), so that the LHS is defined over all bid times.

13.5 SELECTING THE BEST GROWTH MODEL

We develop an automated model selection procedure to choose for each auction the best-fitting growth model among the four models. The procedure uses a specialized proximity metric that measures the distance between bids and the fitted curve in the two dimensions of time and price. This metric is reminiscent of the Mahalanobis distance. Most model selection criteria only measure the residual distance in the y (price, in this case) dimension; however, we are interested in capturing the best fit in both the price and time dimensions because bid times are informative and are random variables. Furthermore, the fit for the logarithmic growth and reflected-logistic growth models are minimized over the x (time) dimension. If we were to choose between models based simply on the price dimension, we would tend to choose the exponential growth and logistic growth models, even though the reflected models may provide a better representation of the price process (as can be visually seen). While our model selection criterion is primarily aimed at choosing among growth models, the metric may also be used to choose among other methods: growth models, p-splines, monotone smoothing, etc.

13.5.1 Model Selection Metrics

For auction i, let $\{\mathbf{t_i}, \mathbf{y_i}\}$ be the vector of live bids (t_{ij}, y_{ij}) where bid y_{ij} is placed at time t_{ij}, and the number of bids in the auction is n_i. Define a new vector with two additional price points $(n_i^* = n_i + 2)$ that also includes the opening and closing prices of the auction as $\{\mathbf{t_i^*}, \mathbf{y_i^*}\} = \{(t = 0,\ y = \min(y_{ij})),\ \{\mathbf{t_i}, \mathbf{y_i}\},\ (t = l,\ y = \max(y_{ij}))\}$, where l is the length of the auction. It is important that we examine the fit at the start and end of the auction because that is where most of the bid activity takes place, because they are conceptually important, and because that is where modeling often falls short.

We propose two measures of fit: the weighted sum-of-squares standardized by the range (WSSER) and the weighted sum-of-squares standardized by the variance (WSSEV). Both metrics are weighted averages of fit in the y-direction and fit in the x-direction, using weights w_y and w_x, such that $w_y + w_x = 1$. The WSSER for an auction i is defined as

$$WSSER_i = \frac{w_y \sum_{j=1}^{n_i^*} (y_{ij}^* - \hat{y}_{ij}^*)^2}{(\max_j (y_{ij}^*) - \min_j (y_{ij}^*))^2} + \frac{w_x \sum_{j=1}^{n_i^*} (x_{ij}^* - \hat{x}_{ij}^*)^2}{(\max_j (x_{ij}^*) - \min_j (x_{ij}^*))^2}. \tag{13.17}$$

Notice that the denominator is the squared price range in the y dimension and the squared time range (in our case the auction length l) in the x dimension. The WSSEV for an auction i is defined as

$$WSSEV_i = \frac{w_y \sum_{j=1}^{n_i^*} (y_{ij}^* - \hat{y}_{ij}^*)^2}{variance_j(y_{ij}^*)} + \frac{w_x \sum_{j=1}^{n_i^*} (x_{ij}^* - \hat{x}_{ij}^*)^2}{variance_j(x_{ij}^*)}. \tag{13.18}$$

13.5.2 Model Selection Procedure

In this section, we describe how to choose automatically between different growth models and choose the best one. The model selection procedure is as follows:

1. Select weights, w_y and w_x, representing the importance of fit in the price and time dimensions, respectively.
2. Fit each of the four growth models to the live bids of an auction.
3. Compute the model selection metric(s).
4. Choose the model with the best fit (minimum WSSE).

For our sample of auctions, we choose equal weights $w_y = w_x = \frac{1}{2}$ since we are equally interested in fit in the price and time dimensions. One may overweight time (large w_x) if capturing bid timing is of special interest. One such case is in studying bid shilling, where a seller may cancel the auction or illegally bid on his or her

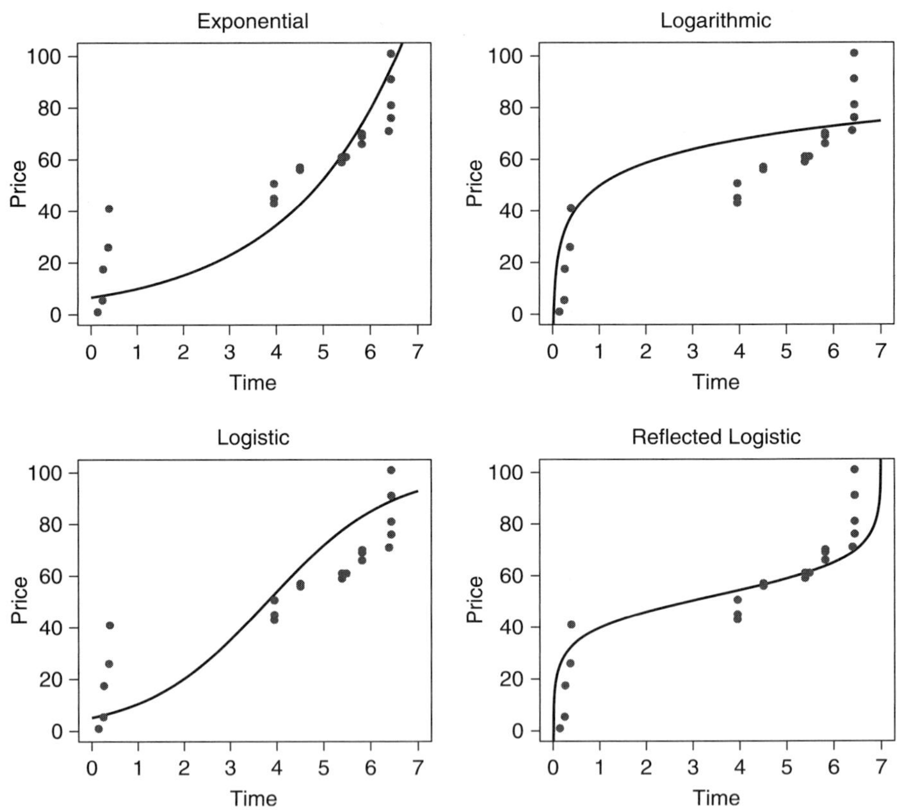

Figure 13.8 Exponential (top left), logarithmic (top right), logistic (bottom left), and reflected-logistic (bottom right) models fit to bids for the same auction. Overweighting in the y-dimension selects exponential growth when reflected-logistic growth should be chosen.

own auction if the price has not reached a certain level by a certain time (Kauffman and Wood 2005). A researcher may overweight price (large w_y) when the focus is on the price level itself (e.g., using this information to make more informed bid decisions.) Note that overweighting price tends to favor exponential and logistic growth models, whereas overweighting time leads to favoring logarithmic and reflected-logistic models. Examples of overweighting are shown in Figures 13.8 and 13.9. In Figure 13.8, overweighting in the price dimension leads to exponential growth selection, whereas reflected-logistic growth would have been selected had the weights been equal. In Figure 13.9, overweighting in the time dimension leads to reflected-logistic growth selection, whereas logistic growth would have been selected had the weights been equal. In addition to the task at hand, visual inspection of a subset of the auctions can provide an appropriate weighting scheme.

From empirical evidence, we find that both WSSE measures provide very similar results, matching on 453 out of 472 (or 95.97%) of the auctions. In cases where the results do not match, visual inspection shows that the models selected by each metric provide reasonably good results, with a slight preference toward WSSER. An

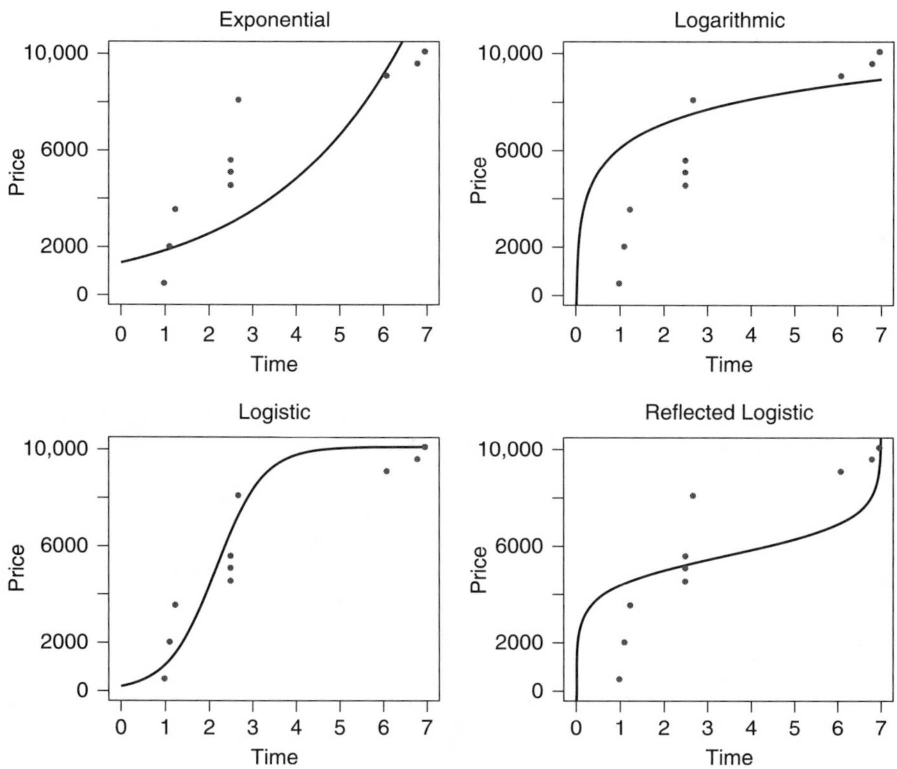

Figure 13.9 Exponential (top left), logarithmic (top right), logistic (bottom left), and reflected-logistic (bottom right) models fit to bids for the same auction. Overweighting in the x-dimension selects reflected-logistic growth when logistic growth should be chosen.

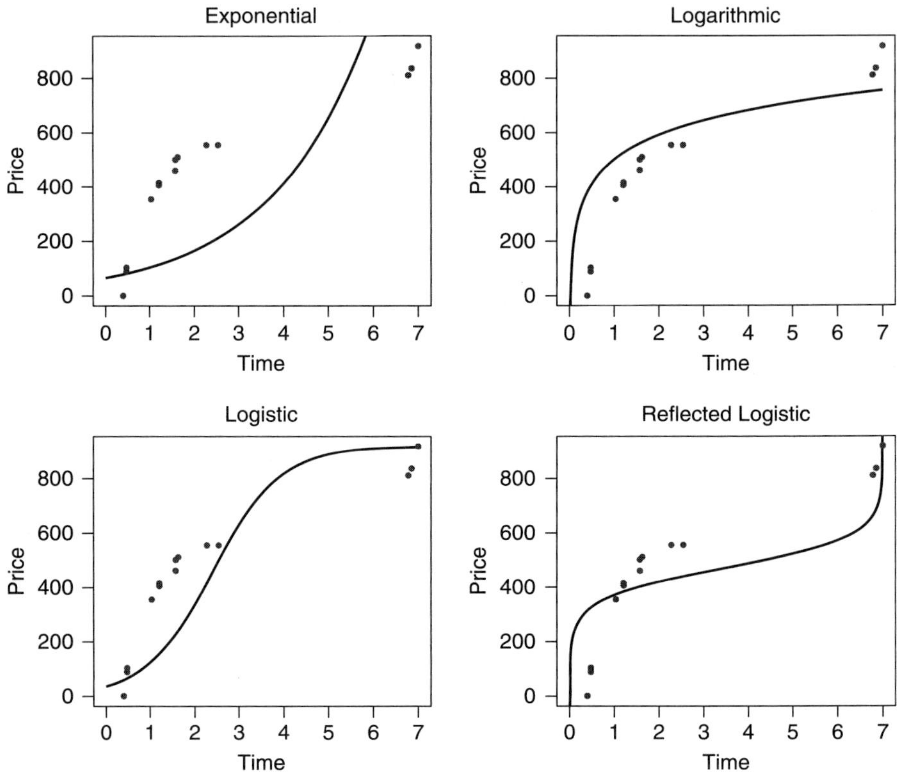

Figure 13.10 Exponential (top left), logarithmic (top right), logistic (bottom left), and reflected-logistic (bottom right) models fit to bids for the same auction. WSSER selects reflected-logistic growth, while WSSEV selects logistic growth.

example is provided in Figure 13.10. WSSEV selects logistic growth, whereas WSSER selects reflected-logistic growth. Both appear to fit the data reasonably well, with reflected-logistic growth capturing the process slightly better. We therefore use WSSER in the following.

Figure 13.11 shows live bids and fitted price curves for four different auctions. The top left model is best fit with exponential growth, the top right with logarithmic growth, the bottom left with logistic growth, and the bottom right with reflected-logistic growth. For model comparison purposes, the selected model is drawn in black and the three models that are not selected (but fitted) are drawn in gray.

Figure 13.12 provides the distribution of auctions across the four models. As expected, and in accordance with previous empirical evidence, exponential growth best fits the majority of the auctions. The next most frequent model is reflected-logistic growth, which captures the common phenomena of early and late bidding. Logistic growth curves are also selected in many cases. In contrast, logarithmic growth is rarely chosen (2.54% of the auctions). This is most likely because, in this set of auctions, early high bidders were rare.

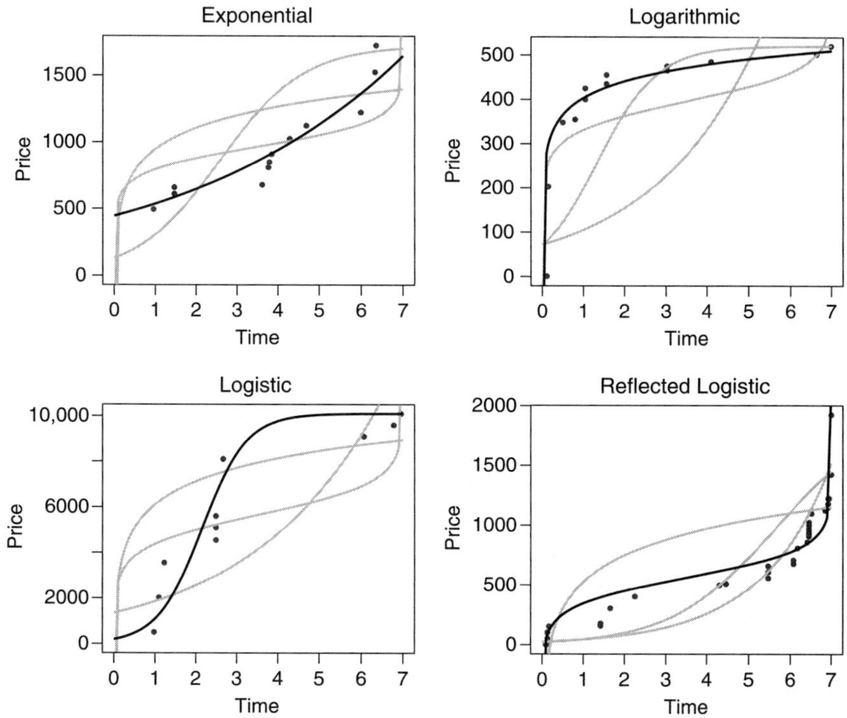

Figure 13.11 Live bids and fitted price curves for four different auctions. The top left is best fit with exponential growth, the top right with logarithmic growth, the bottom left with logistic growth, and the bottom right with reflected-logistic growth.

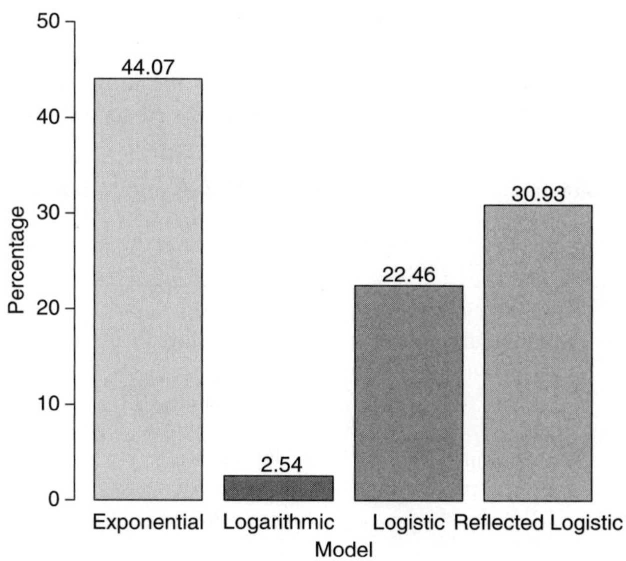

Figure 13.12 Distribution of selected model for 472 luxury wristwatch auctions.

13.6 SMOOTHING METHOD COMPARISON

To assess the differences between our proposed growth models and the nonparametric smoothing methods (p-splines and monotone smoothing splines), we compare them on several dimensions:

Nature of fitted curves: How well does the fitted curve capture the main features of the underlying price process? Specifically, are the estimated curves monotone?

Data fitted: Which data are used for fitting the curve? Can the actual bid data be used or do we have to sample from the "actual price" step function? In addition, what types of auctions, in terms of the number of bids, can be fit?

Overall fit: How well does the curve fit the actual bid data?

Parsimony: What is the level of complication in terms of the number of parameters?

Explanation: How informative are the fitting mechanisms (model driven vs. data driven)?

Automation and computational considerations: What level of user input is needed for curve fitting and the ability to automate the process. In addition, what are the considerations of computational complexity and run time (especially when considering large auction datasets)?

Table 13.3 summarizes the comparison of the parametric and nonparametric curve fitting on all these dimensions.

In terms of fitted curves, growth models have the advantages of fitting monotone curves, fitting directly to the live bids (in some cases, with the addition of the price at the start and end of the auction), and fitting any number of bids, including single-bid auctions (using the additional start and end prices). The resulting curves fit a variety of bid histories and capture the main features of the price dynamics during the auction. In comparison, nonparametric curves are not fit directly to the bid data but rather to the derived step function that conveys the price seen during the auction.

TABLE 13.3 Comparison of Parametric Growth Models and Nonparametric Curve Fitting

	p-splines	Monotone	Growth
Nature of curves	Nonmonotone	Monotone	Monotone
Data fitted	Step function	Step function	Bid data
Overall fit	Good	Variable	Good
Parsimony	Many parameters	Many parameters	Model type + two parameters
Explanation	Unavailable	Unavailable	Available
Automation	User specifies parameters	User specifies parameters	No user input
Computation	Fast	Slow	Fast

Although monotone splines produce monotone curves, p-splines do not guarantee such monotonicity. In fact, there is a balance between monotonicity and data fit, such that a large smoothing parameter might create monotone curves but produce larger deviations between the curve and the data points and vice versa. The wiggliness of the p-splines can be seen in several of the auctions in Figure 13.13. With respect to the minimal number of bids needed for fitting, monotone splines can only be used to fit auctions with at least two bids. p-splines can be fit to single-bid auctions, but the result will be a wiggly horizonal line.

When comparing the goodness of fit of the curve to the bid data, growth models appear to provide a good fit without overfitting. To compare goodness of fit, we fit each of the three methods (growth models, p-splines, and monotone splines) to each of the 472 auctions in the luxury wristwatch dataset. Using the WSSE metric, we find that p-splines provide the best fit roughly 70% of the time, growth models

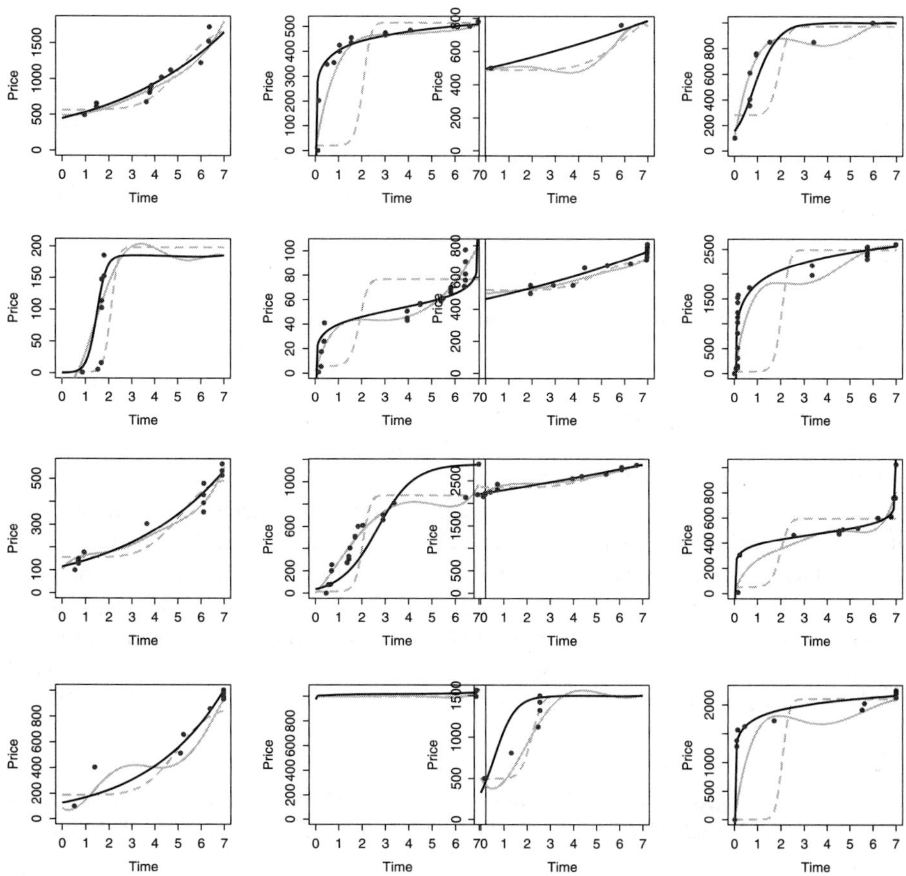

Figure 13.13 Live bids and smoothed price curves for randomly selected seven-day luxury wristwatch auctions. The black line is fit with the growth models, the solid gray line is fit with p-splines, and the gray dashed line is fit with monotone smoothing splines.

TABLE 13.4 Distribution of Chosen Price Model for 472 Completed Seven-Day Luxury Wristwatch Auctions

	p-Splines	Monotone	Growth
Percent chosen when comparing three methods	69.49%	5.08%	25.42%
Percent chosen when comparing two methods	—	11.65%	88.35%

are selected 25% of the time, and monotone smoothing is chosen only 5% of the time (Table 13.4). However, nearly 90% of the curves fit by p-splines are not monotone (which is verified by our sample in Figure 13.13). When comparing only monotone smoothing and growth models, we find that growth models are selected 88% of the time.

With respect to parsimony and explanatory power, the parametric growth models have a clear advantage: They include only two parameters, and the family of four models is able to capture a wide variety of price processes. Furthermore, growth models provide a theoretical basis that describes the price "growth" during the auction and its dynamics. Exponential models are associated with sniping, where the rate of the price increases grows faster and faster. The logistic and reflected-logistic models capture the change in price dynamics associated with early bidding. In contrast, the nonparametric methods are purely data-driven and, as such, do not provide a theoretical model for price growth. While they do capture the price process and its dynamics, they require a large number of parameters (the polynomial coefficients between each pair of consecutive knots; usually each such polynomial is of order 4 to obtain smooth curve derivatives).

Finally, from a computational point of view, fitting the growth models to data is very easy to automate and is reasonably fast, even for a large dataset of auctions. The fitting can be completely automated, and we find that the results of automated fitting are satisfactory. The combination of easy automation and computation time is a major advantage over nonparametric smoothing. When fitting curves nonparametrically, the user is required to specify several parameters in advance: the number and position of knots, the order of the polynomials, and the roughness parameter. The set of knots and the roughness parameter that optimally uncover the price process in one auction may not accurately capture the underlying price process in another auction. However, the same number and position of knots and roughness parameter is often used for all auctions in the dataset of interest in order to avoid confounding the curve fitting from other manipulations (see Jank and Shmueli 2007).

TABLE 13.5 Elapsed Time (in Seconds) Required to Fit 472 and a Subset of 10 Luxury Wristwatch Auctions by p-splines, Monotone Splines, and Growth Models

	p-Spline	Monotone	Growth
10 auctions	2	75	4
472 auctions	6	2082	33

To evaluate computation time, we measure the elapsed time for each of the three smoothing methods on the 472 luxury wristwatch auctions as well as the first 10 auctions (Table 13.5). It is clear that for even moderate datasets monotone splines require very long run times, whereas p-splines and growth models are much faster.

13.7 USING GROWTH CURVES

One of the main advantages of modeling an auction's price process using parametric growth models is that an initial distinction between auctions is directly obtained: Each auction is represented (or labeled) by one of exponential, logarithmic, logistic, or reflected-logistic growth. Knowing the shape of the price curve tells us about the underlying price process. This knowledge is useful in many applications, some of which will be discussed in this section.

13.7.1 Rug Plots

The *rug plot* is a visualization tool, proposed by Hyde et al. (2006), for displaying concurrent processes over a period of time. In the online auction context, the rug plot can display the entire price evolution of all auctions in the dataset over the period of data collection (calendar time). Specifically, the x-axis is calendar time, the y-axis is price, and each auction's price process is plotted as a curve. Rug plots for datasets of Palm Pilot M515 auctions and Xbox auctions (see the Appendix for descriptions of these datasets) are shown in the middle panels of Figure 13.14. The final price of each auction is marked with a dot, and the thick black line and gray band are the daily median closing prices and interquartile ranges (IQRs), respectively.

The rug plot supports visual exploration of temporal groupings of curves. When curves are fit nonparametrically, it is hard to ascertain the types of curves without visual inspection of each curve, which could be a daunting task for even a moderate dataset. Growth models offer an easy solution by using the WSSE measure to choose the best growth model of the four. The model type can then be easily integrated into the rug plot via color coding. To further improve the information contained in a rug plot, we add color-coded dot plots, where a dot represents an auction that closes on that date (the top panels in Figure 13.14). The dots are jittered to visualize periods where many auctions close on the same day. In addition, we create time-grouped stacked bar charts for the volume of auction closings during the time period, as can be seen in the bottom panel of Figure 13.14 (week for Palm data, day for Xbox data). Using this display, we investigate temporal groupings of price processes. For the Palm data, we see a grouping of reflected-logistic price curves over $4/6-4/13$ among very few exponential growth curves, whereas at other times (and especially before $4/6$) most price curves are exponential. Perhaps the large number of exponential growth curves before $4/6$ led bidders in later auctions to believe that most auctions close well over \$150, and they therefore bid early in the auction. In the Xbox data, most curves are either exponential or reflected-logistic with no logarithmic

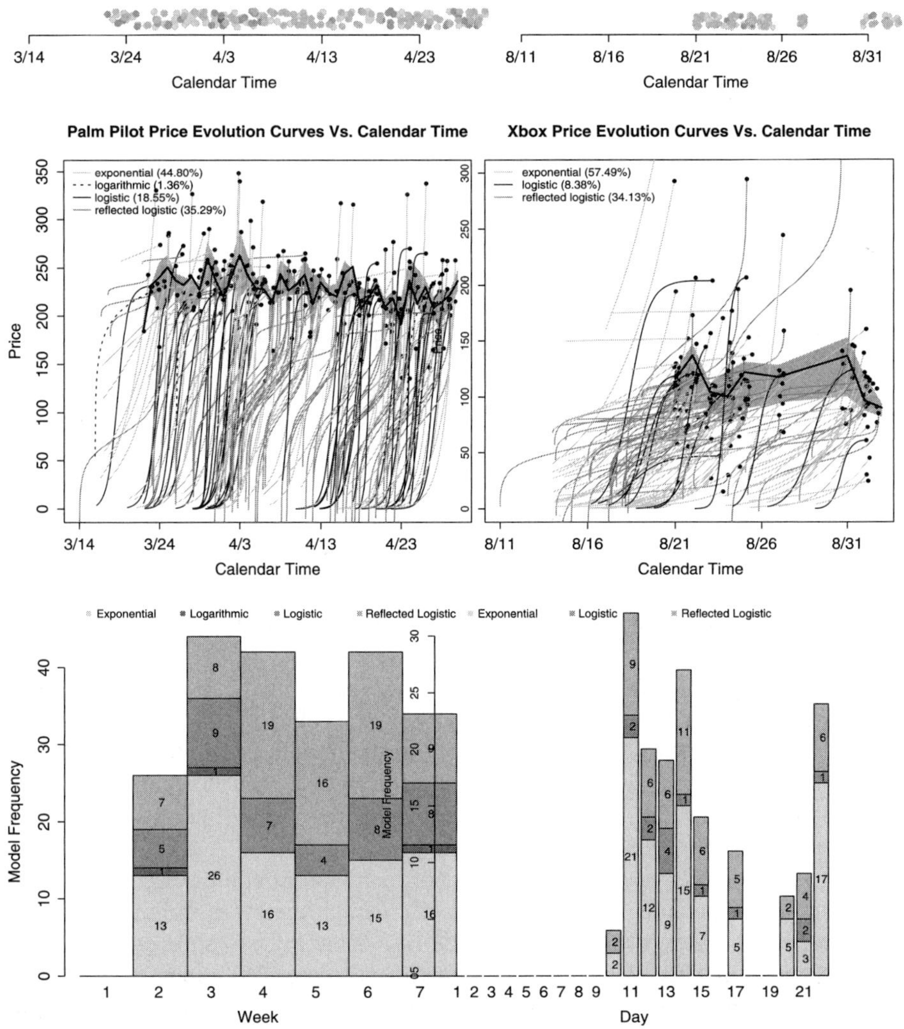

Figure 13.14 Visualizing temporal clustering of price curve types. The left column describes the Palm data; the right column describes the Xbox data. All plots are color-coded by growth model type. The top panels are dot plots (points are jittered for visibility). The middle panels are rug plots. The bottom panels are temporally aggregated, stacked bar charts of auction volume.

price curves. There is also a period with almost no auctions (due to data collection issues). Here we see during the beginning of the period, between 8/14 and 8/18, a clustering of exponential and reflected-logistic curves, with many of the reflected-logistic auctions opening higher than the exponential auctions.

We further explore the relationship of other information, such as auction duration, to the temporal clustering of different price curves. Figure 13.15 is a set of rug plots

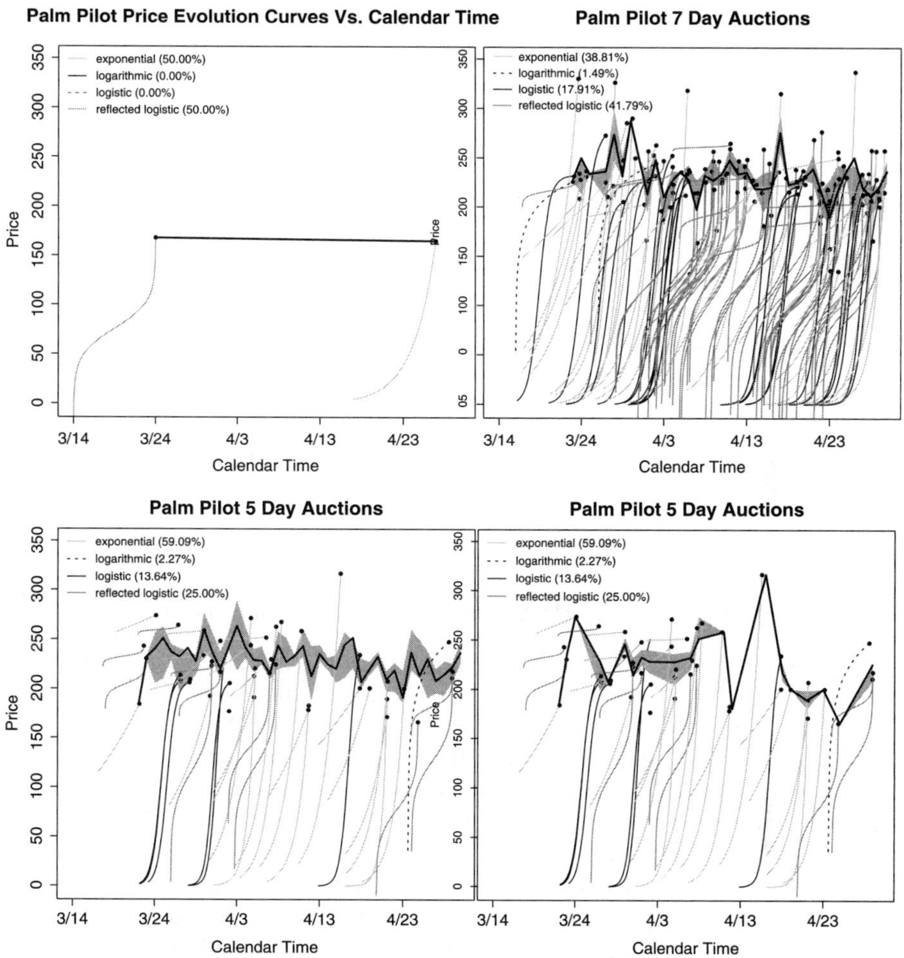

Figure 13.15 Rug plots for Palm Pilot M515 auctions broken down by length: ten-day (top left), seven-day (top right), five-day (bottom left), and three-day (bottom right).

(for the Palm data) separated by auction duration (ten, seven, five, and three days). We see a temporal clustering of reflected-logistic auctions between days 4/6 and 4/13 in the seven-day auctions. Another observation is sporadic exponential price curves in the five- and sometimes seven-day auctions.

Another such exploration is the comparison of new versus used Xbox game consoles (Figure 13.16). We see temporal groupings of logistic curves and reflected-logistic curves for new Xboxes at the beginning of the calendar, whereas for used Xboxes, there does not appear to be such distinct groupings. Because there is still a large volume of auctions in the used dataset, perhaps zooming in on different calendar dates would reveal more temporal groupings.

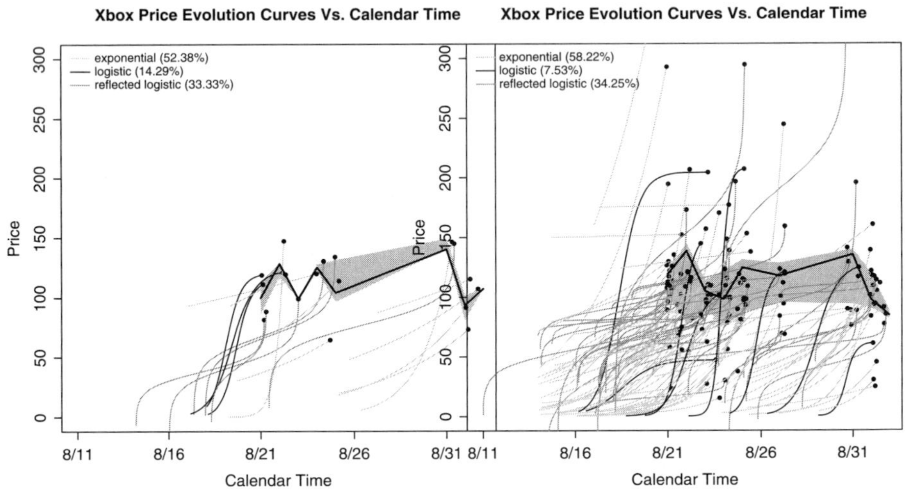

Figure 13.16 Rug plot for Xbox auctions broken down by type: "new" (left) and "used" (right).

13.7.2 Integrating Growth Model Parameters Into Analyses

The parametric growth model representation provides a compact representation of the entire price curve in an auction which is simple and parsimonious—the model type and its two estimated parameters alone. We can then integrate this compact representation into analyses by applying the statistical or data-mining method directly to this representation. This is a classic data-mining approach whereby complicated information is summarized and the summaries are used in the analysis.

One possible application is clustering auctions using the growth model representation and perhaps additional auction-related information. Another direction involves distance-based methods, where the parametric representation can be used for measuring the distance between auctions. A third example is using classification trees, where the model type and the estimated parameters serve either as predictors (for predicting an outcome of interest) or as the outcome variable. In particular, such a tree could be used for predicting the type of price curve of a new auction as a function of information that is given at the auction start (e.g., opening bid, seller rating, presence of a picture, and closing day). Potential bidders can then use the predicted information to decide on which auction to bid, their bid timing, and bid amount.

We describe only a few possible applications here, but obviously the approach is general and parametric representation can be used in almost any type of statistical analysis and/or data-mining technique.

13.8 CONCLUSIONS

In this chapter, we introduce a family of growth models that describe the underlying continuous price process of online auctions: exponential growth, logarithmic growth, logistic growth, and reflected-logistic growth. We also present a metric to choose between models (and, more generally, to choose between any types of fitted curves), which allows automation in the data-fitting stage.

Our parametric approach is parsimonious, the models are easily fitted to bid data, and they capture a variety of price process shapes. They also provide an appealing theoretical explanation of the price process rather than being purely data-driven. The resulting curve is monotone, as expected for price curves in ascending auctions. Our method is computationally fast and can therefore be applied to large auction datasets. All of these reasons give the parametric approach an advantage over nonparametric smoothing methods.

We fit exponential and logistic models in the price dimension but fit logarithmic and reflected-logistic models in the time dimension. For simplicity and computational efficiency, ordinary least square (OLS) fitting is performed in the dimension that is linearizable. However, both metrics (WSSER and WSSEV) for selecting the best growth model evaluate fit in both the price and time dimensions simultaneously to avoid overselection of models fit in a certain dimension. We reason that for online auctions, both the time and magnitude of the bid are random variables, so the fit in both dimensions is important. This suggests that our fitting method should also be based on both dimensions simultaneously. Since we cannot linearize the growth models in both dimensions, we could employ other optimization techniques, such as steepest ascent or Newton-Raphson, to estimate parameters. One of the reasons we initially hesitated to employ optimization methods was the simplicity and computational speed of our method. Further, no parameters needed to be set in advance, as is required of the nonparametric smoothing methods. If we employ iterative fitting in both dimensions simultaneously, we may use the estimates obtained via one-dimensional OLS as the starting values. This line of research should be expanded, and a comparison of parameter estimates as well as computational complexity should be considered.

Our contribution is not limited to the auction setting, but rather proposes the use of parametric functional representations as an alternative to the more popular nonparametric functional objects. We show how the parametric representation provides advantages in data visualization, as well as offering a compact summarization of the price process that can then be used in a variety of statistical analyses and data-mining techniques.

One of the limitations of our family of growth models is in the case of auctions with very sparse activity throughout the auction that changes into very steep price increases at the last moments of the auction. In this case, the exponential growth model, which provides the best fit among the four models, often fails to adequately capture the intense bid activity at the auction's end. One solution is to first transform the data (e.g., by moving to log-scale). Another possibility is to heavily weight the data points toward the auction's end in the fitting process. Yet another option is to include an additional growth model that describes processes that change little until a peak at the end.

There are many other functions that could potentially be used to model growth, and we provide a few examples. The Chapman-Richards growth function is similar to logarithmic growth but places a limit on growth. The Couttsian growth model is similar to exponential growth except that the growth rate is variable. Different models are popular in different disciplines such as biology, ecology, and economics to desribe a variety of phenomena.

We believe that parametric functional representations enhance the field of FDA and provide additional information for statistical analysis. We hope to spur interest in using theoretically relevant parametric models to decribe continuous processes.

ACKNOWLEDGMENTS

The authors would like to thank Shanshan Wang for the Xbox data and Sharad Borle for the luxury wristwatch data.

REFERENCES

Alford, B. and Urimi, L. (2004). An analysis of various spline smoothing techniques for online auctions. Termpaper AMSC Research Interaction Team. Available at http://www.smith.umd.edu/ceme/statistics.

Bajari, P. and Hortascu, A. (2003). Winner's curse, reserve price and endogenous entry: Empirical insights from ebay. *RAND Journal of Economics*, 34: 329–355.

Bapna, R., Goes, P., Gupta, A., and Jin, Y. (2004a). User heterogeneity and its impact on electronic auction market design: An empirical exploration. *MIS Quarterly*, 28(1): 21–43.

Bapna, R., Jank, W., and Shmueli, G. (2004b). Price formation and its dynamics in online auctions. Working Paper, Smith School of Business, University of Maryland. Available at http://ssrn.com/abstract=902887.

Green, P.J. and Silverman, B.W. (1994). *Nonparametric Regression and Generalized Linear Models: A Roughness Penalty Approach*. London: Chapman & Hall.

Hyde, V., Jank, W., and Shmueli, G. (2006). Investigating concurrency in online auctions through visualization. *The American Statistician*, 34(3): 241–250.

Jank, W. and Shmueli, G. (2007). Studying heterogeneity of price evolution in eBay auctions via functional clustering. In *Business Computing* (Adomavicius and Gupta, eds.) Handbook of Information Systems Series. Elsevier.

Jank, W., Shmueli, G., Plaisant, C., and Shneiderman, B. (2007). Visualizing functional data with an Application to eBay's online auctions. In *Handbook on Computational Statistics on Data Visualization* (Chen, Hardie, and Unwin, eds.). Heidelberg: Springer-Verlag.

Kauffman, R.J. and Wood, C.A. (2005). The effects of shiiling on final bid prices in online auctions. *Electronic Commerce Research and Applications*, 4: 18–31.

Ramsay, J.O. (1998). Estimating smooth monotone functions. *Journal of the Royal Statistical Society B*, 60: 365–375.

Ramsay, J.O. and Dalzell, C.J. (1991). Some tools for functional data anaiysis. *Journal of the Royal Statistical Society B*, 53: 539–572.

Ramsay, J.O. and Silverman, B.W. (2002). *Applied Functional Data Analysis: Methods and Case Studies*. New York: Springer-Verlag.

Ramsay, J.O. and Silverman, B.W. (2005). *Functional Data Analysis*. New York: Springer-Verlag.

Roth, A.E. and Ockenfels, A. (2002). Last-minute bidding and the rules for ending second-price auctions: Evidence from ebay and amazon auctions on the internet. *The American Economic Review*, 92.

Shmueli, G., Russo, R., and Jank, W. (2004). The barista: A model for bid arrivals in online auctions. Working Paper, Smith School of Business, University of Maryland. Available at http://ssrn.com/abstract=902868.

Wang, S., Jank, W., and Shmueli, G. (2007). Dynamic forecasting of online auction price using functional data analysis. *Journal of Business and Economic Statistics.*

APPENDIX: ADDITIONAL eBay DATA

Two additional eBay.com datasets are used to illustrate some applications of growth models. They are for Palm Pilot M515 and Xbox auctions.

13.A.1 PALM PILOT M515 DATA

Our data contain information on 221 closed auctions for a brand new Palm Pilot M515. The data were collected between March 11, 2003 and April 20, 2003, roughly a year after the Palm M515 was released to the market. At that time, eBay auctions lasted three, five, seven, or ten days, depending on the length set by the seller. More recently, one-day auctions were introduced on eBay.

Even though the Palm Pilot has a known market value ($250 at the time of the analysis), auctions do not always close near this value. Low prices can result, for instance, if the box is already open or if the seller has a questionable reputation. Auctions can close high if something special is offered with the product, such as an accessory or free shipping. Prices also vary because bidders often get caught up in the excitement of bidding (*auction fever*) and pay more than would be expected. The average selling price for all Palm Pilots in our data is $234 with a median of $232.50 and a standard deviation of $20.86. The least expensive Palm Pilot sold for $172.50 and the most expensive auction closed at $290. Table 13.A.1 provides descriptive statistics for the closing price, opening price, number of bids, number of unique bidders, and unique bidder rating. We also provide descriptive statistics broken down by auction length since we group Palm Pilot auctions by length in Section 13.7. Note that this dataset does not contain any seller information.

TABLE 13.A.1 Descriptive Statistics for 221 Completed eBay Palm Pilot M515 Auctions and Broken Down by Auction Lenth: Three-Day (41), Five-Day (44), Seven-Day (134), and 10-Day (2) Auctions

Variable	Duration	Mean (Std)	Median	Minimum	Maximum
Closing Price	3 Day	\$238.60(\$24.67)	\$232.50	\$177.50	\$290.00
	5 Day	\$233.40(\$23.08)	\$235.00	\$183.50	\$280.00
	7 Day	\$233.60(\$17.78)	\$232.80	\$186.50	\$283.50
	10 Day	\$182.50(\$14.14)	\$182.50	\$172.50	\$192.50
	Total	\$234.00(\$20.86)	\$232.50	\$172.50	\$290.00
Opening Price	3 Day	\$63.46(\$94.87)	\$1.00	\$0.01	\$259.00
	5 Day	\$91.09(\$97.36)	\$35.50	\$0.01	\$259.00
	7 Day	\$41.69(\$72.47)	\$1.00	\$0.01	\$259.00
	10 Day	\$2.51(\$3.53)	\$2.51	\$0.01	\$5.00
	Total	\$63.70(\$84.05)	\$1.00	\$0.01	\$259.00
Number of Bids	3 Day	17.51(9.85)	19.00	2.00	43.00
	5 Day	16.18(9.18)	17.50	2.00	36.00
	7 Day	20.86(10.25)	21.00	2.00	51.00
	10 Day	20.50(6.36)	20.50	16.00	25.00
	Total	19.33(10.09)	19.00	2.00	51.00
Number of Unique Bidders	3 Day	9.22(5.05)	9.00	2.00	23.00
	5 Day	8.25(4.22)	9.00	2.00	19.00
	7 Day	10.85(4.71)	11.50	1.00	23.00
	10 Day	10.50(3.54)	10.50	8.00	13.00
	Total	10.03(4.77)	10.00	1.00	23.00
Unique Bidders Rating	3 Day	91.14(68.43)	86.00	3.00	204.00
	5 Day	89.93(69.01)	84.00	1.00	217.00
	7 Day	84.56(70.29)	74.00	1.00	217.00
	10 Day	73.81(57.02)	74.00	3.00	167.00
	Total	86.46(69.68)	74.00	1.00	217.00

13.A.2 Xbox DATA

Our data contain information on 167 closed auctions for an Xbox game console. The auctions took place between August 11, 2005 and August 30, 2005 and were of fixed duration: one, three, five, seven, or ten days. While the Xbox product is the same, it may be new or used, and the auction may include extras such as additional games and/or controllers. Therefore, bidders will not have the same valuation for each

TABLE 13.A.2 Descriptive Statistics for 167 Completed eBay Xbox Auctions and Broken Down by Condition: New (21) and Used (146)

Variable	Condition	Mean (Std)	Median	Minimum	Maximum
Closing Price	New	\$121.00 (\$14.21)	\$123.20	\$85.00	\$142.50
	Used	\$134.10 (\$66.60)	\$125.50	\$28.00	\$501.80
	Total	\$132.40 (\$62.59)	\$125.00	\$28.00	\$501.80
Opening Price	New	\$26.19 (\$37.33)	\$1.00	\$0.01	\$99.99
	Used	\$40.84 (\$43.40)	\$32.49	\$0.01	\$290.00
	Total	\$39.00 (\$42.86)	\$25.00	\$0.01	\$290.00
Number of Bids	New	20.95 (8.81)	22.00	6.00	38.00
	Used	18.64 (11.75)	18.00	2.00	75.00
	Total	18.93 (11.43)	18.00	2.00	75.00
Number of Unique Bidders	New	9.19 (3.16)	9.00	3.00	14.00
	Used	8.18 (3.80)	8.00	1.00	19.00
	Total	8.31 (3.73)	8.00	1.00	19.00
Unique Bidders Rating	New	234.70 (340.79)	164.00	5.00	1325.00
	Used	299.70 (837.90)	36.00	−1.00	5560.00
	Total	291.50 (792.29)	44.00	−1.00	5560.00
Seller Rating	New	30.16 (69.63)	5.00	0.00	605.00
	Used	42.51 (166.44)	5.00	−1.00	2736.00
	Total	40.79 (156.63)	5.00	−1.00	2736.00

auction. Descriptive statistics for all the auctions as well as broken down by item condition (new or used) are shown in Table 13.A.2.

This game console is no longer sold in stores (as it is the predecessor of the Xbox 360); however, Amazon.com's list price was \$179.98 at the time the data were collected. The average selling price in our sample is \$132.40, with a standard deviation of \$62.59, median of \$125.00, minimum of \$28.00, and maximum of \$501.80. The auction that closed at \$28.00 is for a damaged console, and the auction that closed at \$501.80 is for a used console but includes 84 games.

14

MODELS OF BIDDER ACTIVITY CONSISTENT WITH SELF-SIMILAR BID ARRIVALS

Ralph P. Russo and Nariankadu D. Shyamalkumar
Department of Statistics and Actuarial Science, The University of Iowa, Iowa City, Iowa

Galit Shmueli
Department of Decision and Information Technologies, R.H. Smith School of Business, University of Maryland, College Park, Maryland

14.1 INTRODUCTION

The fast-growing popularity and importance of online auction websites such as eBay has led to a surge in empirical studies of a wide range of online auction phenomena. Using publicly available data from such sites, multiple studies have observed a divergence from classic auction theory. Online auctions differ from offline auctions in several key respects: their length, their lowered barriers of entry for bidders and sellers, and their globalism. Phenomena such as bid *sniping* (last-minute bidding) and bid *revising* (an individual bidder placing multiple bids) are prevalent in online auctions, where, according to the offline theory, they should not exist. Efforts to study online auctions have focused mainly on the identification and quantification of bidding strategies and their justification from a game theoretic perspective (Wilcox 2000; Ockenfels and Roth 2002; Roth and Ockenfels 2002; Bajari and Hortacsu 2003; Bapna et al. 2004; Borle et al. 2006). Our goal is to develop models of bidder activity consisting of bidder arrival and departure, and bid placement, that are consistent with the observable phenomena. The difficulty of creating such models stems from the fact that several aspects of bidder behavior (namely,

Statistical Methods in e-Commerce Research. Edited by W. Jank and G. Shmueli

bidder arrivals, departures, and strategies) are largely unobservable. Bid placements, on the other hand, are fully observable. In a recent paper, Shmueli et al. (2007) proposed their bid arrivals in stages (BARISTA)[1] model, which captures some of the main features of the bid arrival process of online auctions as observed and documented by several authors. Here we introduce a set of bidder behaviors that jointly produce bid arrivals that (under the appropriate conditions) give rise to BARISTA-like bid arrivals, or to bid arrival processes that possess some of the special characteristics that have been observed in empirical studies.

Let $N(s)$, $0 \leq s \leq T$, denote the bid arrival process associated with an online auction having a start time 0 and a deadline (hard close) T. In their study of online second-price auctions,[2] Roth and Ockenfels (2000) noted two interesting characteristics common to the aggregations of such processes:

(P1) An *increasing intensity* of bid arrivals as the auction deadline approaches

(P2) A *striking similarity* in shape among the left truncated bid time distributions on $[s,T]$, as s approaches T

The first phenomenon, also known as *bid sniping*, has been widely observed in fixed-length auctions and has been documented in multiple empirical studies (e.g., Wilcox 2000; Bajari and Hortacsu 2003; Bapna et al. 2004; Borle et al. 2006; Shmueli et al. 2007). The second phenomenon is referred to as *self-similarity*. We illustrate both in Figure 14.1, which displays the empirical cumulative distribution functions (CDFs) of normalized left truncated bid times based on 3643 bid times, placed in 189 online seven-day auctions on eBay.com for new Palm M515 personal digital assistants (PDAs). The graphs are plotted at several different resolution levels (or left truncation points), *zooming in* from the last day, to the last 12 hours, to the last 3 hours, to

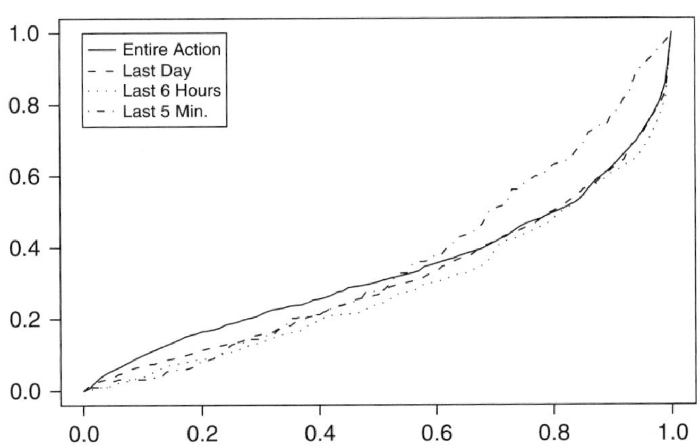

Figure 14.1 Empirical CDFs of truncated bid arrivals—189 Palm auctions.

[1]The BARISTA (bid arrivals in stages) process is a nonhomogeneous Poission process with an intensity function that takes three distinct functional forms over the interval [0, T].

[2]In a second-price auction the highest bidder wins the item and pays the second highest bid.

the last 5 minutes. These functions have a similar shape, independent of scale, until the last few minutes (note the five-minute curve) when the similarity breaks down (see Shmueli et al. 2007).

For $0 \leq s < t \leq T$, define $N(s,t) = N(t) - N(s)$. *Increasing bid intensity* refers to the stochastic monotonicity of $N(t-\delta,\ t)$ as a function of t for any fixed $\delta > 0$, while *self-similarity* refers to the regularity in shape of the distribution functions

$$F_s(\eta) := \frac{N(T - \eta s,\ T)}{N(T - s,\ T)},\quad 0 \leq \eta \leq 1$$

for all s sufficiently close to T (for $s \in [0, T-b]$, some b). Since $N(t-\delta,\ t)$ and $F_s(\eta)$ are determined empirically, to be precise we define these properties in terms of the expected bid counts: We say that the *N-process* has *increasing bid intensity* if $\mathbb{E}(N(t-\delta,\ t))$ is increasing and is *self-similar* over the interval $[b,T]$ if $E_s(\eta)$, defined as

$$E_s(\eta) := \frac{\mathbb{E}(N(T - \eta s,\ T))}{\mathbb{E}(N(T - s,\ T))},\quad \text{for } (s,\ \eta) \in [0,\ T - b] \times [0,\ 1], \tag{14.1}$$

is independent of the value of s in $[0, T-b]$. In words, $E_s(\eta)$ is the same function of $\eta \in [0,1]$ for all $s \in [0, T-b]$. From the Cauchy equation (see p. 41 of Aczel 1966), for (14.1) to hold, we must have for some $\gamma > 0$

$$\frac{\mathbb{E}N(s)}{\mathbb{E}N(T)} = 1 - \left(1 - \frac{s}{T}\right)^{\gamma} b \leq s \leq T, \tag{14.2}$$

in which case $E_s(\eta) = \eta^{\gamma}$ (see footnote 29 of Roth and Ockenfels 2000).

In the following section, we define a *general bid process* (GBP) of bidder arrivals and departures, and bid placement that (under sufficient conditions) yields a bid-arrival sequence that possesses one or both of the above properties. We derive an expression for the probability that an individual bidder is *active* at time s (has not departed the auction as of that time) and an expression for $\mathbb{E}(N(s))$. Moreover, we show that $N(s)/N(T) \to \mathbb{E}(N(s))/\mathbb{E}(N(T))$ uniformly in $s \in [0,\ T]$ as the number of bidders increases. Under a simplifying restriction, the GBP reduces to the *general Poisson bid process*, (GPBP), an aggregation of nonhomogeneous Poisson processes having randomly determined start times and stopping rules. This latter process is related to the BARISTA process of Shmueli, et al. (2007) and is shown to possess the above property **(P1)** under very general conditions. A further simplification results in the *self-similar bid process* (SSBP), which, as its name suggests, possesses the above property **(P2)**.

14.2 THE GENERAL BID PROCESS

Auction theory focuses on bidder behavior and, in particular, on finding optimal bidding strategies for different auction formats. However, the online implementation of auctions has created a different environment where nonoptimal bidding is often observed. Although various empirical studies have documented and quantified these phenomena, there exists a gap in the development of models of bidder behavior that are consistent with them. This is due largely to the fact that one cannot directly observe bidder behavior in publicly available online auction data. Typically, bid placements are completely observable from the auction's bid history, whereas bidder arrivals and departures and bidder strategies are not. On eBay, for example, the temporal sequence of all bids placed over the course of the auction is publicly available. In particular, every time a bid is placed, its exact timestamp is posted. In contrast, the time when bidders first arrive at an auction is unobservable from the bid history. Bidders can browse an auction without placing a bid and thereby not leave a trace or reveal their interest in that auction. That is, they can look at a particular auction, inform themselves about current bid and competition levels in that auction, and make decisions about their bidding strategies without leaving an observable trace in the bid history that the auction site makes public.

Our goal is to establish a model of bidding activity that is consistent with phenomena observable in the bid arrival process.[3] We now define bidder activity more formally.

Suppose that m bidders participate in an online auction starting at time 0 and closing at time T. The parameter m can be fixed or random (see Remark 3 below). With each bidder, associate a random triple $\Theta = (X, \Pi, \mathcal{H})$, which we refer to as the bidder's *type*, that is comprised of an absolutely continuous random variable $X \in [0, T)$ a continuous function Π that maps $[0,T]$ into $(0,1]$, and a family $\mathcal{H} = \{H_s\}$ of real valued distribution functions indexed by a real parameter $s \in [0, T]$, with $H_s(\cdot)$ having support $[s,T]$, where it is differentiable. The variable X represents the arrival time of an individual bidder at the auction, the function Π determines on each of his bid placements whether he will remain in the auction and make a future bid, and the family $\mathcal{H}$ determines the timing of each bid that he makes. Given that the bidder's type is $\theta = (x, \pi, \{H_s\})$, he enters the auction at time x and places an initial bid at time $Y_1 \sim H_x$. If $Y_1 = y_1$, he departs the auction with probability $\pi(y_1)$, or otherwise places a second bid at time $Y_2 \sim H_{y_1}$. If $Y_2 = y_2$, he departs the auction with probability $\pi(y_2)$, or otherwise places a third bid at time $Y_3 \sim H_{y_2}$, etc., ultimately placing a random number of bids during $[0, T]$. All m bidders are assumed to act in a like manner, independently of each other.

In online auctions, bid snipers often attempt to place their final bid close to the deadline to forestall a response from competing bidders. It has been observed that these late bids often fail to transmit for technical reasons such as network congestion.

[3]eBay auctions use a proxy bidding mechanism whereby bidders are advised to bid the highest amount they are willing to pay for the auctioned item. The auction mechanism then automates the bidding process to ensure that the bidder with the highest proxy bid is in the lead at any given time. Thus, when Bidder A places a proxy bid that is lower than the highest proxy bid of (say) Bidder B, the new displayed highest bid and its timestamp will appear with Bidder B's username (although Bidder A is the one who placed the bid). In our discussion we shall consider such a bid as having been placed by A, rather than B, as it is A's action that led to a change in the displayed bid.

Thus, as $s \rightarrow T$, there may be a growing probability that a late bid fails to be recorded. This phenomenon can most efficiently be accommodated within our GBP by building the probability of a failed last bid directly into the π function, as in (14.14).

For a more precise definition of the GBP, let $X_1, X_2, \ldots, X_m$ denote the arrival times of the m bidders. For $1 \leq k \leq m$, let $N_k(s)$ denote the number of bids placed by bidder k during the period $[0, s]$, $0 \leq s \leq T$, and let $Y_{k,j}$ denote the time of the jth bid placed by the kth bidder, $1 \leq j \leq N_k(T)$. For convenience, set $Y_{k,0} = X_k$ and suppose that

(A1) $\Theta_1 = (X_1, \Pi_1, \mathcal{H}_1), \ldots, \Theta_m = (X_m, \Pi_m, \mathcal{H}_m)$ are independent and identically distributed.

(A2) $\Pr\,(Y_{k,j+1} \leq t | Y_{k,j} = y_{k,j}) = H_{k,y_{k,j}}(t)$ for $1 \leq k \leq m$ and $j \geq 0$.

(A3) The sequences $\{Y_{1,j}\}_{j \geq 0}, \ldots, \{Y_{m,j}\}_{j \geq 0}$ are independent.

(A4) $\Pr\,[N_k(T) = r | Y_{k,r} = y_{k,r} \textit{ and } \Pi_k = \pi_k] = \pi_k(y_{k,r})$ for $r \geq 1$ and $1 \leq k \leq m$.

Remark 1. We observe that condition **(A1)** allows dependence among the elements of Θ_k. Thus, the model can accommodate a tendency (say) for infrequent bidders to place their bid(s) late. We observe further that the above model accounts for heterogeneity in bidder probabilities of remaining in the auction after placing a bid, both across bidders and from bid to bid. Empirical evidence (Bapna et al. 2004) suggests that a realistic distribution of Π should reflect the dichotomy of one-time bidders (*opportunists*) versus multibid bidders (*participators*).

Remark 2. Since π is a continuous function on a compact set, it achieves its minimum ($=\pi_{\min}$) on $[0, T]$. Since π maps into $(0, 1]$, we have $\pi_{\min} > 0$; Thus, with each bid, the probability of departure is bounded below by $\pi_{\min}$. The total number of bids placed by the bidder on $[0, T]$ is therefore stochastically bounded above by a geometric($\pi_{\min}$) variable and thus has finite moments of all order.

Remark 3. (Poisson bidder arrivals) The parameter m can be either fixed or random. If random, it is natural to assume that m is Poisson distributed, as would be the case when bidders arrive in accordance with a nonhomogeneous Poisson process having an intensity $\mu g(t)$, $t \in [0, T]$, where $\mu > 0$ and g is a density function on $[0,T]$. Then $X_1, \ldots, X_m$ is a random sample of random size $m \sim$ Poisson (μ) from a fixed distribution with density function $g(t)$, $0 < t < T$.

14.2.1 A Single-Bidder Auction

The following two results pertain to an auction involving a single bidder ($m = 1$) whose *type* is $\theta = (x, \pi, \{H_s\})$. Let N_1 denote the resulting bid counting process. Obviously, a single bidder would not bid against himself. However, it is convenient to study the actions of a single bidder in the context of an m-bidder auction. We say that the bidder is *active* at time s if $N_1(T) > N_1(s)$ (he does not depart the auction during $[0, s]$). In particular, a bidder is active during the time period prior to his arrival. Let $p(s \mid \theta)$ denote the probability that this event occurs, and for $0 \leq s \leq t \leq T$ define

$$G_s(t \mid \theta) = \Pr(Y_{N(s)+1} \leq t \mid \theta), \quad \text{on}\{N_1(T) > N_1(s);\ X < s\}.$$

Given that the bidder arrived at time x (<s) and is still active at time s, his next bid time $Y_{N_1(s)+1}$ has the distribution above. Let $g_s(\theta)$ denote the right derivative of $G_s(t \mid \theta)$ evaluated at s.

To derive the form of $g_s(\theta)$ we define the functional sequence

$$\begin{aligned}
\phi_0(s;\, t) &= h_s(t) \\
\phi_1(s;\, t) &= \int_s^t (1 - \pi(u))\phi_0(s;\, u)\phi_0(u;\, t)du \\
&\vdots \\
\phi_n(s;\, t) &= \int_s^t (1 - \pi(u))\phi_0(s;\, u)\phi_{n-1}(u;\, t)du.
\end{aligned}$$

We now have

$$g_s(\theta) = \frac{\sum_{n=0}^{\infty} \phi_n(x;\, s)}{\sum_{n=0}^{\infty} \int_s^T \phi_n(x;\, u)du}, \qquad x < s.$$

Proposition 1. Suppose that

$$\sup_{s \leq v \leq s+\delta} H_v(s + \delta) := \omega(s,\, \delta) \to 0 \ as\ \delta \to 0,\ all\ s \in [0,\, T). \tag{14.3}$$

Then, for $0 \leq s \leq T$,

$$p(s \mid \theta) = 1_{x<s} \exp\left(-\int_x^s \pi(t)g_t(\theta)dt\right) + 1_{x>s}.$$

Proof. If $x > s$ the result is trivial. Fix $s \in [x, T)$ and define

$$\pi_\delta = \inf\{\pi(t) : s \leq t \leq s + \delta\} \text{ and } \pi^\delta = \sup\{\pi(t) : s \leq t \leq s + \delta\}.$$

Writing $p(s)$ for $p(s \mid \theta)$, we have

$$p(s + \delta) \geq p(s)[1 - G_s(s + \delta \mid \theta) + G_s(s + \delta \mid \theta)(1 - \pi^\delta)(1 - \omega(s,\, \delta))]$$

since an active bidder at time s remains active at time $s+\delta$ if either (i) $Y_{N_1(s)+1} \in (s+\delta, T)$, or (ii) $Y_{N_1(s)+1} < s+\delta$, the bidder stays active, and $Y_{N_1(s)+2} > s+\delta$. It follows that

$$\liminf_{\delta \to 0} \frac{p(s+\delta) - p(s)}{\delta} \geq -\pi(s)p(s)g_s(\theta).$$

Moreover,

$$p(s+\delta) \leq p(s)[1 - G_s(s+\delta \mid \theta) + G_s(s+\delta \mid \theta)(1 - \pi_\delta)]$$

since an active bidder at time s is active at time $s+\delta$ only if (i) $Y_{N_1(s)+1} > s+\delta$ or (ii) $Y_{N_1(s)+1} < s+\delta$, and the bidder stays active. It follows that

$$\limsup_{\delta \to 0} \frac{p(s+\delta) - p(s)}{\delta} \leq -\pi(s)p(s)g_s(\theta).$$

Thus, $p'(s) = -\pi(s)p(s)g_s(\theta)$. Hence the proof. □

Remark 4. Condition (14.3) is a mild condition which will be assumed from here on. For the condition to fail would require a pathological $\mathcal{H}$.

The following is an easy consequence of the above and is hence stated without proof.

Proposition 2. For $0 \leq s \leq T$,

$$\mathbb{E}[N_1(s) \mid \theta] = 1_{x<s} \int_x^s p(t \mid \theta) g_t(\theta) dt.$$

14.2.2 A Multibidder Auction

Returning to the case of m bidders, we state a uniform limit result for $N(s)/N(T)$. It should be noted that many online auctions involve small numbers of bidders. Self-similarity, as studied in Roth and Ockenbels (2000), is not a phenomenon that can be easily observed from an individual auction where the number of bids is small. To observe the self-similarity property, one usually must aggregate many *equivalent* auctions (i.e., auctions for the same item, over the same duration, etc.). Such aggregations often involve hundreds of bidders, hence our interest in $m \to \infty$.

Proposition 3. If $\mathbb{E}(N_1(T)) < \infty$, then as $m \to \infty$,

$$\sup_{0 \leq s \leq T} \left| \frac{N(s)}{N(T)} - \frac{\mathbb{E}(N_1(s))}{\mathbb{E}(N_1(T))} \right| \to 0 \text{ almost surely.}$$

Proof. For fixed $s \in [0, T]$, $N(s)$ is the sum of m independent and identically distributed random variables. By the *strong law of large numbers* we have almost sure pointwise (in s) convergence:

$$\frac{N(s)}{N(T)} \to \frac{\mathbb{E}(N_1(s))}{\mathbb{E}(N_1(T))} \quad \text{almost surely.}$$

By the continuity of the limit as a function of s and by Polya's Theorem, the above convergence is uniform in s. Hence the proof. □

Remark 5. By Remark 2 we have $\mathbb{E}(N_1(T)|\, \theta) \leq 1/\pi_{\min}$, so that by the double expectation formula, $\mathbb{E}(N_1(T)) \leq \mathbb{E}(1/\Pi_{\min})$, where $\Pi_{\min} = \min\{\Pi(s) \colon s \in [0, T]\}$. Thus, the finiteness of $\mathbb{E}(1/\Pi_{\min})$ is sufficient for the convergence in Proposition 3.

Remark 6. (Poisson bidder arrivals). In the case where bidders arrive in accordance with a nonhomogeneous Poisson process with intensity $\mu g(t)$ (as in Remark 3), Proposition 3 and a standard coupling argument yield the following result:

$$\varepsilon > 0 \text{ and } \mathbb{E}N_1(T) < \infty \Rightarrow \lim_{\mu \to \infty} \Pr\left(\sup_{0<s<T} \left| \frac{N(s)}{N(T)} - \frac{\mathbb{E}N_1(s)}{\mathbb{E}N_1(T)} \right| > \varepsilon \right) = 0. \tag{14.4}$$

The practical significance of (14.4) is that in an auction with a large number of participants or in aggregations of many equivalent auctions, the observed distribution of bid times on $[0, T]$ will be uniformly close to the deterministic function $\mathbb{E}N_1(s)/\mathbb{E}N_1(T)$ with high probability.

14.3 THE GENERAL POISSON BID PROCESS

We consider now a simplifying restriction on the form of $\mathcal{H}$. Given $\theta = (x, \pi, \{H_s\}$, suppose that all of the H_s functions are driven by the single (randomly determined) function H_0 as follows:

$$H_s(t) = \frac{H_0(t) - H_0(s)}{1 - H_0(s)} \quad \text{for } 0 \leq s \leq t < T. \tag{14.5}$$

Under condition (14.5) the bidder's jth bid time ($j \geq 1$), conditional on $Y_{j-1} = y_{j-1}$, is distributed as H_0 restricted to $[y_{j-1}, T]$. We note that a bid time chosen from H_s that is not realized by time $t > s$ is probabilistically equivalent to one chosen from H_t.

In this subsection, N_1 denotes the single-bidder process of bid arrivals under condition (14.5). We define an auxiliary bid counting process $M(s)$, $0 \leq s \leq T$, under the additional conditions that $\Pr(X = 0) = 1$ (the bidder enters the auction at time $s = 0$) and $\Pr(\Pi(s) \equiv 0) = 1$ (the bidder never departs the auction). We observe that

$M(0) = 0$ and that the M-process possesses independent increments. Moreover, writing H for H_0, we have

$$
\begin{aligned}
\limsup_{\delta \to 0} & \frac{\Pr(M(s+\delta) - M(s) \geq 2)}{\delta} \\
& \leq \limsup_{\delta \to 0} \left[\frac{H(s+\delta) - H(s)}{1 - H(s)} \right]^2 \frac{1}{\delta} \\
& = \left(\frac{1}{1 - H(s)} \right)^2 h(s) \limsup_{\delta \to 0} \, [H(s+\delta) - H(s)] = 0 \qquad (14.6)
\end{aligned}
$$

and by (14.6)

$$
\begin{aligned}
\lim_{\delta \to 0} \frac{\Pr(M(s+\delta) - M(s) = 1)}{\delta} &= \lim_{\delta \to 0} \left[\frac{\Pr(M(s+\delta) - M(s) \geq 1)}{\delta} \right] \\
&= \lim_{\delta \to 0} \frac{H(s+\delta) - H(s)}{(1 - H(s))\delta} \\
&= \frac{h(s)}{1 - H(s)}.
\end{aligned}
$$

The M-process is thus a nonhomogeneous Poisson process with intensity function

$$\lambda(s) = \frac{h(s)}{1 - H(s)}.$$

Intuitively, when the auction clock reaches time s, no matter how many bids have been placed and no matter when they have been placed, the probability that a bid will be placed during $(s, s+\delta)$ is approximately $\delta h(s)/(1 - H(s))$. Given θ (associated with the N_1-*process*), we may use the function $\pi(\cdot)$ to randomly label an *M-process* arrival at time s as A or B with respective probabilities $1 - \pi(s)$ and $\pi(s)$. The resulting offspring processes M_A and M_B are independent nonhomogeneous Poisson processes with respective intensity functions

$$\lambda_A(s) = \frac{(1 - \pi(s))h(s)}{1 - H(s)} \quad \text{and} \quad \lambda_B(s) = \frac{\pi(s)h(s)}{1 - H(s)}.$$

Note that arrivals from the N_1-process are the M arrivals that occur after the bidder arrival time X, up to and including the first arrival from M_B. That is, N_1 is a nonhomogeneous Poisson process with intensity function λ, restricted to the random interval $[X, T]$, and stopped upon the first arrival from M_B. The N-process (involving all m bidders) is an aggregation of m such independent processes.

Given $\theta = (x, \pi, \{H_s\})$, a bidder is active at time s if and only if either $x > s$ or $x < s$ and there are no arrivals from M_B during the period $[x, s]$. Accordingly,

$$\begin{aligned} p(s \mid \theta) &= 1_{x<s}\Pr[M_B(s) - M_B(x) = 0] + 1_{x\geq s} \\ &= 1_{x<s}\Pr\left[\text{Poisson}\left(\int_x^s \lambda_B(t)dt\right) = 0\right] + 1_{x\geq s} \\ &= 1_{x<s}\exp\left[-\int_x^s \frac{\pi(t)h(t)}{1-H(t)}dt\right] + 1_{x\geq s}. \end{aligned} \tag{14.7}$$

From the above, we obtain the conditional bid intensity $\lambda(\cdot\,|\theta)$ of an individual bidder of type θ:

$$\lambda(s|\theta) = \begin{cases} \exp\left[-\int_x^s \frac{\pi(t)h(t)}{1-H(t)}dt\right]\frac{h(s)}{1-H(s)}, & s > x; \\ 0, & s \leq x. \end{cases}$$

In the case where $\lim\sup_{s\to T}\pi(s) < 1$ and $\lim_{s\to T} h(s) < 0$, the conditional intensity explodes as s approaches T, i.e., $\lim_{s\to T}\lambda(s|\theta) = \infty$. The condition on $\pi(\cdot)$ can be dropped if $h(s)$ increases to infinity sufficiently fast as s approaches T (e.g., at any polynomial rate).

14.3.1 A Constant Probability of Departure

In this subsection, we suppose that (upon each bid placement) the bidder has a randomly determined time-invariant probability of departure:

$$\Pr(\Pi(s) = \Pi(0),\ 0 \leq s \leq T) = 1. \tag{14.8}$$

Under the above assumption, the number of bids placed by an individual bidder is geometrically distributed with a randomly determined parameter. A time-invariant departure probability, while certainly not true for the entire auction duration, should be approximately true over short intervals. Under (14.5) and (14.8), statement (14.7) reduces to

$$p(s \mid \theta) = 1_{x<s}\left[\frac{1-H(s)}{1-H(x)}\right]^{\pi(0)} + 1_{x>s}. \tag{14.9}$$

By Proposition 2 we get

$$\begin{aligned} \mathbb{E}(N_1(s) \mid \theta) &= 1_{x<s}\int_x^s \left[\frac{1-H(t)}{1-H(x)}\right]^{\pi(0)} \frac{h(t)}{1-H(t)}dt \\ &= \frac{1_{x<s}}{\pi(0)}\left(1 - \left[\frac{1-H(s)}{1-H(x)}\right]^{\pi(0)}\right), \end{aligned}$$

so that by the double expectation formula, writing Π for $\Pi(0)$,

$$\mathbb{E}(N_1(s)) = \mathbb{E}\left(\frac{1_{X<s}}{\Pi}\left(1 - \frac{1-H(s)}{1-H(X)}\right)^{\Pi}\right). \tag{14.10}$$

Hence, by Proposition 3, under conditions (14.8) and (14.5), if $\mathbb{E}\Pi^{-1} < \infty$,

$$\sup_{0<s<T}\left|\frac{N(s)}{N(T)} - \mathbb{E}\frac{1_{X<s}}{\Pi}\left(1 - \frac{1-H(s)}{1-H(X)}\right)^{\Pi}\frac{1}{\mathbb{E}\Pi^{-1}}\right| \to 0 \; a.s. \text{ as } m \to \infty$$

and for Poisson arrivals (as in Remarks 3 and 6),

$$\lim_{\mu\to\infty} \Pr\left(\sup_{0<s<T}\left|\frac{N(s)}{N(T)} - \mathbb{E}\frac{1_{X<s}}{\Pi}\left(1 - \frac{1-H(s)}{1-H(X)}\right)^{\Pi}\frac{1}{\mathbb{E}\Pi^{-1}}\right| > \varepsilon\right) = 0.$$

14.3.2 The Self-Similar Bid Process

Suppose that conditions (14.5) and (14.8) hold, and that for some constant $r > 0$ (the same for all bidders) we have

$$H_0(s) = 1 - \left(1 - \frac{s}{T}\right)^{r/\pi(0)}. \tag{14.11}$$

Suppose, in addition, that all bidders arrive by time $b < T$:

$$\Pr(0 \le X < b) = 1. \tag{14.12}$$

Under (14.11), the higher a bidders' likelihood of departure upon the placement of a bid at time s, the stochastically greater the time of his next bid (i.e., the more inclined he is to choose his next bid near the deadline T). Condition (14.11) ties the selection function H_0 directly to the constant departure probability π (0). Again writing Π for Π (0) and assuming that $\mathbb{E}\Pi^{-1} < \infty$, we have by (14.10)

$$\begin{aligned}\mathbb{E}N_1(T-s, T) &= \mathbb{E}\left(\Pi^{-1}\left(\frac{s}{T-X}\right)^r 1_{X<T-s} + \Pi^{-1}1_{X>T-s}\right)\\ &= \mathbb{E}\Pi^{-1}\left(\frac{1}{T-X}\right)^r s^r \text{ for } s < T-b.\end{aligned}$$

Thus, for $(s, \eta) \in [0, T-b] \times [0, 1]$, we have

$$E_s(\eta) = \eta^r, \tag{14.13}$$

where $E_s(\eta)$ is defined in (14.1)

By statement (14.13) and Proposition 3, we have for large m

$$N_s(\eta) = \frac{N(T-\eta s, T)}{N(T-s, T)} \approx \eta^r \text{ for } (\eta, s) \in [0, 1] \times [0, T-b].$$

14.3.3 The BARISTA Process

The efforts of bidders to place bids late in the auction are often thwarted by transmission failures. We build this phenomenon into the GPBP by assuming (14.11) and supposing that each bidder has a randomly determined, time-invariant probability of departure upon the placement of all bids on $[0, T-d]$, for some small $d > 0$, and that this probability is magnified by a constant β for all bids placed after time $T-d$ (β and d being the same for all bidders):

$$\Pi(s) = \Pi(0)1_{s \leq T-d} + \beta\Pi(0)1_{s > T-d}. \tag{14.14}$$

The resulting process of bid arrivals is a GPBP (but not a SSBP) as described in Section 14.3.1, and can thus also be characterized as an aggregation of independent nonhomogeneous Poisson processes with randomly determined start times and stopping rules. At time $s \in (b, T-d]$, the bid intensity associated with an individual bidder of type θ is

$$\begin{aligned}
\lambda_1(s \mid \theta) &= p(s \mid \theta)\frac{h(s)}{1 - H(s)} \\
&= \left(\frac{1 - H(s)}{1 - H(x)}\right)^{\pi(0)} \frac{h(s)}{1 - H(s)} \quad \text{by (1.9)} \\
&= \left(1 - \frac{s}{T}\right)^{r-1} \left(1 - \frac{x}{T}\right)^{-r} \frac{r}{\pi(0)T} \quad \text{by (14.11)}.
\end{aligned}$$

Hence, for a given collection of m bidder types the intensity of the bid arrival sequence at $s \in (b,T-d]$ is given by

$$\lambda(s \mid m, \ \theta_1, \ldots, \theta_m) = \left[\frac{r}{T}\sum_{k=1}^{m} \frac{1}{\pi_k(0)}\left(1 - \frac{x_k}{T}\right)^{-r}\right]\left(1 - \frac{s}{T}\right)^{r-1}.$$

For $s \in (T-d,T]$,

$$\begin{aligned}
\lambda_1(s \mid \theta) &= p(T - d \mid \theta)\left(\frac{1 - H(s)}{1 - H(T - d)}\right)^{\beta\pi(0)} \frac{h(s)}{1 - H(s)} \\
&= \left(\frac{d}{T}\right)^{r - r\beta}\left(1 - \frac{x}{T}\right)^{-r} \frac{r}{\pi(0)T}\left(1 - \frac{s}{T}\right)^{r\beta - 1}
\end{aligned}$$

and hence, for a given collection of m bidder types, the intensity of the bid arrival sequence at $s \in (b,T)$ is given by

$$\lambda(s \mid m, \ \theta_1, \ldots, \theta_m) = \begin{cases} c\left(1 - \frac{s}{T}\right)^{r-1}, & s \in (b, T - d] \\ c\left(\frac{d}{T}\right)^{r - r\beta}\left(1 - \frac{s}{T}\right)^{r\beta - 1}, & s \in (T - d, T], \end{cases}$$

where

$$c = \left[\frac{r}{T} \sum_{k=1}^{m} \frac{1}{\pi_k(0)} \left(1 - \frac{x_k}{T}\right)^{-r} \right]$$

is the result of the realization $(m,\ \theta_1, \ldots, \theta_m)$. This is the form of the two-stage BARISTA (Shmueli et al. 2007) intensity with $d_1 = 0,\ d_2 = d,\ \alpha_2 = r,$ and $\alpha_3 = \beta r$. It can be shown similarly that multiple shifts in the departure probability will lead to an intensity of the multistage BARISTA form. In particular, a double shift yields the three-stage intensity discussed in Shmueli et al. (2007).

Remark 7. We note that in the case of $r\beta < 1$, $\lambda(s \mid m,\ \theta_1, \ldots, \theta)$ as a function of s is increasing to infinity (i.e., increasing and exploding) as s approaches T.

Working with the general Poisson model under condition (14.8), we now derive the distribution of an individual bidder's final bid time Y_{final}. Again, writing Π for $\Pi(0)$, we have by (14.9),

$$P(Y_{\text{final}} > s) = \mathbb{E}\left(\frac{1 - H(s)}{1 - H(X)}\right)^{\Pi} 1_{X<s} + \Pr(X > s) \tag{14.15}$$

and thus

$$\mathbb{E}Y_{\text{final}} = \mathbb{E}X + \int_0^T \mathbb{E}\left(\frac{1 - H(s)}{1 - H(X)}\right)^{\Pi} 1_{X<s} \mathrm{d}s.$$

For the SSBP, the above statements simplify to

$$P(Y_{\text{final}} > s) = \mathbb{E}\left(\frac{T - s}{T - X}\right)^{r} 1_{X<s} + \Pr(X > s)$$

and

$$\mathbb{E}Y_{\text{final}} = \mathbb{E}X + \int_0^T \mathbb{E}\left[\left(\frac{T-s}{T-X}\right)^{r} 1_{X<s}\right] \mathrm{d}s.$$

In the case of m-bidders, $Y_{m,\text{final}}$ is the sample maximum from a random sample of size m from the distribution of Y_{final}. Hence, properties of $Y_{m,\text{final}}$ are easily derived from those of Y_{final} given above. Also, by using conditioning, we can extend the above results to the case of random m.

14.3.4 Examples

We end this section with two simulations.

Example 1 (SSBP). We set $m = 1000$, $T = 7$, $X \sim U(0,\ 5.6)$, $\Pr(\Pi(0) \leq t) = 4t^2$ for $t \in (0,\ 1/2)$, and $r = 2/5$. In our simulation, 1000 bidders arrive uniformly during

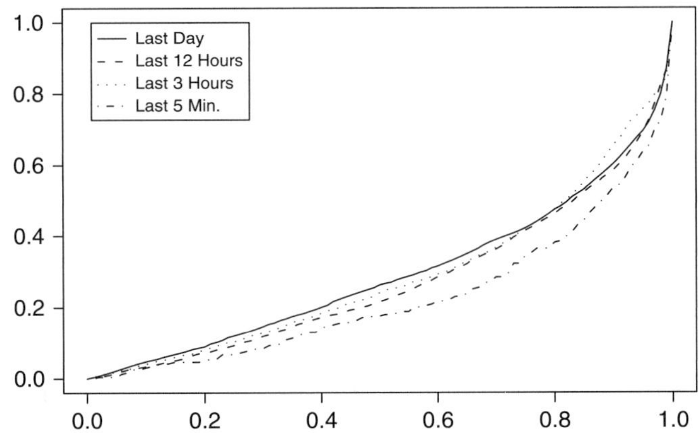

Figure 14.2 Empirical CDFs of truncated bid arrivals—simulated data.

the first 5.6 days (the first 80%) of a 7-day auction, and their departure probabilities constitute a random sample from a distribution having a triangular shape on (0, 1/2), so that each bidder is expected to place 4 bids (4619 were placed in our simulation). Moreover, any bidder placing a bid at time t and opting to remain active has

$$\Pr(\text{time of placement of next bid} > s) = \left[\frac{1-s}{1-t}\right]^{\frac{5}{2\pi}}.$$

Uniform bidder arrivals during a fixed time span is a natural assumption. The exact distribution of the arrival times over (0, 5.6) determines the distribution of the number of active bidders at time point 5.6 but is otherwise irrelevant. In the case of U(0, 5.6), we expect $.678m = 678$ bidders to be active at time $t = 5.6$. In our simulation, the

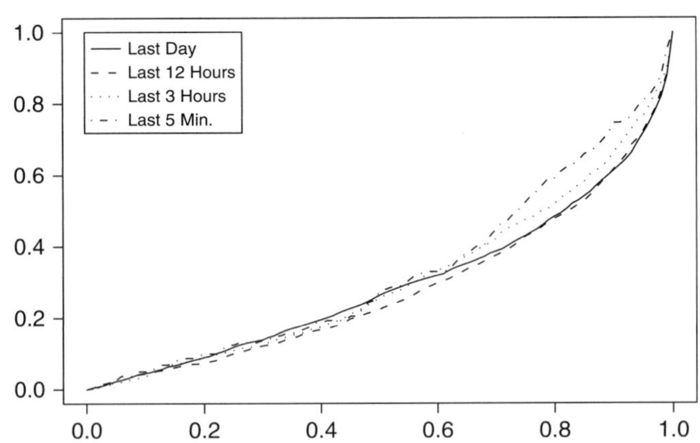

Figure 14.3 Empirical CDFs of truncated bid arrivals—simulated data.

observed number of such bidders was slightly lower (665). The function H is of the form required in the SSBP model. Our value of $r = 2/5$ is in the range (0, 1), which guarantees that h increases as t approaches T. Figure 14.2 displays the empirical CDFs of normalized left truncated bid times. Note the similarity (excluding the five-minute graph) of Figures 14.1 and 14.2.

While the simple model of the above example is able to capture the self-similarity displayed in the Palm data, it fails to capture its breakdown, as indicated by the five-minute curve of Figure 14.1. In the following example we capture this breakdown by magnifying, during the final minute, each bidder's probability of departure (as in (14.14)). This magnification incorporates the greater likelihood of nontransmittal of bids during the final minute.

Example 2 (GPBP, BARISTA). We maintain the same setup used in the above example, except that $\pi(0)$ is doubled during the final minute ($\beta = 2$ in (14.14)). Figure 14.3 displays the empirical cumulative distribution functions of normalized left truncated bid times. Note the similarity of Figures 14.1 and 14.3.

REFERENCES

Aczel, J. (1966). *Lectures on Functional Equations and Their Applications.* New York: Academic Press.

Bajari, P. and Hortacsu, A. (2003). Cyberspace auctions and pricing issues: A review of empirical findings. In *New Economy Handbook* (Derek C. Jones, ed.). Academic Press.

Bapna, R., Goes, P., Gupta, A., and Jin, Y. (2004). User heterogeneity and its impact on electronic auction market design: An empirical exploration. *MIS Quarterly*, 28: 1:21–43.

Borle, S., Boatwright, P., and Kadane, J.B. (2006). The timing of bid placement and extent of multiple bidding: An empirical investigation using ebay online auctions. *Statistical Science*, 21(2): 194–205.

Ockenfels, A. and Roth, A.E. (2002). The timing of bidding in internet auctions: Market design, bidder behavior and artificial agents. *AI Magazine*, 23(3): 79–87.

Roth, A.E. and Ockenfels, A. (2000). Last-minute bidding and the rules for ending second-price auctions: Theory and evidence from a natural experiment on the internet. Technical report, NBER Working Paper No. 7729.

Roth, A.E. and Ockenfels, O. (2002). Last-minute bidding and the rules for ending second-price auctions: Evidence from ebay and amazon auctions on the internet. *The American Economic Review*, 92(4): 1093–1103.

Shmueli, G., Russo, R.P., and Jank, W. (2007). The BARISTA: A model for bid arrivals in online auctions. *Annals of Applied Statistics*, 1(2): 412–441.

Varian, H. (2000). Online auctions as a laboratory for economists to test their theories. *New York Times*, Nov. 16.

Wilcox, R.T. (2000). Experts and amateurs: The role of experience in internet auctions. *Marketing Letters*, 11(4): 363–374.

15

DYNAMIC SPATIAL MODELS FOR ONLINE MARKETS

WOLFGANG JANK

Department of Decision and Information Technologies, R.H. Smith School of Business, University of Maryland, College Park, Maryland

P.K. KANNAN

Department of Marketing, R.H. Smith School of Business, University of Maryland, College Park, Maryland

15.1 INTRODUCTION

Spatial models are statistical models that take into account the spatial proximity of observations. Until recently, the notion of space has always been viewed as *geographical space*, and with the availability of spatial information associated with traditional data sources (e.g., ZIP code information, GIS systems) spatial modeling has become much more popular. Spatial models can be thought of as an extension of classical statistical models: They have all the features of classical (e.g., regression) models; additionally, they use covariate information hidden in the geographical data. Spatial models are often based on the notion of distance-based similarity: If two observations lie in close proximity to one another in space, then the typical assumption in spatial modeling is that they are more similar than another observation that is farther away. Thus, the two observations are correlated in such a way that the closer they are, the more similar they are (Cressie 1993; Anselin 1988; Ripley 1998).

Given the initial interpretation of space as geographical space, traditionally spatial models have found applications in the earth sciences (e.g., mining: Journal and Huijbregts 1978). Spatial models are especially popular in biology or medicine

Statistical Methods in e-Commerce Research. Edited by W. Jank and G. Shmueli (e.g., to model the outbreak of diseases or the spread of viruses across geographical regions: Marshall 1991), they have found applications in social sciences (e.g., to model migration across parts of the country). The geographical interpretation of space has also seen applications in the area of marketing to explain retailer promotions (Bronnenberg and Mahajan 2001), customer satisfaction across geographical markets (Mittal et al. 2004), market rollout and retailer adoption of new brands (Bronnenberg and Mela 2004), and global diffusion phenomena (e.g., Albuquerque et al. 2007). More recently, the notion of space has also been interpreted more generally, e.g., as economic space (Slade 2004; Beck et al. 2006) or as the feature space associated with a product category (Jank and Shmueli 2005) or as demographic and psychometric space (e.g., Gao and Kannan 2007).

A natural extension of the marketing applications has been to the online markets. The widespread use of the Internet has led to online markets being a very popular channel for business transactions for many consumers all over the world. Online market players such as eBay, Amazon, or more traditional retailers (e.g., Best Buy, Circuit City) that also have a presence online, are posting rapid growth in their online sales. One of the big (perceived) advantages of online markets is that they are "on" day and night and are free of geographical boundaries. In particular, the independence of geographical boundaries has often led to the (wrong) conventional wisdom that online markets do not depend on geography. However, in reality, the shopping behavior of consumers online depends strongly on where they live. For example, Bell and Song (2008) show that a consumer's decision to adopt a new Internet service is affected by his interactions with other consumers who live near him, thus capturing the social contagion effect. Jank and Kannan (2005) argue that spatial models can be very effective in capturing variations in demand-side factors such as geographical characteristics and customer characteristics impacting preferences, product/service adoption, and consumption patterns resulting from variations in physical and psychological landscapes. They show that geographical data can act as a proxy for many such demand-side factors in explaining customers' differences in purchasing different forms of content online.

In this chapter, we demonstrate how spatial models can be quite useful in online contexts as e-commerce gathers steam, as more customers shop online, and as more data become available and are gathered online. We provide an overview of emerging applications where spatial models can be used very effectively to solve managerial problems. We then focus on a specific online application—online mortgage leads qualification—and apply both static and dynamic spatial models and empirically demonstrate the value of such models in lead targeting. The chapter is organized as follows. In Section 15.2 we discuss the emerging applications in the e-commerce realm where spatial models can play an important role. We then describe the specific application of targeting mortgage leads online, which will be the focus of our analysis, along with the data description. In Section 15.3 we discuss the model and estimation. Section 15.4 describes the results and empirically compares our approach with competing approaches, e.g., classical regression models/choice models.

We conclude in Section 15.5 with a consideration of directions for future research in the realm of spatial model applications for e-commerce.

15.2 SPATIAL MODELS IN E-COMMERCE

15.2.1 The Need

Much has been made about the "boundaryless" nature of online markets and how online markets break all types of geographical constraints restraining commerce. While some of this is true, especially in the context of supply-side factors (same products and services available at all locations, uniformity in information, and so on), there is much variation in the demand-side factors, which are still relevant even though the transactions are performed online. These demand-side factors are very important in many retail applications. For example, time-starved consumers who live in urban areas with significant traffic congestion problems are prime targets for online grocery shopping. Thus, online grocery ordering may occur mostly in high-income ZIP codes where the values of time and convenience may be more salient factors driving online purchasing. Even within large metropolitan areas, online purchasing may be more prevalent in higher-income, more educated neighborhoods than in inner cities. Geographical spatial data can be used as a proxy for many such variables—income, age, education, house size, age of the neighborhood, property values, and other such demographic differences, and will be able to capture some of the variances in online purchasing due to these variables. Thus, we see that spatial data can capture many of the customer taste variations exhibited online and are highly correlated with geography due to the factors mentioned above.

In addition to the above, geographical data may capture differences due to infrastructural variation that can impact e-commerce. For example, businesses selling products online (videos and online movies) might see more of a demand from customers with high-speed Internet connections, which can vary significantly across geographical markets. A recent Federal Communications Commission Section 706 report (March 2004) shows that nationwide in the (United States, about 10% of the ZIP codes do not have broadband availability. At the state level, this percentage varies from 0 to 32%. Also, these data deal only with availability (and not penetration—that is, the percentage of households subscribing to broadband), and there are reasons to believe that low-income ZIP codes will have lower penetration and higher-income neighborhoods higher penetration with higher affordability. This will certainly have an impact on customers' ability to transact business online. This is critical for emerging e-commerce applications such as online video streaming. It not only has an impact on the demand side but also has implications for online traffic management of content.

It is well known that geographical regions vary significantly in terms of percentage of highly educated persons, percentage of professionals, income, age, etc. People of similar background (lifestyle, life stage, ethnic background) also tend to colocate in terms of geography. Proprietary data services such as Spectra (www.spectramarketing.com), which sell market intelligence to retailers, classify consumer segments as upscale suburbs, traditional families, metro elite, working class, downscale urban, etc., and provide segment breakup in each ZIP code and market area, and these typically show high variance across ZIP codes. Such factors could contribute to variations of tech savvyness or technology readiness of consumers across geographical regions. In fact, the psychographic profiles of consumers residing in different ZIP codes, according to GeoVALS (which estimates the percentage of each VALS segment within each five-digit U.S. residential ZIP code and is marketed by a proprietary firm, SRI Consulting Business Intelligence, www.sric-bi.com), show a significant degree of variance in the segments that make up each ZIP code. For example, ZIP code 89510 has 10.5% of residents as Innovators, while ZIP code 20741 has 14.2% of residents as Innovators and ZIP code 20745 has only 8%. These data suggest that tech savvy-ness and technical readiness could well vary across geographical areas, and spatial data can serve as a useful proxy for capturing such variation.

While the demand variations that spatial models can capture (as discussed above) are very useful for existing e-commerce applications, there are many emerging applications where such data will be indispensable. These are geo-targeting applications that are making e-commerce increasingly *local* in a geographical sense. There are billions of dollars at stake in local advertising, specifically in local search. Search engines such as Google, Yahoo, and others are increasingly using their sites as local Yellow Pages to provide users with locally relevant content—and spatial models can be very useful in this context. Targeting content (be it search information, advertising, or product/service information) that is different for different consumers is the holy grail of e-commerce applications that can be made a reality with the help of spatial data and models. Thus, the trend in the online world is to move toward more and more localized features. Google, AOL, and Yahoo offer local search together with geographically varying sponsored links. Similarly, Overture offers a service called *local match* linking businesses with local customers. LiveDeal.Com also offers a local marketplace to consumers.

Another emerging application, apart from the search realm, is geo-targeting customers/users on the basis on their location, with the location established by GPS systems and/or mobile devices that customers use. For example, if a customer drives by a mall, advertisements from the retailers in the mall could be pushed to customers' GPS and mobile devices. Some GPS systems are already providing information/advertisement on Zagat-rated restaurants to their customers based on their locations. With 3G mobile devices, customers can surf online seamlessly using their mobile devices, with their geographic location captured easily. These applications will not only help online businesses touch customers wherever they are to provide information, advertisements, and promotion, but will also help them relate location data to customers' choice and buying behavior. Using spatial modeling techniques can help retailers improve the predictive power of their forecasting, targeting, and customer scoring models.

In the next subsection, we describe the geo-targeting application that we focus on in this chapter, which highlights the potential of spatial models for many emerging e-commerce applications.

15.2.2 Online Mortgage Leads

The geo-targeting application we focus on involves an online mortgage firm that has it own website, where potential customers visit and fill out preapplication forms, which capture basic data about the customers and their property on which the financing or refinancing is to be done. During the refinancing boom, the website attracted a large number of visitors who filled out in the pre-applications. Once this was done, the online mortgage firm would "qualify" each application based on the information provided. The qualified applications were then followed up by the mortgage agents through a phone call to the prospective customer to obtain more information and determine whether the customer was qualified to receive a mortgage finance/refinance loan. This process was a time- and effort-intensive process. Since the firm did not have a good process for screening the application for the time-consuming qualification process, agents spent a significant amount of time on leads that were not successful in terms of mortgage sales in the end. Thus, even though the firm had a large number of applications and spent a significant amount of time processing them, the number of loans ultimately granted was fairly low, making the customer acquisition process very expensive. The firm needed a customer *scoring* model that could predict the successful leads with a reasonable degree of accuracy to help to identify good applications from the pool of initial applications and select them for further processing based on the data collected in the initial application (see the next subsection). It was hypothesized that the online firm would have greater success in converting leads to actual sales when they incorporated spatial factors in their selection criteria and scoring model—those geographical areas where property values have risen significantly and/or houses at the appropriate age provide good prospects for debt consolidation through refinancing of homes. In this application, spatial data could act as a proxy for many geography related variables—income, education, age, house size, property values, etc.—and capture some of the variances in probability of a successful sale due to these variables, thus predicting the most promising leads.[1] The next subsection describes the data used for building the scoring model using a spatial formulation.

[1]The spatial data may also capture information about competition in a specific ZIP code. Typically, applicants are in their prequalification stage when submitting their application information. As a result, they may be comparing rates and shopping around. Even if an applicant is later qualified, he or she may go with another lender if the rate and terms are more competitive there. Since competition is related to geographical areas, the extent of competition may be captured by the ZIP code.

15.2.3 Data Description

The data consist of 5000 mortgage leads from 95 metropolitan statistical areas (MSAs) along the U.S. East Coast. Figure 15.1 displays the geographical scatter of these MSA's. Table 15.1 provides a list of MSA names and locations.

For each mortgage lead, the company records the following information: type of loan (TOL), amount the customer wishes to borrow (AWB), type of home (TOH), estimated market value (EMV), amount customer currently owes (ACO), gross household income (GHI), overall consumer credit (OCC), overall mortgage history (OMH), and the way in which the customer wishes to be contacted (CON). Table 15.2 summarizes these nine variables.

As a first step in investigating the spatial variation of the data, we break up the data by geographical region (see Table 15.4). In particular, the table shows the rate of success and the four numerical variables (amount a customer wishes to borrow, estimated market value, amount a customer currently owes, and gross household income), each broken up by region. We focus on the four numerical variable for simplicity only. The remaining categorical variables behave similarly. For each region,

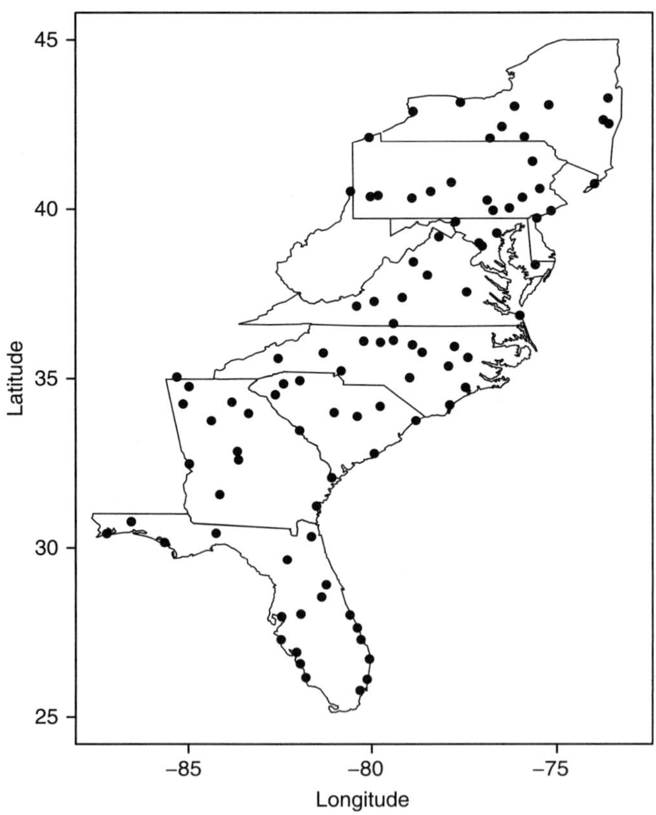

Figure 15.1 Ninety-five MSAs along the U.S. East Coast.

TABLE 15.1 Names and Locations of 95 MSAs Along the U.S. East Coast

Region	MSA
DC-VA-MD-WV	WASHINGTON-ARLINGTON-ALEXANDRIA
DE-MD-NJ	WILMINGTON
FL	CAPE CORAL-FORT MYERS
FL	DELTONA-DAYTONA BEACH-ORMOND BEACH
FL	FORT LAUDERDALE-POMPANO BEACH-DEERFIELD BEACH
FL	FORT WALTON BEACH-CRESTVIEW-DESTIN
FL	GAINESVILLE
FL	JACKSONVILLE
FL	LAKELAND
FL	MIAMI-MIAMI BEACH-KENDALL
FL	NAPLES-MARCO ISLAND
FL	ORLANDO
FL	PALM BAY-MELBOURNE-TITUSVILLE
FL	PANAMA CITY-LYNN HAVEN
FL	PENSACOLA-FERRY PASS-BRENT
FL	PORT ST.LUCIE-FORT PIERCE
FL	PUNTA GORDA
FL	SARASOTA-BRADENTON-VENICE
FL	TALLAHASSEE
FL	TAMPA-ST.PETERSBURG-CLEARWATER
FL	VERO BEACH
FL	WEST PALM BEACH-BOCA RATON-BOYNTON BEACH
GA	ALBANY
GA	ATHENS-CLARKE COUNTY
GA	ATLANTA-SANDY SPRINGS-MARIETTA
GA	BRUNSWICK
GA	DALTON
GA	GAINESVILLE
GA	MACON
GA	ROME
GA	SAVANNAH
GA	WARNER ROBINS
GA-AL	COLUMBUS
GA-SC	AUGUSTA-RICHMOND COUNTY
MD	BALTIMORE-TOWSON
MD	BETHESDA-FREDERICK-GAITHERSBURG
MD	SALISBURY
MD-WV	CUMBERLAND
MD-WV	HAGERSTOWN-MARTINSBURG
NC	ASHEVILLE
NC	BURLINGTON
NC	DURHAM
NC	FAYETTEVILLE
NC	GOLDSBORO

(*Continued*)

TABLE 15.1 *Continued*

Region	MSA
NC	GREENSBORO-HIGH POINT
NC	GREENVILLE
NC	HICKORY-LENOIR-MORGANTON
NC	JACKSONVILLE
NC	RALEIGH-CARY
NC	ROCKY MOUNT
NC	WILMINGTON
NC	WINSTON-SALEM
NC-SC	CHARLOTTE-GASTONIA-CONCORD
NY	ALBANY-SCHENECTADY-TROY
NY	BINGHAMTON
NY	BUFFALO-NIAGARA FALLS
NY	ELMIRA
NY	GLENS FALLS
NY	ITHACA
NY	NASSAU-SUFFOLK
NY	ROCHESTER
NY	SYRACUSE
NY	UTICA-ROME
NY-NJ	NEW YORK-WAYNE-WHITE PLAINS
PA	ALTOONA
PA	ERIE
PA	HARRISBURG-CARLISLE
PA	JOHNSTOWN
PA	LANCASTER
PA	LEBANON
PA	PHILADELPHIA
PA	PITTSBURGH
PA	READING
PA	SCRANTON-WILKES-BARRE
PA	STATE COLLEGE
PA	YORK-HANOVER
PA-NJ	ALLENTOWN-BETHLEHEM-EASTON
SC	ANDERSON
SC	CHARLESTON-NORTH CHARLESTON
SC	COLUMBIA
SC	FLORENCE
SC	GREENVILLE
SC	MYRTLE BEACH-CONWAY-NORTH MYRTLE BEACH
SC	SPARTANBURG
SC	SUMTER
TN-GA	CHATTANOOGA
VA	BLACKSBURG-CHRISTIANSBURG-RADFORD
VA	CHARLOTTESVILLE
VA	DANVILLE
VA	HARRISONBURG

(*Continued*)

TABLE 15.1 *Continued*

Region	MSA
VA	LYNCHBURG
VA	RICHMOND
VA	ROANOKE
VA-NC	VIRGINIA BEACH-NORFOLK-NEWPORT NEWS
VA-WV	WINCHESTER

TABLE 15.2 Numerical and Categorical Mortgage Lead Predictors

Variable	Avg	Median	St Dev	Min	Max
AWB	$188.14	$176.02	$90.54	$25.15	$449.96
EMV	$347.52	$324.44	$172.03	$37.73	$844.85
ACO	$294.05	$299.21	$174.84	$0.00	$598.90
GHI	$122.87	$118.46	$54.35	$25.00	$234.98

Variable	Code	Value	Count	Percent
TOL	1	New Home Purchase	383	7.66%
	2	Refinance	2320	46.40%
	3	Home Equity	1365	27.30%
	0	Help Me	932	18.64%
TOH	1	Single Family	2991	59.82%
	2	Mobile Home	369	7.38%
	0	Other	1640	32.80%
OCC	1	Excellent	2023	40.46%
	2	Good	1148	22.96%
	3	Fair	1389	27.78%
	0	Poor	440	8.80%
OMH	1	I do currently not have a mortgage	385	7.70%
	2	I have never been 30 days late	2709	54.18%
	3	I have been 30 days late	1031	20.62%
	4	I have been 60 days late	247	4.94%
	5	I have been 90 days or more late	367	7.34%
	0	I have been in foreclosure	261	5.22%
CON	1	Contact by phone	4104	82.08%
	0	Contact by email	896	17.92%

Note: The table lists all numerical (top) and categorical (bottom) predictor variables. The variables are type of loan (TOL), amount customer wishes to borrow (AWB), type of home (TOH), estimated market value (EMV), amount customer currently owes (ACO), gross household income (GHI), overall consumer credit (OCC), overall mortgage history (OMH), and the way in which the customer wishes to be contacted (CON). We report average, median, standard deviation, minimum, and maximum values for numerical variables and the variable distribution for categorical variables. The field "Code" denotes the variable coding in our subsequent analyses. Notice that the coding "0" corresponds to the baseline category in our regression models.

TABLE 15.3 Geographical Variation

Region	Success Rate	AWB	EMV	ACO	GHI
DC-VA-MD-WV	1.2094	0.0127	0.0129	−0.0022	0.1879
DE-MD-NJ	1.1296	−0.1799	−0.1800	0.1492	0.1342
FL	1.1350	0.0028	0.0028	−0.0073	0.0335
GA	−0.4292	−0.0670	−0.0670	0.0546	0.0061
GA-AL	−0.8267	0.0024	0.0022	−0.0241	−0.1340
GA-SC	−0.8267	0.0403	0.0403	−0.0313	−0.2583
MD	1.0939	0.0320	0.0322	−0.0051	0.1435
MD-WV	0.0677	0.0900	0.0900	−0.0772	−0.0566
NC	−0.6599	0.0505	0.0505	−0.0432	−0.0360
NC-SC	−0.7650	−0.0679	−0.0678	0.0995	0.0790
NY	−0.1555	−0.0161	−0.0162	0.0220	0.0079
NY-NJ	1.0605	0.3848	0.3849	−0.2870	0.1480
PA	−0.6348	0.0264	0.0264	−0.0267	−0.0586
PA-NJ	1.0200	−0.2179	−0.2179	0.1920	0.3983
SC	−0.5286	0.0055	0.0055	−0.0086	−0.0210
TN-GA	−0.8267	−0.2076	−0.2076	0.2454	−0.3284
VA	−0.1443	0.0030	0.0030	−0.0131	−0.0216
VA-NC	1.0666	−0.0413	−0.0412	0.0569	0.3025
VA-WV	1.1687	−0.1845	−0.1847	0.1215	−0.0286

Note: The table shows data summaries by geographical regions (as defined in Table 15.1). In particular, the table shows five variables: the success rate, the amount wished to borrow (AWB), the estimated market value (EMV), the amount customer currently owes (ACO), and the gross household income (GHI), each broken up by region. For each variable, we report the z-score relative to the mean and standard deviation across all regions.

we report the *z-score*, that is, the average value in that region, standardized by the mean and standard deviation across all regions. In that sense, each z-score value gives an idea of how much a particular region varies relative to all regions. For instance, the value 1.2094 in the first row indicates that for the DC-VA-MD-WV area, the rate of success is 120% higher than the average. The interpretation is similar for the remaining values.

Table 15.3 shows that variation in the success rate is quite large, much larger than the variation in the explanatory variables. We take this as evidence that the explanatory variables do not capture all of the information hidden in the data. Rather, knowledge of a customer's location captures additional important information and should be accounted for.

15.3 MODEL AND ESTIMATION

We propose a semiparametric approach to modeling the binary outcomes (success/failure). For the sake of exposition, we term this a *choice model* in keeping with the terminology used in extant literature. A semiparametric model has the advantage of making only minimal distributional assumptions. It also handles large databases with greater ease than computationally more involved parametric models. We estimate our model within the context of the expectation/maximization (EM) algorithm. The EM algorithm is a very popular tool for model-fitting. It is numerically stable and results in an improvement of the objective function in every iteration, thus making very efficient use of computing resources. Moreover, we can use EM to dynamically update the model, as we will demonstrate at the end of this section.

15.3.1 Semiparametric Spatial Choice Model

First, we introduce the basic principles of semiparametric models based on a simple example. We then apply these principles to derive a semiparametric spatial choice model. After that, we discuss modeling alternatives for the spatial component of that model.

15.3.1.1 Semiparametric Models. We consider the very flexible class of semiparametric models. These models combine the advantages of parametric and nonparametric models. Parametric models postulate a strict functional relationship between the predictors and the response. The functional form often allows for new insight into the relation between variables, such as in growth models, where new insight could be derived from the observation that the response grows at an exponential rate rather than at a polynomial one. The strict functional relationship imposed by parametric models, though, can sometimes be overly restrictive. This restriction can be alleviated using nonparametric approaches. Nonparametric models are very flexible and posit only minimal functional assumptions. As a consequence, they often provide a much better fit to the data. On the downside, nonparametric models typically do not provide much insight into the functional relationship between variables. In the above example, nonparametric models would not allow us to distinguish between exponential and polynomial growth. Semiparametric models can be thought of as a wedding between, parametric and nonparametric models, and they combine the advantages of both approaches.

Semiparametric models also provide computational advantages. They can handle large databases without much extra computing effort. This is in contrast to, say, Bayesian or hierarchical models, which often rely heavily on computer-intensive estimation procedures such as Markov chain Monte Carlo (MCMC) techniques. In fact, while semiparametric models can deliver answers in seconds or minutes, it may take hours or days for MCMC to terminate.

We describe the general principles of semiparametric models next. For illustrative purposes, we consider a simple scenario with one response variable, y, and only two predictor variables, x_1 and x_2. More details, also for more complex models, can be found in e.g., Ruppert et al. (2003). Similar to classical regression, we model the response in additive fashion, say as

$$y = \beta_0 + \beta_1 x_1 + f(x_2) + \varepsilon. \tag{15.1}$$

The function *f* describes the relationship between x_2 and y, and it is completely unspecified. It corresponds to the nonparametric part of the model. On the other hand, the relationship between x_1 and y is parametric. It is given by $\beta_0 + \beta_1 x_1$, which specifies a linear relationship between x_1 and y.

In order to estimate the function *f* from the data, one assumes a suitable set of basis functions. While many different choices are possible, a popular approach is to use *truncated line bases* (Wand 2003), which yields

$$f(x) = \vartheta_0 + \vartheta_1 x + \sum_{k=1}^{K} u_k (x - \tau_k)_+, \tag{15.2}$$

where $\tau_1, \ldots, \tau_K$ denotes a suitable set of knots and the function $(x - \tau_k)_+$, equals $x - \tau_k$ if and only if $x > \tau_k$, and it is zero otherwise. Instead of the truncated line basis one could also use the truncated polynomial basis, which leads to a smoother fit, or radial basis functions, which are often useful for higher-dimensional smoothing (Ruppert et al. 2003). Notice that the set of all parameters that needs to be estimated from the data is then given by $(\vartheta_0, \vartheta_1, u_1, u_2, \ldots, u_K)$. Using (15.2), the model in (15.1) becomes

$$y = \beta_0^* + \beta_1 x_1 + \vartheta_1 x_2 + \sum_{k=1}^{K} u_k (x_2 - \tau_k)_+ + \varepsilon, \tag{15.3}$$

where we set $\beta_0^* := (\beta_0 + \vartheta_0)$ to avoid two confounded intercepts in the model.

15.3.1.2 Spatial Choice Model. We propose a semiparametric approach to modeling spatially varying choice decisions or the success/failure rates in the context of our application. The model is semiparametric in that we model the dependence of choice on a customer's *observable* characteristics in the traditional, parametric form. We combine this with a flexible nonparametric model for the *unobservable*, spatially varying effects that also affect choice decisions.

Alternatives to the semiparametric approach exist. The most common alternative is to model both the observed and the unobserved data in a parametric fashion. An example is the work of Jank and Kannan (2005), who propose a parametric model for spatially varying choice decisions. In that model, the unobserved data are modeled with the help of random effects for which a suitable parametric distribution (typically the multivariate normal distribution) has to be specified. This parametric approach, though, leads to estimation difficulties. Indeed, the resulting model no longer has a closed-form solution, and simulation-based methods have to be used for parameter estimation. While in principle this is no problem, it can lead to long run-times, which our present approach avoids.

Let $\mathbf{z}_i = (z_{i1}, z_{i2})$, $i = 1, \ldots, N$, denote the spatial coordinate or location of the observed response y_i. Our model assumes that the response variable takes on only one of J values, $y_i \in \{1, 2, \ldots, J\}$ and that the y_i's are independent realizations of a multinomial random variable; that is,

$$y_i \sim \text{Multinomial}(\pi_{i1}, \pi_{i2}, \ldots, \pi_{iJ}), \tag{15.4}$$

where $\pi_{ij} = \text{Prob}(y_i = j)$ is the probability of choosing category j, $(j = 1, \ldots, J)$. Let J denote the *baseline* category. We then model the logit of π_{ij} as

$$\log\left(\frac{\pi_{ij}}{\pi_{iJ}}\right) = \mathbf{x}_i\beta_j + f_j(\mathbf{z}_i), \quad j = 1, \ldots, J - 1. \tag{15.5}$$

Equation (15.5) deserves some discussion. Notice that $\mathbf{x}_i$ is a $p \times 1$ vector of known covariates and β_j is a $p\times 1$ vector of unknown logit parameters associated with category j. In that sense, $\mathbf{x}_i$ and β_j are the traditional components of (standard) choice models. They also form the parametric part of the model. The spatial aspect of the model is captured in the (nonparametric) function $f_j(\cdot)$. This function is completely unspecified and depends only on the spatial location $\mathbf{z}$. It captures spatial dependencies in that for two locations $\mathbf{z}$ and $\mathbf{z}'$, if the locations are in close geographical proximity to one another (i.e., $\|\mathbf{z} - \mathbf{z}'\| \approx 0$), then the effect of these two locations on the choice j will be similar (i.e., $f_j(\mathbf{z}) \approx f_j(\mathbf{z}')$). Operationally, this is accomplished by choosing an appropriate spatial smoothing function. In our application we use the popular *radial smoothing splines* for $f_j(\cdot)$, but we are quick to point out that alternatives are also possible.

15.3.1.3 Spatial Smoothing. We model the dependence on the geographical location using a nonparametric approach based on radial smoothing splines. Let $\vartheta = (\vartheta_0, \vartheta_1, \vartheta_2)$ denote a parameter-vector and let $\{\tau_k\}_k$, $1 \leq k \leq K$ denote a set of knots in the two-dimensional space spanned by the geographical locations $\mathbf{z} = (z_1, z_2)$. A radial smoothing spline (Ruppert et al. 2003) is defined as

$$f(\mathbf{z}) = \vartheta_0 + \vartheta_1 z_1 + \vartheta_2 z_2 + \sum_{k=1}^{K} u_k C(r_k), \tag{15.6}$$

where, similar to equation (15.2), the set of all unknown parameters is $(\vartheta, u_1, u_2, \ldots, u_K)$. We let $r_k = d(\mathbf{z}, \tau_k)$, where $d(\cdot, \cdot)$ denotes the Euclidian distance. Notice that $C(\cdot)$ denotes the covariance function. Many different choices exist for the covariance function such as the family of Matérn or power covariance functions (Cressie 1993). We use $C(r) = r^2 \log|r|$, which corresponds to low-rank thin-plate splines with smoothness parameters set to 2 (French et al. 2001). For more on spatial modeling alternatives see, e.g., Cressie (1993), Olea (1999), or Ruppert et al. (2003).

15.3.2 Model Estimation and Dynamic Updating

In this section we discuss estimation of our semiparametric spatial choice model. First, we discuss the relationship between semiparametric models and mixed models. Then we use this relationship to derive an estimation approach based on the powerful EM paradigm. And lastly, we discuss how the EM algorithm can be used to produce updated predictions in a dynamic, forward-looking fashion.

15.3.2.1 Semiparametric and Mixed Models. Before going into details about the model estimation, we point to a clever and very useful connection between semiparametric models and mixed models. Specifically, one can estimate the semiparametric model within the very familiar mixed model setting. Mixed models (McCullogh and Searle 2000) have become a very popular toolset in the statistics literature, and methods to fit mixed models are readily available. Going back to the illustrative example, we can embed (15.3) within the mixed models framework by rewriting it suitably. Define the vectors

$$\beta^* = [\beta_0^*, \beta_1, \vartheta_1]^T \quad \mathbf{u} = [u_1, \ldots, u_K]^T. \tag{15.7}$$

In the mixed model context, β^* and $\mathbf{u}$ would be referred to as *fixed* and *random effects*, respectively. We define the design matrices for these effects as

$$\mathbf{x}^* = [1, x_1, x_2], \quad \mathbf{z}^* = [(x_2 - \tau_1), \ldots, (x_2 - \tau_K)]. \tag{15.8}$$

Then we can write the model in (15.3) in the familiar mixed model notation

$$y = \mathbf{x}^* \beta^* + \mathbf{z}^* \mathbf{u} + \varepsilon, \tag{15.9}$$

with the standard mixed model assumptions on the random effect $\mathbf{u} \sim$ MVN $(0, \sigma_u^2 \mathbf{I})$.

We can apply the above principles to write model (15.5) in mixed model notation. Let $\mathbf{y} = (y_1, \ldots, y_N)^T$ denote the vector of observed choice responses. Let $\mathbf{z}_i$ denote the location associated with response y_i, and let $\mathbf{x}_i$ be the corresponding vector of known covariates. Write

$$\mathbf{x}_i^* = [1, \mathbf{x}_i, \mathbf{z}_i]_{1\times(p+3)}, \quad \mathbf{z}_i^* = [C(r_k)]_{1\times K}. \tag{15.10}$$

Then (15.5) can be expressed as

$$\log\left(\frac{\pi_{ij}}{\pi_{iJ}}\right) = \mathbf{x}_i^* \beta_j^* + \mathbf{z}_i^* \mathbf{u}, \tag{15.11}$$

with $\beta_j^* = (\vartheta_j, \beta_j)$ and $\mathbf{u} = (u_1, \ldots, u_K)$. Notice that (15.11) is a special case of the well-known generalized linear mixed model (GLMM). We describe a maximum likelihood estimation procedure for this model next.

15.3.2.2 Parameter Estimation Via the EM Algorithm. The EM algorithm is a powerful tool in the context of unobserved or missing information. Let $\mathbf{y}$ denote the observed (or incomplete) data; let $\mathbf{u}$ denote the unobserved (or missing) data; and let $\boldsymbol{\theta}$ be the parameter of interest. EM finds the best estimate of $\boldsymbol{\theta}$ based on the *complete* data $(\mathbf{y}, \mathbf{u})$ using the principles of imputation. In addition, it produces an estimate $\hat{\mathrm{u}}$ of the unobserved data. In our context, the parameter of interest is $\boldsymbol{\theta} = (\beta_1^*, \ldots, \beta_J^*)$. On the other hand, the parameters u_k of the radial basis functions $C(\tau_k)$ in (15.6) are considered unobserved in our estimation framework. Using the powerful paradigm of EM, we obtain estimates for both, $\boldsymbol{\theta}$ and $\mathbf{u}$.

More formally, let $f(\mathbf{y}, \mathbf{u}; \boldsymbol{\theta})$ denote the joint distribution of the observed and unobserved data, $\mathbf{y}$ and $\mathbf{u}$, and indexed by $\boldsymbol{\theta}$. The goal is to find the maximum likelihood estimate of $\boldsymbol{\theta}$, that is, the value of $\boldsymbol{\theta}$ that maximizes the marginal likelihood of the observed data. Notice that the marginal likelihood is obtained by integrating out the random effects from the joint likelihood of the observed and unobserved data:

$$L(\boldsymbol{\theta}; \mathbf{y}) = \int f(\mathbf{y}, \mathbf{u}; \boldsymbol{\theta}) d\mathbf{u}. \tag{15.12}$$

Let $\hat{\boldsymbol{\theta}}$ denote the maximizer of (15.12). We obtain $\hat{\boldsymbol{\theta}}$ using the EM algorithm.

The EM algorithm (Dempster et al. 1977) is an iterative procedure to find the maximum of likelihood functions in incomplete data problems. Let $\boldsymbol{\theta}^{(l-1)}$ denote the current parameter value. Then, in the lth iteration of the algorithm, the E-step computes the conditional expectation of the complete data log-likelihood, conditional on the observed data and the current parameter value,

$$Q(\boldsymbol{\theta} | \boldsymbol{\theta}^{(l-1)}) = E[\log f(\mathbf{y}, \mathbf{u}; \boldsymbol{\theta}) | \mathbf{y}; \boldsymbol{\theta}^{(l-1)}]. \tag{15.13}$$

This conditional expectation is often referred to as the *Q-function* since it plays a central role in the EM algorithm. The lth EM update $\boldsymbol{\theta}^{(l)}$ maximizes the Q-function. That is $\boldsymbol{\theta}^{(l)}$ satisfies

$$Q(\boldsymbol{\theta}^{(l)} | \boldsymbol{\theta}^{(l-1)}) \geq Q(\boldsymbol{\theta} | \boldsymbol{\theta}^{(l-1)}) \tag{15.14}$$

for all $\boldsymbol{\theta}$ in the parameter space. This is the M-step. Given an initial value $\boldsymbol{\theta}^{(0)}$, the EM algorithm produces a sequence $\{\boldsymbol{\theta}^{(0)}, \boldsymbol{\theta}^{(1)}, \boldsymbol{\theta}^{(2)}, \ldots\}$ that, under regularity conditions (Boyles 1983; Wu 1983), converges to $\hat{\boldsymbol{\theta}}$.

The EM algorithm is a very popular tool due to its many unique properties. For instance, in contrast to many other optimization methods, it guarantees an increase in the likelihood function in every iteration of the algorithm. Also, since it operates on the log scale, it allows for significant analytical and numerical simplifications, especially for models in the exponential family. The only downside is that for GLMMs, the conditional expectation in (15.13) typically has no closed-form solution. In fact, in many applications this expectation yields an intractable integral of high dimension. Solutions to overcome an intractable E-step are plentiful. If the integral dimension is large, then Monte Carlo approaches are typically the solution of choice (Booth and Hobert 1999; Levine and Casella 2001; Jank 2004; Caffo et al. 2005). Fortunately in our case, the components of $\mathbf{u}$ are uncorrelated with one another, which yields only a one-dimensional integration problem. For this type of problem, analytical approximation is a fast and reliable solution. Analytical approximation is based on a Laplace approximation of the intractable integral. The Laplace method approximates an integral of the form $\int f(\mathbf{w}) d\mathbf{w}$ by fitting a Gaussian at the maximum $\hat{\mathbf{w}}$ of $f(\mathbf{w})$ and computing the volume under the Gaussian. For more details on this type of approximation see, e.g., Breslow and Clayton (1993). In our application we will use the Laplace approximation to the E-step in (15.13).

Using the above method, we obtain an estimate $\hat{\boldsymbol{\theta}}$ which maximizes (15.12). Using this estimate, we now obtain an estimate for $\mathbf{u}$ in the following way. The minimum mean squared error estimate of $\mathbf{u}$ is given by the conditional expectation $\hat{\boldsymbol{u}} = \mathrm{E}[\mathbf{u} | \mathbf{y}; \hat{\boldsymbol{\theta}}]$, which we evaluate using the Laplace approximation described above. This yields the pair $(\hat{\boldsymbol{\theta}}, \hat{\mathbf{u}})$, which completes the estimation process.

15.3.2.3 Dynamic Updating of Parameter Estimates. In many situations, information becomes available only successively. For instance, new customers arrive at an online business over the course of weeks and months. In order to make up-to-date predictions, every new piece of information has to be incorporated into the choice model as soon as it becomes available. In the following, we propose an online version of the EM algorithm to update the spatial model in real time.

Let $\mathbf{y}_{t-1} = (y_1, \ldots, y_{t-1})$ denote the training data observed over the previous $t-1$ time periods. We initialize the model by applying Monte-Carlo expectation maximization (MCEM) algorithm to $\mathbf{y}_{t-1}$ and obtain an initial estimate $\boldsymbol{\theta}_{t-1}$. The goal is to update $\boldsymbol{\theta}_{t-1}$ as soon as new information arrives.

Let $t, t+1, t+2, t+3, \ldots,$ denote the times at which new information becomes available and let y_t denote the information at time t. If we assume that in the short term the model parameters remain constant, statistical reasoning suggests that we obtain a more accurate model by incorporating all of the available information. Let $\mathbf{y}_t = (\mathbf{y}_{t-1}, y_t)$ denote all of the data available at time t, and let $\boldsymbol{\theta}_t$ denote the corresponding parameter estimate via MCEM.

We use this information to predict the next observation in the following way. A new customer arrives at time $t+1$. We predict this customer's choice using spatial prediction based on $\hat{\boldsymbol{\theta}} = \boldsymbol{\theta}_t$. Once we observe the customer's true choice y_{t+1}, we incorporate it into the training data and learn the model parameters all over again.

The procedure described above is a rather straightforward (yet effective) way of updating the model parameters. In other situations, more intricate updating schemes may lead to better performance. For instance, Jank and Kannan (2006) propose to update the model only on a smaller, more recent subset of all the data. Let $(y_1, \ldots, y_t)$ denote all of the information available until time t. Then, if there is reason to believe that earlier data carry less valuable information about the future, one may want to use a shorter sequence, say, $(y_{t0}, \ldots, y_t)$, $t_0 > 1$, to update the model parameters. While this approach may get rid of patterns that vanish over time, it is not obvious at all how to choose the value of t_0. An alternate (or additional) way of updating the model parameters is via a convex combination of earlier and more recent model parameters. Let $\boldsymbol{\theta}_{t-1}$ denote an estimate based on the earlier information, and let $\boldsymbol{\theta}^*$ denote the estimate based on the most recent information. Then, rather than taking $\boldsymbol{\theta}_t = \boldsymbol{\theta}^*$, one can use a weighing scheme that places less importance on the earlier information and up-weights the more recent information, e.g., $\boldsymbol{\theta}_t = \gamma\boldsymbol{\theta}^* + (1-\gamma)\theta_{t-1}$, where $\gamma \in (0,1)$. However, while this approach looks appealing and has some similarity to the *stochastic approximation* procedure (e.g., Kesten 1958), again it is not obvious how to optimally choose γ. (See also Jank (2006) for more on the difficulty of best choosing the weighing parameter γ.) In that sense, while our approach is very straightforward, it is easy to implement and, as we will see, results in an impressive performance.

In general, the specific feature of online learning can be very useful in contexts where customer preferences and behavior are quite dynamic, changing depending on factors that are related to their location and time of interaction, as highlighted in Section 15.2.1 in the chapter. Online learning allows us to update these changes almost in real time and thus be in sync with the dynamic environment.

15.4 EMPIRICAL APPLICATION AND RESULTS

In this section, we investigate the performance of the semiparametric spatial choice model applied to the online mortgage leads data described earlier. We first focus on the model fit and compare our model to the traditional nonspatial choice model. Then we investigate the predictive performance on a holdout sample. In particular, we investigate the predictive performance of the "static" spatial model (i.e., the model that use all the data at once and does not update its parameters) against the dynamic implementation via the online version of the EM algorithm.

15.4.1 Model Estimation

We first inspect our semiparametric spatial choice model and compare it to a traditional non-spatial choice model. Table 15.4 shows the model fit and the parameter estimates on the training data. We can see that the spatial model formulation results in a much better fit to the data. Moreover, we can see that the model's effects are, on average, much more significant in the spatial model. For instance, it is interesting to see that in the traditional choice model, only the second type of loan (TOL2) is

TABLE 15.4 Parameter Estimates on the Training Data

	Regular Choice Model			Spatial Choice Model		
Variable	Estimate	SE	P-Val	Estimate	SE	P-Val
(Intercept)	7.40	3.53	0.04	−0.08	0.00	0.00
TOL1	0.14	0.15	0.35	0.27	0.08	0.00
TOL2	0.23	0.10	0.02	0.28	0.03	0.00
TOL3	0.09	0.11	0.37	0.23	0.04	0.00
AWB	−1492.15	666.63	0.03	0.34	0.24	0.16
TOH1	−0.01	0.08	0.90	0.25	0.02	0.00
TOH2	−0.12	0.14	0.40	0.02	0.06	0.72
EMV	785.08	350.85	0.03	0.65	0.87	0.46
ACO	−0.92	0.64	0.15	−1.17	1.41	0.41
GHI	1.23	0.63	0.05	1.45	1.35	0.28
OCC1	0.05	0.13	0.70	−0.04	0.06	0.49
OCC2	0.01	0.14	0.96	−0.08	0.07	0.26
OCC3	−0.03	0.13	0.85	−0.15	0.07	0.03
OMH1	0.32	0.19	0.09	0.73	0.14	0.00
OMH2	0.08	0.15	0.59	0.08	0.09	0.37
OMH3	0.06	0.17	0.73	0.18	0.11	0.10
OMH4	0.29	0.21	0.17	−0.03	0.17	0.86
OMH5	0.10	0.20	0.63	−0.33	0.15	0.02
CON	0.04	0.09	0.65	0.38	0.03	0.00
Model fit						
Deviance		652.07			321.82	
AIC		690.07			487.35	

significant, while in the spatial model all four types of loans result in different success rates (or conversion rates), where we define "success" as a successful conversion from an initial application to a final loan.

More specifically, we can see from Table 15.4 that in the spatial model, refinances (TOL2) result in the highest conversion rates, followed closely by new home purchases (TOL1). Moreover, of the three different home types, the single-family home (TOH1) results in the highest conversion rates. Interestingly, different types of consumer credit (OCC) barely make a difference. On the other hand, mortgage history (OMH) does matter: Consumers with no existing mortgage have significantly higher conversion rates than those with existing mortgages. It is also interesting that the medium through which the consumer would like to be contacted matters, and phone follow-ups result in significantly higher conversions. And lastly, we note that none of the remaining variables result in a significantly different conversion rate: amount a customer wishes to borrow (AWB), estimated market value (EMV), gross household income (GHI), amount a customer currently owes (ACO).

15.4.2 Predictive Performance

While the insights from the estimated model are revealing, they do not convey how well the model *predicts* a future conversion. To do so, we conduct an analysis of the model's predictive performance. To that end, we divide the data randomly into a training sample (50% of the data) and a holdout sample (the remaining 50%). We calibrate all models on the training sample and check their predictive performance on the holdout sample. Both the static spatial and nonspatial models are checked in *batch mode*; that is, after calibrating the model on the training sample, we use it to predict on the *entire* holdout sample. For the dynamic model, we operate *sequentially*: We first predict the first observation on the holdout sample and check this prediction against the true value. We then augment the training data with the true observation and recalibrate the model. We use the new model to predict the second observation, check its prediction and move the true value back into the training data, and so on.

Table 15.5 shows the overall performance for the three models.[2] Notice that the traditional choice model (i.e., the nonspatial model) has a horrendous misclassification rate of almost 50%. Basically, the traditional choice model mimics a coin flip and predicts with almost equal probability a conversion as well as a nonconversion. This indicates that there is a lot of information hidden in the unobserved spatial dependencies. Consequently, the spatial model (operating in batch mode) performs much better: Its misclassification rate is only 33%. The dynamic version of the spatial model further improves upon the batch-mode model and unveils the power of sequential information updating: Its misclassification rate is only 13%. In other words, it improves upon the batch-mode model by over 60%!

[2]One may be able to further improve the predictive performance of each of these models by deriving new features from the data. However, the point here is to show the capabilities of a spatial approach based on the identical information.

TABLE 15.5 Misclassification Rates for Nonspatial, Static Spatial, and Dynamic Spatial Models

Model	Correct	Incorrect
Nonspatial	51%	49%
Static spatial	77%	33%
Dynamic spatial	87%	13%

Figure 15.2 further illustrates the performance of the dynamic spatial model. In that graph, we show the cumulative performance over different time periods. In fact, for each time period, we show the cumulative percentage of correctly predicted conversions at that time. We can see that as time progresses, the dynamic model learns and its predictions improve. In fact, its predictions improve from an early rather poor 74% correct classification rate to an excellent 87% at the end of

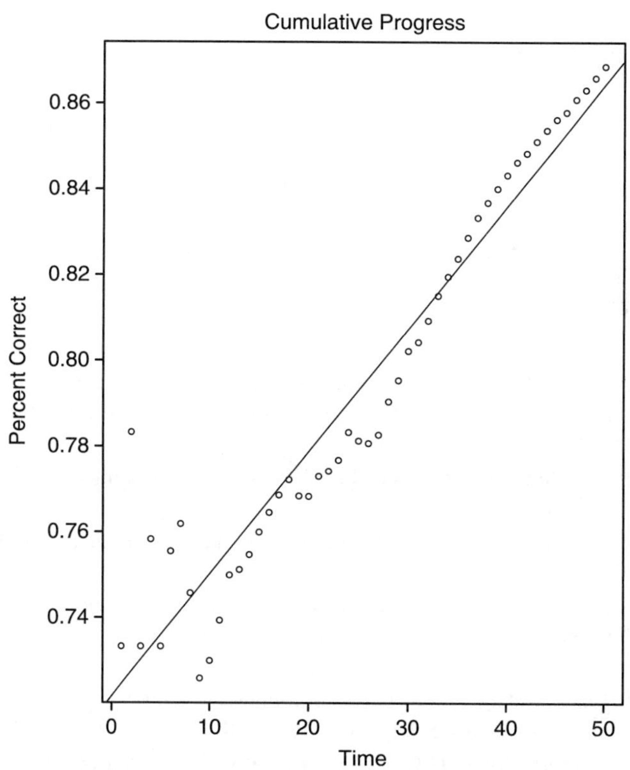

Figure 15.2 Cumulative progress of the dynamic model over time. The x-axis denotes time; the y-axis shows the cumulative number of correctly predicted conversions at that time. The solid line corresponds to a linear fit. The regression equation for this fit is Pct.Correct = 0.72 + 0.003 Time.

our experiment. The solid line illustrates the rate of improvement. In fact, for each additional time period, the dynamic spatial model learns, on average, at a rate of 0.03%.

15.5 CONCLUSIONS

In this chapter, we have highlighted the power of spatial models for online market applications. We have shown how the emerging online applications have the potential to generate gigabytes of spatial data, which can help explain to a significant extent customer online behavior, which otherwise cannot be captured easily. We have also illustrated the power of spatial models through their application to our online mortgage loans case. As the example, showed the dynamic versions of the model hold the greatest potential for explaining conversion rates, giving us a glimpse of the spatial models' potential in location-based geo-targeting applications.

REFERENCES

Albuquerque, P., Bronnenberg, B.J., and Corbett, C. (2007). A spatio-temporal analysis of global diffusion of iso certification. *Management Science*, 53: 451–468.

Anselin, L. (1988). *Spatial Econometrics: Methods and Models*. Kluwer Academic, Boston.

Beck, N., Gleditsch, K.S., and Beardsley, K. (2006). Space is more than geography: Econometrics in the study of political economy. *International Studies Quarterly*, 50(1): 27–44.

Bell, D.R. and Song, S. (2008). Neighborhood effects and trial on the internet: Evidence from online grocery retailing. *Quantitative Marketing and Economics*, forthcoming.

Booth, J.B. and Hobert, J.B. (1999). Maximizing generalized linear mixed model likelihoods with an automated Monte Carlo EM algorithm. *Journal of the Royal Statistical Society B*, 61: 265–285

Boyles, R.A. (1983). On the convergence of the EM algorithm. *Journal of the Royal Statistical Society B*, 45: 47–50.

Breslow, N.E. and Clayton, D.G. (1993). Approximate inference in generalized linear mixed models. *Journal of the American Statistical Association*, 88: 9–25.

Bronnenberg, B.J. and Mahajan, V. (2001). Unobserved retailer behavior in multimarket data: Joint spatial dependence in market shares and promotion variables. *Marketing Science*, 20: 284–299.

Bronnenberg, B.J. and Mela, C. (2004). Market rollout and retail adoption for new brands of non-durable goods. *Marketing Science*, 23: 500–518.

Caffo, B.S., Jank, W., and Jones, G.L. (2005). Ascent-based Monte Carlo EM. *Journal of the Royal Statistical Society, Series B*, 67: 235–252.

Cressie, N.A.C. (1993). *Statistics for Spatial Data*. New York: Wiley.

Dempster, A.P., Laird, N.M., and Rubin, D.B. (1977). Maximum likelihood from incomplete data via the EM algorithm. *Journal of the Royal Statistical Society B*, 39: 1–22.

French, J.L., Kammann, E.E., and Wand, M.P. (2001). Comment on ke and wang. *Journal of the American Statistical Association*, 96: 1285–1288.

Gao, J. and Kannan, P.K. (2007). Comparing apples with oranges: A spatial framework to control for biases in cross-cultural customer satisfaction measures. Technical report, Working Paper, Robert H. Smith School of Business, University of Maryland, College Park, MD 20742.

Jank, W. (2004). Quasi-Monte Carlo sampling to improve the efficiency of Monte Carlo EM. *Computational Statistics and Data Analysis*, 48: 685–701.

Jank, W. (2006). Implementing and diagnosing the stochastic approximation em algorithm. *Journal of Computational and Graphical Statistics*, 15: 1–27.

Jank, W. and Kannan, P.K. (2005). Understanding geographical markets of online firms using spatial models of customer choice. *Marketing Science*, 24: 623–634.

Jank, W. and Kannan, P.K. (2006). Dynamic e-targeting using learning spatial choice models. *Journal of Interactive Marketing*, 20: 30–42.

Jank, W. and Shmueli, G. (2005). Modeling concurrency of events in online auctions via spatio-temporal semiparametric models. *Journal of the Royal Statistical Society Series C*, 56: 1–27.

Journel, A.G. and Huijbregts, C.J. (1978). *Mining Geostatistics*. New York: Academic Press, London.

Kesten, H. (1958). Accelerated stochastic approximation. *The Annals of Mathematical Statistics*, 29: 41–59.

Levine, R.A. and Casella, G. (2001). Implementations of the Monte Carlo EM algorithm. *Journal of Computational and Graphical Statistics*, 10: 422–439.

Marshall, R.J. (1991). A review of methods for the statistical analysis of spatial patterns of disease. *Journal of the Royal Statistical Society Series C*, 154: 421–441.

McCulloch, C.E. and Searle, S.R. (2000). *Generalized, Linear, and Mixed Models*. Wiley.

Mittal, V., Kamakura, W.A., and Govind, R. (2004). Geographic patterns in customer service and satisfaction: An empirical investigation. *Journal of Marketing*, 68: 48–62.

Olea, R.A. (1999). *Geostatistics for Engineers and Earth Scientists*. Kluwer Academic Publishers, Boston.

Ripley, B.D. (1998). *Statistical Inference for Spatial Processes*. Cambridge University Press.

Ruppert, D., Wand, M.P., and Carroll, R.J. (2003). *Semiparametric Regression*. Cambridge: Cambridge University Press.

Slade, M.E. (2004). The role of economic space in decision making. *Annales d'Economie et de Statistique*, 77: 1–21.

Wand, M.P. (2003). Smoothing and mixed models. *Computational Statistics*, 18: 223–249.

Wu, C.F.J. (1983). On the convergence properties of the EM algorithm. *The Annals of Statistics*, 11: 95–103.

16

DIFFERENTIAL EQUATION TREES TO MODEL PRICE DYNAMICS IN ONLINE AUCTIONS

WOLFGANG JANK AND GALIT SHMUELI
Department of Decision and Information Technologies, R.H. Smith School of Business, University of Maryland, College Park, Mayrland

SHANSHAN WANG
Modeling and Analytical Services, DemandTec Inc., San Carlos, California

16.1 INTRODUCTION

Empirical research of online auctions has been growing steadily in the past several years. Online auctions are different from offline auctions in several important ways: They are usually much longer, bidders and sellers are anonymous, and the barriers of entry are much lower for both bidders and sellers. These differences lead to auction dynamics that can be very different from those in offline auctions. One important aspect of these dynamics is their effect on the auction price. In this chapter, we are particularly interested in the price path of an online auction and its dynamics, that is, the speed of the price increases and how it changes over the duration of the auction.

Within the rich empirical online auction literature, there has been very little study of price dynamics. Most price-related studies have focused on the final price alone. However, a series of recent papers by Jank and Shmueli and co-authors (Bapna et al. 2005; Hyde et al. 2006, 2008; Jank and Shmueli 2006, 2008; Shmueli and Jank 2006; Wang et al. in press) show that dynamics matter, that even auctions for the same product can have very different price paths and dynamics (Hyde et al.

Statistical Methods in e-Commerce Research. Edited by W. Jank and G. Shmueli

2008), and that incorporating the information contained in the price dynamics of an ongoing auction greatly improves the ability to forecast its final price (Wang et al. in press). In particular, Wang et al. (in press) find that the availability of dynamics greatly improves the forecasting error compared to powerful competitors such as double exponential smoothing. Furthermore, the relationship between the price path and other auction-related information (e.g., seller rating, auction duration, opening bid, and item properties) changes during the auction (Bapna et al. 2005). One example is the effect of the opening bid on the price at different times during the auctions. Bapna et al. (2005) and Shmueli and Jank (2006) find that although there is a positive relationship between the price and the opening bid at any point in the auction, the strength of this relationship declines as the auction progresses, implying that bidders derive less and less information from the opening bid.

In order to estimate the price path and its dynamics from the discrete observed bids, Jank and Shmueli take a functional data analytic approach. In that approach, the price path of each auction is represented by a smooth, continuous curve. The derivatives of this curve capture price dynamics: The first derivative captures the price velocity, indicating when the price increases fast and when the increase slows down. The second derivative captures price acceleration.

The estimation of smooth, continuous price curves is achieved via smoothing methods, as is customary in functional data analysis (FDA) (Ramsay and Silverman 2005). Wang et al. (in press) build upon this and identify a family of differential equation models (DEM) that parsimoniously capture auction price dynamics. However, it is not clear how to incorporate external auction-related information into the DEM. In this chapter, we build upon the work of Wang et al. (in press) and show how regression trees can be extended to model the relationship between price dynamics and other

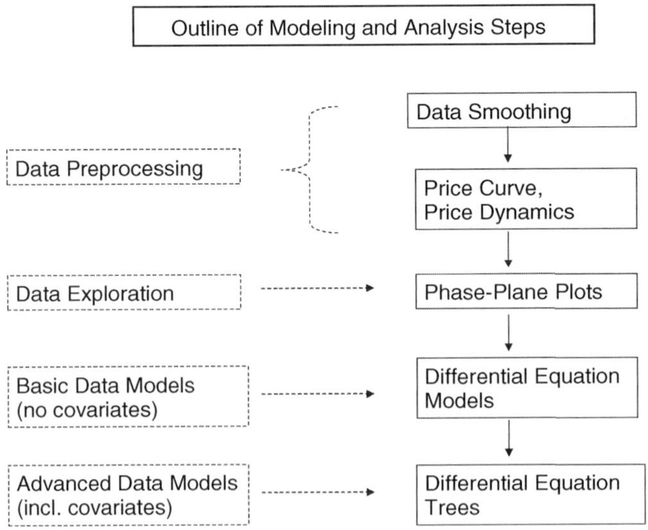

Figure 16.1 Road map for the remainder of this chapter.

auction-related variables. Specifically, we propose a novel tree-based approach for DEM based on recursive partitioning. We want to point out that this chapter is rather technical in that its primary focus is on developing methodology for differential equation trees.

The road map for the remainder of the chapter is as follows (see also Figure 16.1). We first describe online auction data and the features that make them amenable to FDA models. We then describe the steps necessary for identifying and fitting a DEM to auction data. These include data smoothing to create a smooth functional object in the first step. The smooth functional object allows for a gauging of the price dynamics. We describe a rather novel approach of exploring price dynamics via phase-plane plots. The observed price dynamics are our main motivation for studying DEMs. We describe the general nature of DEMs and their shortcomings in that they do not allow for easy incorporation of external, non-price-related covariates. This shortcoming is addressed via our new methodology of differential equation trees. We describe our method and illustrate it on a dataset of eBay auctions.

16.2 DATA

The data used in this chapter consist of closed seven-day auctions for two different products, Microsoft Xbox gaming systems and the book *Harry Potter and the Half-Blood Prince*. All auctions were transacted and included at least two bids. Xbox systems were popular items on eBay at the time of data collection and had a market value of \$179.98 (based on Amazon.com). The Harry Potter books were also very popular items and sold for \$27.99 on Amazon.com. In that sense, we can

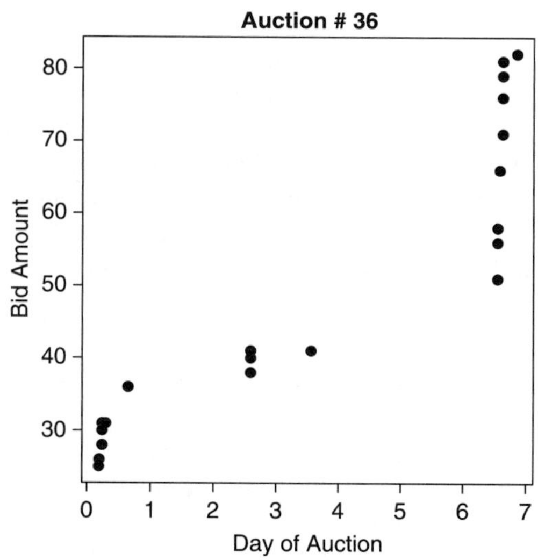

Figure 16.2 The bids placed in auction number 36 of a Microsoft Xbox auction. The horizontal axis denotes time (in days); the vertical axis denotes the bid amount (in £).

TABLE 16.1 Summary Statistics for All Continuous (Top) and Categorical (Bottom) Variables

Variable	Mean	Median	Min	Max	StDev.
Opening bid	19.84	5.99	0.01	175.00	31.07
Winning bid	71.78	17.75	7.00	405.00	76.99
Number of bids	14.12	11.00	2.00	75.00	11.05
Seller rating	280.00	85.5	0	9515.00	829.96
Bidder rating	61.72	34.23	0	758.00	85.31

Variable	Case	Count	Proportion (%)
Value	High	93	48.95
	Low	97	51.05
Reserve price	Yes	5	2.63
	No	185	97.37
Condition	New	60	31.58
	Used	130	68.42
Early bidding	Yes	81	42.63
	No	109	57.37
Jump bidding	Yes	34	17.89
	No	156	82.11

consider Xbox systems high-valued items and can contrast them with the lower-valued Harry Potter books.

For each auction, we collected the bid history, which reveals the temporal order and magnitude of bids and forms the basis of the DEM. Figure 16.2 shows an example of a typical bid history for an Xbox auction. In addition to the bid history, we collected information on a wide variety of other auction characteristics, such as the opening bid and the final price, the number of bids, and the seller and bidder ratings (summary statistics of these continuous variables are given at the top of Table 16.1). We also recorded item condition (used vs. new), whether or not the seller set a secret reserve price, and whether or not the auction exhibited early bidding or jump bidding (see the summary statistics at the bottom of Table 16.1). For further details on these data, see Wang et al. (in press).

16.3 PRICE CURVES AND DIFFERENTIAL EQUATION MODELS

Although observed bidding data are discrete, we prefer a smooth, continuous representation. The reason for this is that, from a conceptual point of view, price during an auction is a continuous process. Moreover, from a practical point of view, we want to estimate price dynamics, which can be done by calculating derivatives of the price process. For that reason, we also prefer a smooth representation. This lends itself to the use of FDA. In FDA the object of analysis is a continuous (functional) object rather than a scalar or vector. In recent years, there has been a surge of research on FDA (mainly due to the monographs by Ramsay and Silverman 2002, 2005), both in application areas and in theoretical research. FDA is especially suitable in the online auction context, since our object of interest is the price path and we have many replications of the same (or similar) path (i.e., a sample of auctions for the same or a similar product). The standard approach for estimating a functional object from discrete data is via smoothing, which we describe next.

16.3.1 Fitting Price Curves Via Smoothing

Our goal is to measure, for each auction, the dynamics of its associated price process. To that end, we first need a smooth representation of the process itself. We refer to this object as the *smooth functional object*. After creating the functional object, we obtain estimates for its dynamics via the first and second derivatives.

More specifically, let $\tilde{y}_i^{(j)}$ denote the jth ($j = 1, \ldots, n_i$) bid in auction $i (i = 1, \ldots, N)$ on day t_{ij} $(0 \leq t_{ij} \leq 7)$.[1] Note that because the bids arrive at irregularly spaced times, the t_{ij}'s (and n_i's) vary from one auction to another. To account for the irregular spacing, we sample the current price step function at a common set of time points t_j, $0 \leq t_j \leq 7$, $j = 1, \ldots, n$. Thus, the observed price path for auction i can be represented by a vector of fixed length n:

$$y_i(t) = (y_i^{(1)}, \ldots, y_i^{(n)}), \tag{16.1}$$

where $t = (t_1, \ldots, t_n)$ and $y_i^{(j)} = y_i(t_j)$ denotes the value of the bid sampled at time t_j.

In order to arrive at a smooth representation, we approximate y_i using basis functions. We write

$$y_i(t) = f_i(t) + \epsilon_i(t), \tag{16.2}$$

where the error term $\epsilon_i(t)$ is assumed to be the only cause of roughness for an otherwise smooth object. Using an appropriate basis functions expansion, we can represent $f_i(t)$ as

$$f_i(t) = \sum_{k=1}^{K} c_{ik}\phi_k(t) \tag{16.3}$$

for a set of known basis functions $\phi = (\phi_1(t), \ldots, \phi_K(t))$ and a coefficient vector $c_i = (c_{i1}, \ldots, c_{iK})^T$. Then the $K \times N$ estimated coefficient matrix $\hat{c} = (\hat{c}_1, \ldots, \hat{c}_N)$

[1]In our application, bid values are transformed into log scores to better capture common price surges, especially toward the end of the auction.

minimizes the penalized sum of squares

$$PENSSE_{\lambda}(c) = \sum_{i=1}^{N}\sum_{j=1}^{n}(y_i(t_j) - f_i(t_j))^2 + \lambda\int(Lt)^2 dt. \tag{16.4}$$

Using $Lt = f''(t)$, $\hat{c}$ is given by

$$\hat{c} = (c^T c + \lambda K)^{-1} c^T Y(t), \tag{16.5}$$

where c is the $n \times K$ basis matrix, $Y(t)$ is the $n \times N$ matrix of responses, and λ is a smoothing parameter that controls the trade-off between data fit and smoothness. Note that the elements of K are given by $K_{kl} = \int c_k''(t)c_l''(t)dt$. In this work, we use B-splines of order 6 to allow for a reliable estimation of at least the first three derivatives of f. The selection of the knots and the smoothing parameter is driven by visual inspection of the resulting functional objects.[2]

The left panel in Figure 16.3 shows the smooth price curves for the 190 auctions in our dataset. As mentioned earlier, an advantage of having smooth price curves is that we can readily obtain estimates for their dynamics via their derivatives. First and second derivatives of the price curve correspond to the price velocity and price acceleration, respectively. The middle panel in Figure 16.3 shows that most price velocities are close to zero, especially during midauction, implying a process with linear growth. In contrast, velocities are often very high at the start of the auction and especially at the end. However, the magnitude of the dynamics differs quite significantly from auction to auction. (This can also be seen from the range of the price accelerations in the right panel of Figure 16.3.) Thus, auction dynamics can be quite heterogeneous.

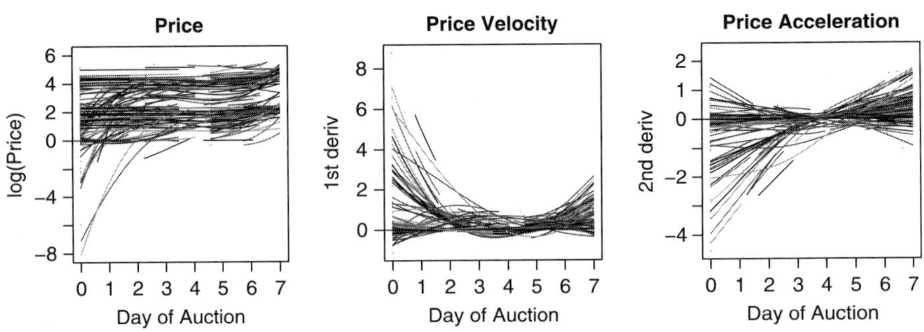

Figure 16.3 Price curves for the 190 seven-day auctions on Xbox play stations and Harry Potter books, together with their estimated first two derivatives.

[2]For more on the quality of fit of the smooth curves to the observed data, see Wang et al. (in press).

In the following, we use *phase-plane plots* to further investigate auction dynamics. In particular, we use these plots to investigate the relationships between dynamics of different order. This investigation will motivate our subsequent efforts to model dynamics via differential equations.

16.3.2 Exploring Auction Dynamics Via Phase Plane Plots

At the heart of differential equations are models that relate the function and its derivatives to one another. A preliminary step to choosing an appropriate differential equation model is to examine plots of pairs of derivatives, one versus the other. Such plots are called *phase-plane plots* (PPPs). In the functional context, where one has repeat observations at each derivative level, we plot the *averages* versus one another.

Figure 16.4 shows a PPP for average price acceleration versus the average price velocity. The numbers along the curve indicate the day of the auction (for seven-day auctions). We see that price velocity is high at the beginning of the auction and then the dynamics slow down: Price acceleration becomes negative, and there is a slowdown in price velocity. This continues until about day 4, after which the dynamics reverse: Acceleration becomes positive and velocity increases rapidly until the end of the auction.

There are several interesting aspects that appear in Figure 16.4. First, the C shape of the PPP is typical of an online auction: a phase of decrease in dynamics followed by a transitional phase of change and finally a phase of increase in dynamics. Second, the magnitude/importance of each phase varies as a function of different auction

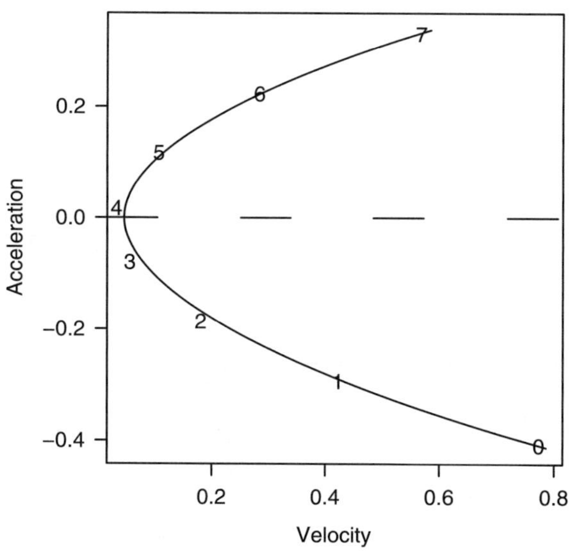

Figure 16.4 PPP for the average price curve of the data: the second derivative (acceleration) versus the first derivative (velocity).

characteristics. This can be seen in the series of *conditional* PPPs in Figure 16.5, where the average derivatives are conditional on the auction characteristics described in Table 16.1. Here, pairs of plots can be compared to see the effect of the two different levels (e.g., new vs. used items in the top left or high vs. low bidder ratings in the

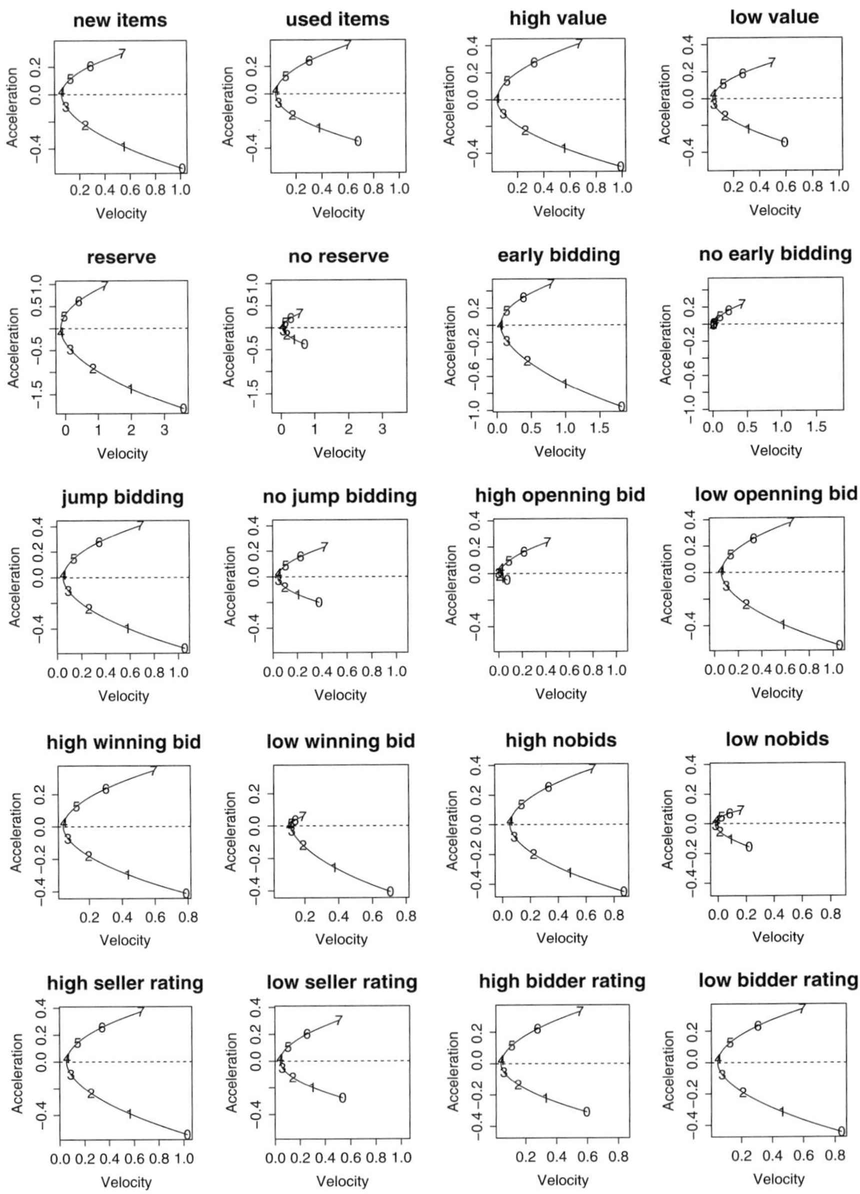

Figure 16.5 Conditional PPP for the average price curve of the data, conditional on 10 auction characteristics from Table 16.1.

bottom right). While the general C shape persists in all PPPs, the *magnitude* of the dynamics varies. Moreover, the different size of the C shapes also indicates that the magnitude of the *relationship* between velocity and acceleration differs.

To summarize the findings on the relationship between price dynamics and individual auction-related variables, see Figure 16.5. For item value, high-valued items appear to have a larger range of dynamics compared to lower-valued items. Early bidding seems to have a large effect on the dynamics not only at the start of the auction but throughout the entire auction. For jump bidding, we see that auctions that experience jump bids have a different relationship between velocity and acceleration than auctions without jump bids. Additional observations are that the opening bid and the number of bidders both have an impact on the dynamics. Interestingly, while different bidder ratings do not appear to make much of a difference, seller ratings do. Note, however, that several of these variables are correlated (e.g., closing price and item value). Our goal is therefore to look at the effect of the combined information on price dynamics. For this purpose, we formulate a differential equation model that relates price dynamics to a set of predictor variables.

The previous analysis shows that dynamics exist and that heterogeneity among auction dynamics is due to different auction subpopulations determined by the product, the auction format, the seller, or the bidders. Moreover, while the dynamics vary, we see a persistent C shape for the relationship between acceleration and velocity. We take this as motivation that online auction dynamics can be captured by a single family of DEMs. In the following, we derive such a class of DEMs.

16.3.3 DEMs for Price

Differential equations are widely used in engineering and physics. A differential equation describes a process with changing dynamics by specifying relationships among the function and its derivatives. In the context of online auctions, we view the price process as a dynamic system with many observed and unobserved factors acting upon it. Our exploratory analysis using PPPs indicates a general structure of price dynamics, and we now set out to find a DEM that can capture this structure.

Let y_i be the price function for auction i ($i = 1, \ldots, N$) and let $D^m y_i$ be the mth derivative of y_i. Our goal is the identification of a linear differential operator (LDO) of the form

$$L = \omega_0 I + \omega_1 D + \cdots + \omega_{m-1} D^{m-1} + D^m \tag{16.6}$$

that satisfies the homogeneous linear differential equation $Ly_i = 0$ for each observation y_i. In other words, we seek a linear differential equation model so that our data satisfy

$$D^m y_i = -\omega_0 - \omega_1 D y_i - \cdots - \omega_{m-1} D^{m-1} y_i. \tag{16.7}$$

In practice, due to the prevalence of noise, it is impossible to find a model that satisfies (16.7) *exactly*. Hence, principal differential analysis adopts a least squares approach to the fitting of the DEM. For details of parameter estimation, see Wang et al. (in press).

Next, we focus on a second-order differential equation, since the PPPs indicated varying relationships between the first and second derivatives of price. The general second-order differential equation is of the form

$$Ly_i = \omega_0 y_i + \omega_1 D y_i + D^2 y_i = 0. \tag{16.8}$$

However, to obtain a strictly monotone, twice-differentiable function as required from a price process in online auctions, we must set $\omega_0 = 0$ (Ramsay 1998). This leads to the differential equation

$$Ly_i = \omega D y_i + D^2 y_i = 0. \tag{16.9}$$

Note that coefficient function $\omega^* = -D^2 y/Dy$ measures the relative curvature of the monotone function in the sense that it assesses the size of the curvature of $D^2 y$ relative to the slope Dy. A constant value of ω^* implies an exponential process of the form $y(t) = C_0 + C_1 \exp(\omega t)$, with $\omega^* = 0$ the special case of a linear function. Thus, small or zero values of ω^* correspond to locally linear functions, whereas very large values correspond to regions of sharp curvature. In mechanical systems, the latter type is generally caused by internal or external frictional forces or viscosity. In the context of online auctions, sharp curvature in the price process can be related to jump bids caused by bidders attempting to apply external force to the bidding process, and locally linear motion is observable during the middle of the auction.

Next, we estimate the second-order differential equations model for our 190 auctions (for details on model fit, see Wang et al. in press). Figure 16.6 shows the estimated coefficient curve ω^*. We can see that ω^* has three phases: negative, zero, and finally positive. These correspond to the three typical bidding phases during an auction: early activity, little midauction activity, and high late activity. Recall that a value of zero indicates linear motion of the price process (i.e., no dynamics), whereas large positive or negative values are indicative of changes in the dynamics (oppressing them or increasing them, respectively). The first phase (up to day 3) is characterized by a negative ω^*, with a dip on day 2. This negative dip marks the change from early bidding to *bidding draught*, when velocity decreases. Then we see that $\omega^* = 0$ during the bidding draught, until price starts to increase again with a peak on day 6, in transition to high-intensity last-moment bidding.

In summary, we find that a second-order differential equation model fits the data reasonably well. It captures the three typical phases of bidding and the interplay of dynamics that change over the course of the auction. However, Wang et al. (in press) also find that the degree of model fit varies at different periods of the auction and that a considerable amount of variation is remained unexplained. The next step is therefore to incorporate external auction-related information.

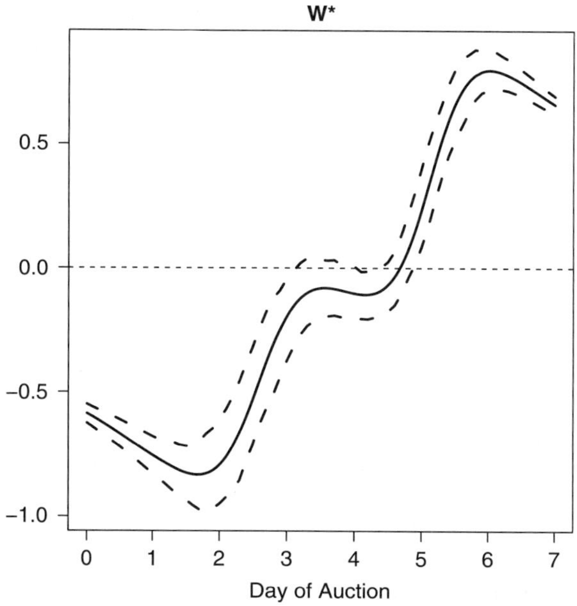

Figure 16.6 Estimated coefficient function of the monotone second-order differential equation fitted to online auction data.

Wang (in press) fit separate models to different levels of the categorical covariates and find differences between pairs of models. However, we strive for a more general approach that allows for the incorporation of covariate information directly into the dynamic model. For that purpose, we develop a novel approach based on regression trees. More specifically, we propose a new modeling approach for dynamic data via functional differential equations trees.

16.4 FUNCTIONAL DIFFERENTIAL EQUATION TREES

The previous section shows that differential equation models can capture the changing dynamics in online auctions but also that a significant amount of the variation in the dynamics is left unexplained. The road map for the rest of this chapter is thus as follows. In the following, we explain some of the residual variation using covariate information. In the auction context, plenty of covariate information is available (see, e.g., Table 16.1). However, incorporating covariate information into DEMs is not straightforward. Ramsay and Silverman (2005) suggest a way of incorporating covariate information via the forcing function, thereby creating a nonhomogeneous differential equation, but this approach has not received much traction to date. We thus propose an alternative and innovative approach, borrowing ideas from recursive partitioning. In particular, we propose a novel tree-based approach for DEMs. We discuss how to estimate differential equation trees, and we subsequently illustrate their performance on our online auction data. We are quick to point out that our description is rather technical and that the data example is for illustrative purposes only. The full power of our approach still remains to be investigated.

Tree models give simple descriptions of often complex nonlinear relationships between several predictors and a univariate or multivariate response. A classical reference is the monograph *Classification and Regression Trees* by Breiman et al. (1984). While tree models are generally very powerful, they encounter problems when the response is high dimensional. Yu and Lambert (1999) explore two ways of fitting trees to high-dimensional data. Both approaches proceed by first reducing the dimensionality of the data and then fitting a standard multivariate tree to the reduced response. In the first approach, the dimensionality is reduced by representing the response as a linear combination of spline basis functions. In the second one, the dimensionality is reduced using principal component analysis, retaining only the first several principal components.

Note that classical trees fit a scalar in each node of the tree. Since this tends to produce rather large and complex trees (see, e.g., Chan and Loh 2004), research on incorporating (simple) parametric models into trees has recently received considerable attention. Such approaches are often referred to as *functional trees* (Gamma 2004), with the most notable being *M5* (Quinlan 1993). (See also Kim and Loh 2001; Loh 2002; Chan and Loh 2004 for related approaches.) One particularly noteworthy approach is that of Zeileis et al. (2005), who take the integration of parametric models into trees one step further by embedding recursive partitioning into the model estimation and variable selection framework. Within that framework, every leaf is associated with a conventionally fitted model, such as a maximum likelihood model or linear regression. The model's objective function is used for estimating the parameters as well as the split points. The appeal of this approach is that the same objective function is used for partitioning and for parameter estimation. Building upon these ideas, we propose a novel model-based functional differential equation tree which allows the incorporation of covariate information into dynamic models. We first briefly review the main ideas of model-based recursive partitioning. We then use these ideas to derive our functional differential equation tree methodology.

16.4.1 Model-Based Recursive Partitioning

Let $M(y, \theta)$ be a parametric model where $y = (y_1, \ldots, y_N)$ are (possibly vector-valued) observations and $\theta \in \Theta$ is a k-dimensional vector of parameters. We assume that $M(\cdot)$ can be estimated by minimizing some objective function, say, $\Psi(y, \theta)$, yielding the parameter estimate $\hat{\theta}$, where

$$\hat{\theta} = \text{argmin}_{\theta \in \Theta} \Psi(y, \theta). \tag{16.10}$$

Estimators of this type are based on many well-known estimation techniques, the most popular ones being ordinary least squares (OLS) and maximum likelihood

(ML). In the case of OLS, Ψ is given by the error sum of squares, and in the case of ML it is the negative log-likelihood.

Let $(Z_1, \ldots, Z_L)$ be a set of partitioning variables (i.e., covariates). We assume that there exists a partition $\{\beta_b\}_{(b=1,\ldots,B)}$ of the space $Z = Z_1 \times \cdots \times Z_L$ into B cells (or segments) such that in each cell β_b, a model $M(y, \theta_b)$ with a cell-specific parameter θ_b holds.

The basic idea of model-based recursive partitioning is now that each node is associated with a single model. In the first step, the designated model $M(y, \theta)$ is fit to all observations by estimating $\hat{\theta}$ via minimization of the objective function Ψ; this yields a model in the top node. Then in the second step, a fluctuation test for parameter instability is performed to assess whether splitting of the node is necessary. Generally speaking, determining the necessity of a split based on a partitioning variable Z_l is based on comparing the estimated parameters of the post-split resulting daughters to determine whether they come from the same mean (and thus the split is unnecessary) or not. If there is significant parameter instability with respect to any of the partitioning variables Z_l $(1 \leq l \leq L)$, then we select the variable Z_l that is associated with the highest parameter instability. In the third step, we compute the split point(s) of Z_l that locally optimize Ψ. Finally, we split the node into B locally optimal segments and repeat the procedure. If no more significant instabilities can be found, the recursion stops and returns a tree where each terminal node is associated with a model of type $M(y, \theta_b)$.

The steps of the algorithm are as follows:

1. Fit the model $M(y, \theta)$ to all observations in the current node by estimating $\hat{\theta}$ via minimization of the objective function Ψ.
2. Assess the stability of the parameters w.r.t. every ordering $Z_1, \ldots, Z_L$. If there is some overall instability, choose the variable Z_l associated with the highest parameter instability for partitioning; otherwise, stop.
3. Search for the locally optimal split point(s) in Z_l by minimizing the objective function of the model ψ.
4. Split the node into daughter nodes and repeat the procedure.

The details for steps 1–3 are described below. To keep notation simple, dependence on the current segment $b \in \{1, \ldots, B\}$ is suppressed.

16.4.1.1 Testing for Parameter Instability. We assess parameter instability adopting the method of Zeileis et al. (2005). The basic idea is to check whether the score functions $\hat{\psi}_i$ ($\hat{\psi}_i = \hat{\psi}(y_i, \hat{\theta})$, where $\psi = \frac{\partial \Psi(y, \theta)}{\partial \theta}$) fluctuate randomly around their mean of 0 or exhibit systematic deviations from 0 over Z_l. These deviations can be captured by the empirical fluctuation process

$$W_l(t) = \hat{J}^{-1/2} n^{-1/2} \sum_{i=1}^{\lfloor nt \rfloor} \hat{\psi}_{\sigma(z_{il})}, \; (0 \leq t \leq 1), \tag{16.11}$$

where $\sigma(Z_{il})$ is the ordering permutation which gives the antirank of the observation Z_{il} in the vector $Z_l = (Z_{1l}, \ldots, Z_{nl})^T$. Thus, $W_l(t)$ is simply the partial sum process of the scores ordered by the variable Z_l, scaled by the number of observations n and a suitable estimate $\hat{J}$ of the covariance matrix $COV(\psi(y,\hat{\theta}))$, e.g., $\hat{J} = n^{-1}\sum_{i=1}^{n} \psi(y_i, \hat{\theta})\psi(y_i, \hat{\theta})^T$. This empirical fluctuation process is governed by a functional central limit theorem under the null hypothesis of parameter stability: It converges to a Brownian bridge. We can derive a test statistic by applying a scalar functional $\lambda(\cdot)$ capturing the fluctuation in the empirical process to the fluctuation process $\lambda(W_l(\cdot))$ and determine its limiting distribution. Then, in order to test whether there is instability in the current node, we check whether the observed significance level falls below a certain threshold α (e.g., $\alpha = 5\%$), and we consequently split the variable Z_l with the smallest p-value. In the following, we describe this process in more detail for assessing numerical and categorical variables.

Assessing Numerical Variables. For capturing instabilities of a numerical variable Z_l, we use the sup*LM* statistic proposed by Andrews (1993):

$$\lambda_{\sup LM}(W_l) = max_{i=\underline{i},\ldots,\bar{i}} \left(\frac{i}{N}\frac{N-i}{N}\right)^{-1} \left\| W_l\left(\frac{i}{N}\right) \right\|_2^2. \tag{16.12}$$

This statistic returns the maximum of the squared L_2 norm of the empirical fluctuation process scaled by its variance function. This type of statistic first appeared in Andrews (1993); it can be interpreted as the *LM* statistic[3] against a single change point alternative where the potential change point is shifted over the interval $[\underline{i}, \bar{i}]$. The interval is defined by requiring some minimal segment size $\underline{i}$ and then $\bar{i} = N - \underline{i}$. The limiting distribution of (16.12), as shown in Andrews (1993), is given by the supremum of a squared, m-dimensional, tied-down Bessel process $\sup_t(t(1-t))^{-1}||W^0(t)||_2^2$, where W^0 denotes a Brownian bridge.

Assessing Categorical Variables. For capturing the instabilities of a categorical variable, we need a different statistic. A categorical variable Z_l with G levels (or categories) has ties, and a total ordering of the observations is not available. We therefore need a different statistic. An appropriate statistic is one that is insensitive to the ordering of the G levels and of the ordering of observations within each level. One such statistic is given by Hjort and Koning (2002):

$$\lambda_{\chi^2}(W_l) = \sum_{g=1}^{G} \frac{|I_g|^{-1}}{N} \left\| \Delta_{I_g} W_l\left(\frac{i}{N}\right) \right\|_2^2, \tag{16.13}$$

where $\Delta_{I_g} W_l$ is the increment of the empirical fluctuation process over the observations in category $g = 1, \ldots, G$ (i.e., essentially the sum of the scores in category g).

[3]The *LM* (Lagrange multiplier) statistic is based on the generalized method of moments (GMM) estimators (see Hansen 1982).

The test statistic is then the weighted sum of the squared L_2 norm of the increments which has an asymptotic χ^2 distribution with $m \cdot (G-1)$ degrees of freedom. For more details, see Hjort and Koning (2002) and Zeileis et al. (2005).

One last problem remains. Recall that the empirical fluctuation process depends on the score function $\hat{\psi}$, which is given by the derivative of $\Psi(y, \theta)$ w.r.t. θ. Differentiating with respect to the infinite dimensional parameter function $\theta = \theta(t)$ is not straightforward. We solve it via basis expansion of the form $\theta \approx \sum_k c_k \phi_k = c'\phi$, where $\phi = (\phi_1, \ldots, \phi_K)^T$ is a K-vector of basis functions and $c = (c_1, \ldots, c_K)^T$ is a vector of associated coefficients.

16.4.1.2 Splitting. In the third step, the model is split with respect to the variable Z_l into B segments (typically, $B = 2$). Two rival segmentations are compared by comparing the segmented objective function $\sum_{b=1}^{B} \sum_{i \in I_b} \Psi(y_i, \theta_b)$. The optimal partition is found by performing an exhaustive search over all conceivable partitions into B segments. See Zeileis et al. (2005) for more details.

16.4.2 Model-Based Functional Differential Equation Trees

We now adopt this methodology to differential equation trees. We apply recursive model-based partitioning to the differential equation context by suitably defining the objective functions $M(y, \theta)$ and $\Psi(y, \theta)$. Consider again a DEM of the form

$$D^m y_i = -\omega_0 y_i - \omega_1 D y_i - \cdots - \omega_{m-1} D^{m-1} y_i. \tag{16.14}$$

Following the notation from the previous section, we will denote this model by $M(y, \omega)$. As pointed out in Wang et al. (in press), we can estimate this model by minimizing the sum of squared norms

$$\Psi(y, \omega) = \sum_{i=1}^{N} \int \left[\sum_{j=0}^{m} \omega_j(t)(D^j y_i)(t) \right]^2 dt \tag{16.15}$$

over the weight function $\omega = (\omega_0, \ldots, \omega_m)$.

Parameter estimation can be accomplished via basis expansion. First, approximate the coefficients ω_j via a linear combination of basis functions:

$$\omega_j \approx \sum_k c_{jk} \phi_k. \tag{16.16}$$

Using this approximation, we can write $\Psi(y, \omega)$ as a quadratic form in c,

$$\Psi(y, \omega) \approx C + c'Rc + 2c's, \tag{16.17}$$

where the constant C does not depend on c, and hence the estimate $\hat{c}$ is given by the solution of the equation

$$\hat{c} = -R's, \tag{16.18}$$

where the symmetric matrix R consists of an $m \times m$ array of $K \times K$ submatrices R_{jk} of the form

$$R_{jk} = N^{-1} \int \phi(t)\phi(t)' \sum_i D^j y_i(t) D^k y_i(t) dt \tag{16.19}$$

for $j = 0, \ldots, m-1$. For more details on parameter estimation, see, e.g., Ramsay and Silverman (2005).

16.4.3 Application to Online Auction Data

Figure 16.7 shows the fitted functional differential equation tree for the data described in Section 16.2. This tree was obtained by employing a stopping criterion of a statistical significance level of $\alpha = 0.01$ and a minimum number of at least 10 observations within each final node. The resulting tree uses three different splitting variables to arrive at eight different price curves: the opening bid (Obid), the winning bid (Wbid), and the number of bids (Numbid). Returning to Figure 16.4 (the conditional PPP plots), the tree confirms the differences between price dynamics observed for each of the variables from Table 16.1. Moreover, as noted earlier, many of the variables are correlated, and therefore it is not surprising that some of them are absent from the tree. For instance, item value and item condition are associated with the winning bid; reserve price and early bidding are associated with the opening bid. As in the PPP plots, the average bidder rating does not appear to lead to different price dynamics and is absent from the tree. In contrast, seller rating, which did appear to yield different PPP plots, is conspicuously missing from the tree. One possible reason for this is that seller rating (which proxies for experience) is indirectly associated with the opening bid, as the seller determines the opening bid.[4]

With respect to the resulting eight leaf nodes, the estimated price curves cover three main shapes: Shape 1 is characterized by a fast initial price increase followed by an end-of-auction slowdown (L1–3); shape 2 is characterized by an almost linear increase (L4); and shape 3 shows little to moderate initial activity followed by late price spurts (L5–8). We also note that these three main shapes differ by the opening bid only: For auctions with an opening bid lower than \$3.99, the price curves follow the first shape; for those with an opening bid higher than \$6.5, the curves follow the third shape; and for auctions with an opening bid between \$3.99

[4]Using a more lenient stopping criterion results in a larger tree. For example, using $\alpha = 0.1$ and a minimum number of at least five observations within a final node results in a tree that also incorporates the seller rating as a splitting variable. Note that this tree is the same as the earlier except for the addition of seller rating. We leave it up to future researchers to determine the optimal settings for the differential equation tree.

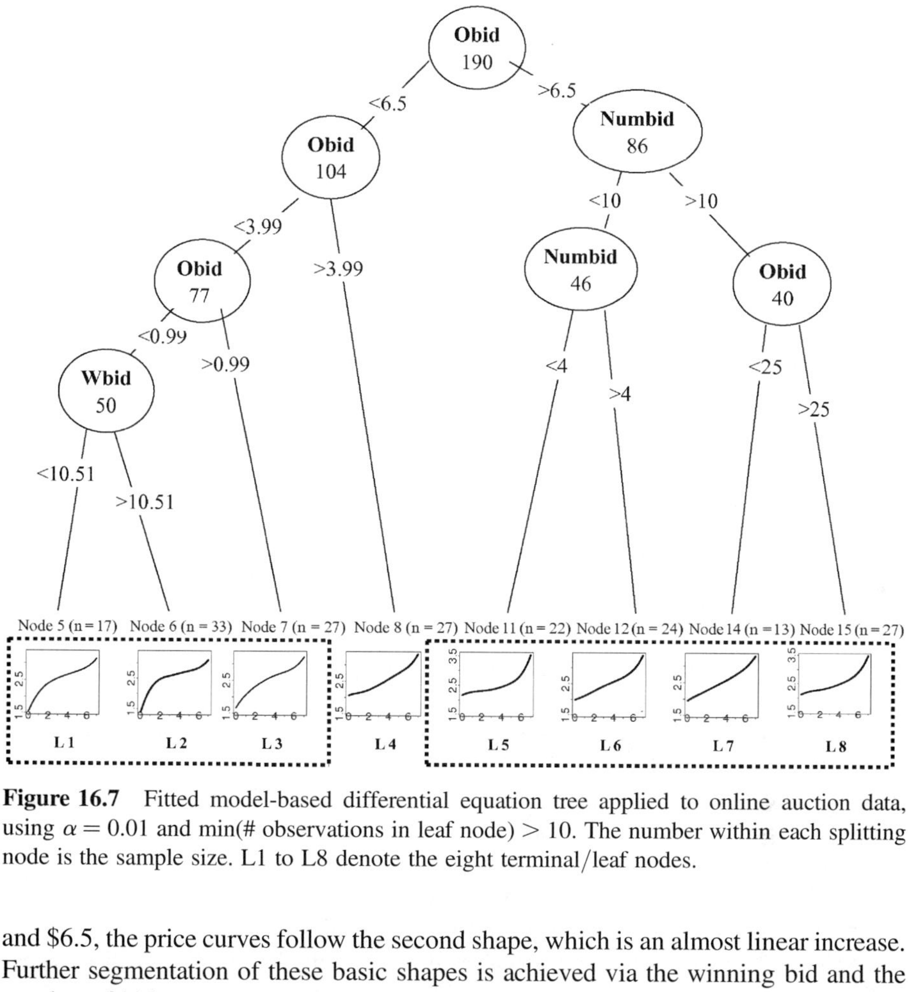

Figure 16.7 Fitted model-based differential equation tree applied to online auction data, using $\alpha = 0.01$ and min(# observations in leaf node) > 10. The number within each splitting node is the sample size. L1 to L8 denote the eight terminal/leaf nodes.

and \$6.5, the price curves follow the second shape, which is an almost linear increase. Further segmentation of these basic shapes is achieved via the winning bid and the number of bids.

16.5 CONCLUSIONS

We propose a novel tree-based approach for incorporating covariate information into differential equation models. Our recursive model-based approach defines an objective function that is also used for determining the splits of the tree.

This work fills a void in the literature in that it offers a direct and practical method for modeling the relationship between process dynamics and external factors. By applying our methodology to online auction data, we show that dynamic models can capture the changing price dynamics in an online auction. In particular, we find that the opening bid, the number of bids, and the winning bids are the major factors determining the shape of the price curve and its dynamics.

REFERENCES

Andrews, D.W.K. (1993). Tests for parameter instability and structural change with unkown change point. *Econometrica*, 61: 821–856.

Bapna, R., Jank, W., and Shmueli, G. (2005). Price formation and its dynamics in online auctions. Technical report, R.M. Smith School of Business, University of Maryland, College Park.

Breiman, L., Friedman, J.H., Olshen, R.A., and Stone, C.J. (1984). *Classification and Regression Trees*. Belmont, CA: Wadsworth.

Chan, K.Y. and Loh, W.Y. (2004). Lotus: An algorithm for building accurate and comprehensible logistic regression trees. *Journal of Computational and Graphical Statistics*, 13(4): 826–852.

Gamma, J. (2004). Functional trees. *Machine Learning*, 55: 219–250.

Hansen, L.P. (1982). Large sample properties of generalized method of moments estimators. *Econometrica*, 50: 1029–1054.

Hjort, N.L. and Koning, A. (2002). Tests for constancy of model parameters over time. *Nonparametric Statistics*, 14: 113–132.

Hyde, V., Jank, W., and Shmueli, G. (2006). Investigating concurrency in online auctions through visualization. *The American Statistician*, 60(3): 241–250.

Hyde, V., Jank, W., and Shmueli, G. (2008). A family of growth models for representing the price process in online auctions. In *Statistical Methods in eCommerce Research* (Jank and Shmueli, eds.). Hoboken, NJ: Wiley.

Jank, W. and Shmueli, G. (2006). Functional data analysis in electronic commerce research. *Statistical Science*, 21(2): 155–166.

Jank, W. and Shmueli, G. (in press). Studying heterogeneity of price evolution in eBay auctions via functional clustering. In *Handbook on Information Series: Business Computing* (Adomavicius and Gupta, eds.). Elsevier, Amsterdam, Netherlands.

Kim, H. and Loh, W.Y. (2001). Classification trees with unbiased multiway splits. *Journal of the American Statistical Association*, 96(454): 589–604.

Loh, W.Y. (2002). Regression trees with unbiased variable selection and interaction detection. *Statistica Sinica*, 12: 361–386.

Quinlan, J.R. (1993). *C4.5: Programs for Machine Learning*. San Mateo, CA: Morgan Kaufmann.

Ramsay, J.O. (1998). Estimating smooth monotone functions. *Journal of the Royal Statistical Society, Series B*, 60: 365–375.

Ramsay, J.O. and Silverman, B.W. (2002). *Applied Functional Data Analysis: Methods and Case Studies*. New York: Springer-Verlag.

Ramsay, J.O. and Silverman, B.W. (2005). *Functional Data Analysis* (2nd ed.). New York: Springer-Verlag.

Shmueli, G. and Jank, W. (2006). Modeling the dynamics of online auctions: A modern statistical approach. In *Economics, Information Systems, and Ecommerce Research II: Advanced Empirical Methods* (Kauffman and Tallon, eds.). Advances in Management Information Systems. Armonk, NY: M.E. Sharpe.

Wang, S., Jank, W., and Shmueli, G. (in press). Explaining and forecasting online auction prices and their dynamics using functional data analysis. *Journal of Business and Economic Statistics*.

Wang, S., Jank, W., Shmueli, G., and Smith, P. (in press). Modeling price dynamics in ebay auctions using principal differential analysis. *Journal of the American Statistical Association*.

Yu, Y. and Lambert, D. (1999). Fitting trees to functional data, with an application to time-of-day patterns. *Journal of Computational and Graphical Statistics*, 8(4): 749–762.

Zeileis, A., Hothorn, T., and Hornik, K. (2005). Model-based recursive partitioning. Technical Report 19, Department of Statistics and Mathematics, Wirtschaftsuniversität Wien.

17

QUANTILE MODELING FOR WALLET ESTIMATION

CLAUDIA PERLICH AND SAHARON ROSSET
IBM T.J. Watson Research Center, Yorktown Heights, New York

17.1 INTRODUCTION

E-commerce provides a wide range of new opportunities for customer relationship management (CRM) and direct marketing including the personalization of products and offers, the analysis of customer behavior from web logs, and new communication channels such as email and personalized web content. A vital component of truly personalized marketing is an understanding of the customer's purchasing power. The total amount of money a customer can spend on a certain product category is a vital piece of information for planning and managing sales and marketing efforts. This amount is usually referred to as the customer's *wallet* (also called *opportunity*) for this product category. There are many possible uses for wallet estimates, including straightforward targeting of sales force and marketing actions toward large-wallet customers and prospects. In a more sophisticated sales and marketing environment, the combination of classical propensity models for a particular product category with the knowledge of the wallet for that category can direct the sales efforts: It allows a company to market not only to customers or potential customers with large wallets, but also to those with a high probability of buying specific products in the category.

By combining customer wallet estimates with the data on how much customers spend with a particular seller, we can calculate the share-of-wallet that the seller has of each customer for a given product category. This information allows the seller to target customers based on their growth potential, a combination of total wallet and share-of-wallet. The classical approach of targeting customers who have

Statistical Methods in e-Commerce Research. Edited by W. Jank and G. Shmueli

historically generated large amounts of revenue for the company (known as *lifetime value modeling*; see, e.g., Rosset et al. 2003) does not give enough importance to customers with a large wallet but a small share-of-wallet, who are the ones with presumably the highest potential for revenue growth.

Share-of-wallet is also important for detecting partial defection or silent attrition, which occurs when customers increase their spending in a given category without increasing the amount purchased from a particular company (Malthouse and Wang 1998). In certain industries, customer wallets can be easily obtained from public data. For example, in the credit card industry, the card-issuing companies can calculate the wallet size and respective share-of-wallet using credit records from the three major credit bureaus (Epsilon 2001). For most industries, however, no public wallet information is available at the customer level. In this case, two approaches are used in practice for obtaining wallet estimates:

1. *Top-down:* starts with a public aggregate estimate for the overall industry opportunity in a given country and splits this estimate across the individual customers using heuristics based on customer characteristics. For example, if the customers are companies, the overall opportunity could be divided among the companies in proportion to their number of employees.
2. *Bottom-up:* estimates the wallet directly at the customer level, using heuristics or predictive models based on customer information. A common approach is to obtain wallet information for a random subset of customers/prospects through primary research. A model is then developed based on these data to predict the wallet for the other customers/prospects.

Although customer wallet and share-of-wallet have been recognized as important customer value metrics in the marketing and services literature for a number of years (Pompa et al. 2001; Keiningham et al. 2003; Garland 2004; Du and KamaKura 2005), there is not much published work on actual wallet modeling. The few references available are limited to commercial white papers from marketing analytics consulting companies (Epsilon 2001), and two recent conference presentations (Du and Kamakura 2005; Zadrozny et al. 2005).

The white paper from Epsilon (2001) describes at a very high level the methodology that the company uses to estimate wallets for both customers and prospects in a given product category. Epsilon uses a bottom-up approach in which a survey provides the self-reported category wallet for a sample of customers. The self-reported wallet is used to develop a multivariate linear regression that can be applied to the whole market or customer base. They do not describe which variables are used in the model and do not provide experimental results. In Du (2005) and Du and Karmakura (2005), the authors propose a multivariate latent factor model that a bank can use to impute a customer's holdings of financial products outside the bank based on a combination of internally available data and survey information about customers' holdings with competitors. In Zadroany et al. (2005), the authors discuss the value of wallet estimation, review methodologies, and compare a top-down approach with two bottom-up approaches, where again, a survey provides the self-reported category wallet for a sample of customers.

In this chapter, we address the issue of predicting the wallet for IBM customers that purchase information technology (IT) products such as servers, software, and services. Our models are bottom-up, with explanatory features drawn from historical transaction data as well as firmographic data taken from external sources like Dun & Bradstreet. We take a predictive modeling approach and attempt to minimize the dependence of our models on primary research data. While we concentrate here on the problem of predicting customer wallets for IT products, the developed methodology can be applied much more widely, to predicting wallets for other durable, as well as nondurable products (provided appropriate data are available) and also to setting prices of services (see Section 17.6).

In Section 17.2, we address the exact meaning of *wallet*, surveying several different definitions and their implications. We then devote the next two sections to developing and analyzing methodologies to address the challenging problem of building prediction models for wallets. The main difficulty stems from the fact that the wallet is an unobserved quantity in the larger customer population. We concentrate on *quantile modeling* as our main solution. In nontechnical terms, this approach allows us to estimate and predict *what we can hope for* rather than *what we expect.* We argue that this matches well at least one definition of *wallet*—a highly optimistic expectation of what the specific customer may spend on the specific product. We discuss evaluation of wallet models in Section 17.3, and in Section 17.4 we discuss and develop various modeling methodologies that can be used to implement our proposed quantile modeling approach.

Finally, in Section 17.6, we describe implementation of a wallet estimation system within IBM and how we evaluated the performance of different wallet estimation approaches using expert input in this system. The conclusion is that the quantile modeling approaches do a good job of approximating the expert decision. This conclusion has led to the selection of our models as the wallet prediction tool used in this application.

17.2 DEFINITIONS OF CUSTOMER WALLET

The first question we need to address is, what exactly is meant by a customer's wallet? We discuss this in the context of IBM as a seller of IT products to a large collection of customers for whom we wish to estimate the wallet. We have considered three nested definitions:

1. The total spending by this customer in the relevant area. In IBM's case, this would correspond to the total IT spending by the customer. We denote this wallet definition by TOTAL.
2. The total attainable (or served) opportunity for the customer. In IBM's case, this would correspond to the total spent by the customer in IT areas covered by IBM's products and services. While IBM serves all areas of IT spending—software, hardware, and services—its products do not necessarily cover all needs of companies in each of these areas. Thus, the served opportunity is smaller than the total IT spending. We denote this definition by SERVED.

3. The *realistically attainable* wallet, as defined by what the best customers spend. This may be different from SERVED, because it may not be realistic to get individual customers to buy every single served product from IBM. Defining the best customers is key in correctly applying this definition. In what follows, we define best customers in a relative sense as ones who are spending as much as we can hope, given a stochastic model of spending. Thus, a good customer is one whose spending is a high percentile of its *spending distribution*. Below, we describe some approaches that can allow us to build models that predict such a high percentile and, perhaps more importantly, evaluate models with regard to the goal of predicting percentiles of individual spending distributions. We denote this definition by REALISTIC.

A key question is, which definition do marketing executives refer to when discussing customer wallet? When discussed in the context of *share of wallet*, it seems likely that SERVED is the most appropriate definition (how much of their spending is done with us). When wallet is discussed as *opportunity*, which is the context in which we have most often encountered it at IBM, it seems more consistent with REALISTIC. As we will see below, our experiments show that REALISTIC seems to line up well with the numbers that IBM sales executives associate with their customers' wallets.

In principle, the three definitions should line up as REALISTIC $\leq$ SERVED $\leq$ TOTAL. An important caveat is that all the three definitions could be affected by marketing actions. Thus, a company could theoretically be convinced by diligent marketing activity to buy an IT product they were not planning to spend any money on. This could affect the value of all three wallet definitions. In the rest of this chapter, we ignore this possible effect of such marketing actions and essentially assume that our marketing actions can affect only wallet share, not the actual wallet. Our current challenge is to model these fixed wallet values. We concentrate on REALISTIC and SERVED, which we view as the two more operational definitions. The modeling approaches we discuss below are for the REALISTIC wallet, but we are also developing an approach to modeling SERVED.

We usually know the total company sales (revenue) of potential customers and the total amount of historical sales made by IBM to these customers (see Section 17.5 for details). In principle, the relation IBM SALES $\leq$ WALLET $\leq$ COMPANY REVENUE should hold for every company and all three measures (for the REALISTIC definition of wallet, we actually expect IBM SALES $\approx$ REALISTIC WALLET for a small percentage of companies).

17.3 EVALUATION APPROACHES

A key issue in any wallet estimation problem is, how can we evaluate the performance of candidate models? Since wallet is usually an unobservable quantity, it is critical to have a clear idea of how model performance can be measured—at least approximately—to give us an idea of future performance.

17.3.1 Evaluation Using Survey Values

The first and most obvious approach is to obtain actual values for customer wallets, typically through a survey, where customers (or potential customers) are called up and asked about their total IT spend for the previous year. We have one such survey, encompassing 2000 IBM customers or potential customers. However, the attempts by us, as well as other groups, to use this survey for model evaluation have not been particularly successful. We attribute this to three main reasons:

1. Of our three wallet definitions, these self-reported wallets correspond to TOTAL, which is the least relevant definition for marketing purposes.
2. Companies have no obligation to carefully consider their answers to such surveys. Exact IT spending may be difficult to calculate; thus, even companies that do not intend to misreport their wallet may give extremely erroneous numbers.
3. Getting surveys which represent unbiased samples of the customer universe is often extremely difficult due to sampling issues, response bias issues, etc.

The prohibitive cost of surveys compounds these difficulties and increases our motivation to design methodologies that do not depend on primary research.

17.3.2 Evaluation of High-Level Indicators

The second common approach is to evaluate wallet models by comparing high-level summaries of model predictions to known numbers, such as those on industry-level IT spending or total spending with IBM, and thus evaluating the high-level performance of models. Such approaches include comparing the total wallet in an industry to a known *industry opportunity* based on econometric models; comparing the total wallet to the total spending with IBM for large segments of companies and comparing the findings to some reasonable numbers; and comparing the total wallet in a segment to the total sales of companies in that segment. These evaluation approaches suffer from lack of specificity, as they do not evaluate the performance of the models on individual companies, but rather compare aggregated measures. A more specific approach in the same spirit is based on the statistics of ordering between model predictions and known quantities such as COMPANY REVENUE and IBM SALES. If many of the wallet predictions are smaller than IBM SALES for the same companies or bigger than COMPANY REVENUE, this is evidence of the inadequacy of the model being evaluated.

17.3.3 Quantile Loss-Based Evaluation

A different approach is to search for evaluation loss functions which evaluate the performance of the model in a way that is directly related to the goal of wallet estimation. Despite the fact that the wallet is unobservable, this evaluation turns out to be possible when the REALISTIC definition of wallet is used, using the quantile regression loss function (Koenker 2005): Given an observed IBM SALES number y and a predicted REALISTIC wallet $\hat{y}$, we define the quantile loss function for the pth quantile to be

$$L_p(y, \hat{y}) = \begin{cases} p \cdot (y - \hat{y}) & \text{if } y \geq \hat{y} \\ (1 - p) \cdot (\hat{y} - y) & \text{otherwise} \end{cases}. \tag{17.1}$$

In Figure 17.1, we plot the quantile loss function for $p \in \{0.2, 0.5, 0.8\}$. With $p = 0.5$ this is just absolute error loss. Expected quantile loss is minimized by correctly predicting the (conditional) pth quantile of the residual distribution. That is, if we fix a prediction point x, and define $c_p(x)$ to be the pth quantile of the conditional distribution of y given x

$$P(y \leq c_p(x)|x) = p, \ \forall x,$$

then the loss function is optimized by correctly predicting $c_p(x)$:

$$\arg\min_c E(L_p(y, c)|x) = c_p(x).$$

With $p = 0.5$, the expected absolute loss is minimized by predicting the median, while when $p = 0.8$ we are evaluating a model's ability to correctly predict the 80th percentile of the distribution of y given x—exactly the definition of the REALISTIC wallet! Thus, we have a loss function which allows us to directly evaluate model performance in estimating REALISTIC wallets. We can now use this loss function $L_p(x)$ both for model training and for model evaluation.

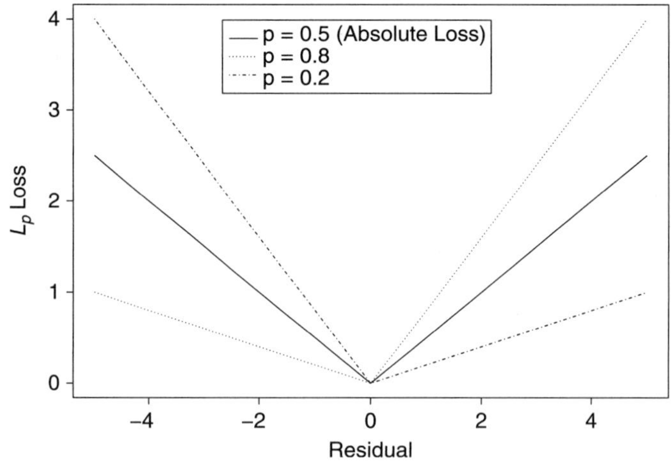

Figure 17.1 Quantile loss functions for various quantiles.

17.3.4 Using Expert Input

If we have access to experts who know what the customer's potential is likely to be, we can use their best guesses as noisy approximations of the true wallet number and evaluate our models' ability to make predictions that agree with the experts' opinions. We employ this approach in Section 17.5.

17.4 ADJUSTING MODELING APPROACHES TO QUANTILE PREDICTION

Recently, the modeling of quantiles has received increasing attention. The modeling objectives were either prediction or the need to gain insights to how the statistical dependencies for quantiles differ from those for expected value models. We review some of these efforts in Section 17.4.1. Two of the best-studied and most common standard regression approaches in machine learning and data mining are k-nearest neighbors and regression trees. We discuss in some detail how these methods can be adjusted to modeling quantiles in Sections 17.4.2 and 17.4.3.

17.4.1 Existing Approaches

We are aware of a number of such quantile estimation methods, including linear quantile regression (Koenker 2005), kernel quantile regression (Takeuchi et al. 2006), Quanting (Langford et al. 2006), quantile regression forests (Meinshausen 2006), and polynomial regression trees (Chaudhuri and Loh 2002). Many of these methods suffer from intractable computational behavior for larger quantile estimation tasks, as in the case of our IBM wallet. Next we discuss the two approaches most relevant to our work in some detail.

17.4.1.1 Linear Quantile Regression. A standard technique for quantile regression that has been developed and extensively applied in the econometrics community is linear quantile regression (Koenker 2005). Linear quantile regression assumes that the conditional quantile function is a linear function of the explanatory variables of the form $\beta\mathbf{x}$, and we estimate the parameters $\hat{\beta}$ that minimize the quantile loss function (equation 17.1). It can be shown that this minimization is a linear programming problem and that it can be efficiently solved using interior point techniques (Koenker 2005). Implementations of linear quantile regression are available in standard statistical analysis packages such as R and SAS. The obvious limitation of linear quantile regression is that the assumption of a linear relationship between the explanatory variables and the conditional quantile function may not be true. To circumvent this problem, Koenker (2005) suggests using nonlinear spline models of the explanatory variables.

17.4.1.2 Quanting. Recently, a reduction from quantile regression to classification has been proposed (Langford et al. 2006). The *quanting* reduction transforms a quantile regression problem into a series of classification problems such that a small average error rate on these problems leads to a provably accurate estimate of the conditional quantile. This allows us to apply any existing classifier learning algorithm to solve quantile regression problems. The essential idea of quanting is that each classifier c_t attempts to answer the question "Is the q-quantile above or below t?" In the (idealized) scenario where A is perfect, one would have $c_t(\mathbf{x}) = 1$ if and only if $q(\mathbf{x}) > t$ for a q-quantile $q(\mathbf{x})$; hence, the algorithm would output $\int_0^{q(x)} dt = q(x)$ exactly. The quanting analysis (Langford et al. 2006) shows that if the error of A is small on average over t, the quantile estimate is accurate.

17.4.2 Quantile k-Nearest Neighbor

The traditional k-nearest neighbor model (kNN) is defined as

$$\hat{y}(\mathbf{x}) = \frac{1}{k} \sum_{\mathbf{x}_i \in N_k(\mathbf{x})} y_i, \tag{17.2}$$

where $N_k(x)$ is the neighborhood of $\mathbf{x}$ defined by the k closest points $\mathbf{x}_i$ in the training sample for a given distance measure (e.g., Eucledian). From a statistical perspective, we can view the set $y_j \in N_k(\mathbf{x})$ as a sample from and approximated conditional distribution of $P(Y|\mathbf{x})$. The standard kNN estimator of $\hat{y}$ is simply the expected value of this conditional distribution approximated by a local neighborhood. For quantile estimation, we are not interested in the expected value (i.e., an estimate of $E(Y|\mathbf{x})$) but rather in a particular quantile $c(\mathbf{x})$ of the conditional distribution $P(Y|\mathbf{x})$ such that $P(Y \leq c(\mathbf{x})|\mathbf{x}) = q$. Accordingly, we can estimate $\hat{c}(\mathbf{x})$ in a kNN setting as the qth quantile of the empirical distribution of $\{y_j : \mathbf{x}_j \in N_k(x)\}$. If we denote that empirical distribution by

$$\hat{G}_{\mathbf{x}}(c) = 1/k \sum_{\mathbf{x}_j \in N_k(\mathbf{x})} \mathbf{I}\{y_j \leq c\}, \tag{17.3}$$

then our kNN estimate of the qth quantile of $P(Y|\mathbf{x})$ would be $\hat{G}_{\mathbf{x}}^{-1}(q)$.

Similarly the interpretation is that the values of Y in the neighborhood $N_k(\mathbf{x})$ are a sample from the conditional distribution $P(Y|\mathbf{x})$, and we are estimating its qth quantile.

An important practical aspect of this estimate is that, in contrast to the standard kNN estimates, it imposes a constraint on k. While $k = 1$ produces an unbiased (while high-variance) estimate of the expected value, the choice of k has to be at least $1/(1 - q)$ to provide an upper bound for the estimate of the qth "high" quantile (more generally, we have $k \geq \max(1/q, 1/(1 - q))$). The issue of how exactly to estimate the qth quantile when q/k is not an integer (and hence the quantile of the empirical distribution falls between observed y_j values) also has to be addressed. In our experiments below, we select the integer closest to q/k as the index for the estimate, although interpolation approaches may be considered as well.

The definition *of neighborhood* is determined based on the set of variables, the distance function, and implicit properties such as scaling of the variables. The performance of a *k*NN model depends strongly on the ability of the neighborhood definition to provide a good approximation of the true conditional distribution. This is true both for the standard problem of estimating the conditional mean and for estimating conditional quantiles.

17.4.3 Quantile Regression Tree

Tree induction algorithms are very popular in predictive modeling, and are known for their simplicity and efficiency when dealing with domains with large number of variables and cases. Regression trees are obtained using a fast divide-and-conquer greedy algorithm that recursively partitions the training data into subsets. Therefore, the definition of the neighborhood that is used to approximate the conditional distribution is not predetermined, as in the *k*NN model, but optimized locally by the choice of the subsets. Work on tree-based regression models traces back to Morgan and Sonquist (1963) but the major reference is the book on classification and regression trees (CART) by Breiman et al. (1984). We will limit our discussion to this particular algorithm. Additional regression tree implementation include RETIS (Karalic 1992), M5 (Quinlan 1993), and RT (Torgo 1997).

A tree-based modeling approach is determined predominantly by three components:

- the *splitting criterion* that selects the next split in the recursive partitioning
- the *pruning method* that shrinks the overly large tree to an optimal size after partitioning has finished in order to reduce variance
- the *estimation method* that determines the prediction within a given leaf

The most common choice for the splitting criterion is least squares error (LSE). While this criterion is consistent with the objective of finding the conditional expectation, it can also be interpreted as a measure of the improvement of the approximation quality of the conditional distribution estimate. Tree induction searches for local neighborhood definitions that provide good approximations for the true conditional distribution $P(Y|\mathbf{x})$. So, an alternative interpretation of the LSE splitting criterion is to understand it as a measure of dependency between Y and an x_i variable by evaluating the decrease of uncertainty (as measured by variance) through conditioning. In addition, the use of LSE leads to implementations with high computational efficiency based on incremental estimates of the errors for all possible splits.

Pruning is the most common strategy to avoid overfitting in tree-based models. The objective is to obtain a smaller subtree of the initial overly large tree, excluding those lower-level branches that are unreliable. CART uses the error complexity pruning approach, which finds an optimal sequence of pruned trees by sequentially eliminating the subtree (i.e., the node and all the its ancestors) that minimizes the increase in error weighted by the number of leaves in the eliminated subtree:

$$g(t, T) = \frac{E(t) - E(T_t)}{S(T_t) - 1}, \tag{17.4}$$

where $E(T_t)$ is the error of the subtree T_t containing t and all its ancestors, and $E(t)$ is the error if it was replaced by a single leaf, and $S(T_t)$ is the number of leaves in the subtree. $E(\cdot)$ is measured in terms of the splitting criterion (i.e., for standard CART, it is squared error loss). Given an optimal pruning sequence, one still needs to determine the optimal level of pruning, and Breiman et al. (1984) suggest cross-validation on a holdout set.

Finally, CART estimates the prediction for a new case that falls into leaf node l similarly to the kNN algorithm as the mean over the set of training responses D_l in the leaf:

$$\hat{y}_l(x) = \frac{1}{n_l} \sum_{y_i \in D_l} y_i, \tag{17.5}$$

where n_l is the cardinality of the set D_l of training cases in the leaf.

Given our objective of quantile estimation, the most obvious adjustment to CART is to replace the sample mean estimate in the leaves with the quantile estimate using the empirical local estimate $\hat{G}_{D_l}(c)$ of $P(Y|\mathbf{x})$, as in (17.3).

A more interesting question is whether the LSE splitting (and pruning) criterion should be replaced by a quantile loss. On the one hand, finding splits that minimize the quantile loss on the training sample in the leaves corresponds directly to our prediction objective. On the other hand, having the best possible approximation of the conditional distribution can be expected to result in the best quantile estimates of the distribution, and minimizing the distribution variance could lead to a better approximation than the direct optimization of quantile loss, in particular for very high quantiles. In addition, changing the splitting criterion to quantile loss causes severe computational problems. The evaluation of a split now requires the construction of two sets of predictions in each leaf, sorting both of them of to find the correct quantile, and calculation of the loss. We will not consider the issue of efficiency any further, as there is already some related work by Torgo (1997) on efficient implementations of trees that minimize mean absolute deviation (MAD), i.e., quantile loss for $q = 0.5$. In our experiments below, we investigate the success of the two splitting criteria in terms of predictive performance only.

17.5 WALLET ESTIMATION AT IBM: THE MAP PROJECT

Our framing of wallet modeling as a quantile estimation task leads to a well-defined machine learning problem. While we can assess the relative model performance for different modeling approaches in terms of quantile, the fundamental question of how well our models perform as wallet estimators remains open. And in particular, we have no strong indication about the appropriate choice of the quantile for the

IBM customer wallets. In this section we describe the Market Alignment Program, which demonstrates a major use for wallet estimates within IBM, and which supplied us with a unique opportunity to evaluate the success of wallet estimation models in capturing experts' notion of IBM customers' wallets.

17.5.1 Market Alignment Program

In 2005 IBM started an initiative called the Market Alignment Program (MAP) (Lawrence et al. 2007) to address the major challenge of aligning sales resources with the best revenue-generating opportunities. The main objective of MAP is to drive the sales resources allocation process based on field-validated analytical estimates of future revenue opportunity. While expert knowledge is crucial to facilitate the sales process, sole reliance on expert knowledge may lead to an overly strong focus on existing and large customers with limited growth opportunities. In order to facilitate the discussion with sales experts, initial model-based wallet estimates are used as a starting point in the assessment of future revenue opportunity. An integral part of the MAP process is validation of the analytical estimates via an extensive set of workshops conducted with sales leaders. These interviews rely on a Web-based tool to convey the relevant information and to capture experts' feedback on the analytical models. The tool allows the sales team to input their estimates of revenue opportunity, as well as their reasons for recommending a change to the model results. As a side effect of this interview process, we now have validated customer-wallet estimates against which to evaluate our models. Figure 17.2 shows a high-level view of the MAP Web-based tool.

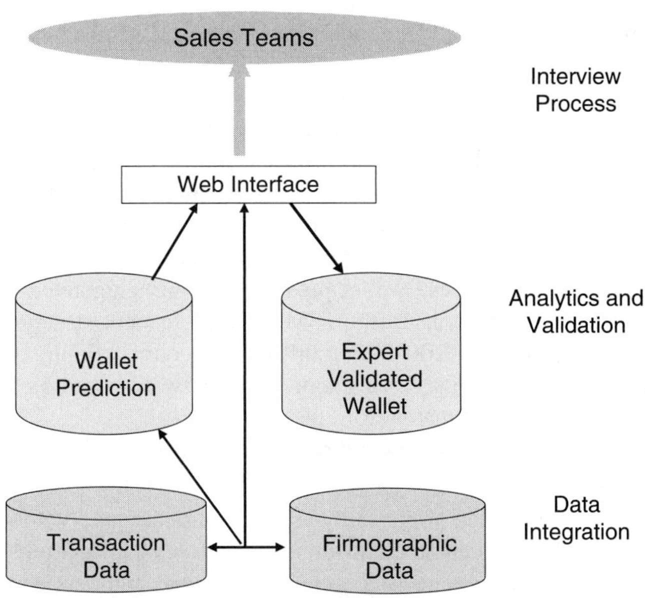

Figure 17.2 Overview of the MAP tool.

The complete MAP process consists of:

1. Developing a consistent data model incorporating all relevant information for each customer, including historical transactions with IBM, company firmographics (e.g., annual revenue, number of employees), and IBM sales coverage information.
2. Estimating the wallets for all IBM customers within each major product group using a simple quantile *k*NN wallet model.
3. Developing a Web-based tool designed to display historical revenue, along with the model-estimated revenue opportunities for each IBM sales account (an account consists of one or more IBM customers), and to capture experts' feedback on these estimates.
4. Conducting workshops with sales leaders to validate the model-estimated future revenue opportunities for each sales account.
5. Shifting sales resources to accounts with large validated revenue opportunities.

17.5.2 Analytical Details

For the initial round of workshops conducted in late 2005, the wallet estimates displayed in the MAP tool were generated using a simple, intuitive quantile *k*NN approach that follows our definition of a REALISTIC wallet. For each of the approximately 100,000 customers available for this study, we identified a set of 20 similar companies where similarity is based on the industry and a measure of size (either sales or number of employees, depending on the availability of the data). From this set of 20 firms, we discarded all companies with zero IBM revenue in the particular product group and reported the median of the IBM revenues of the remaining companies as the wallet estimate for this product group. The choice of the median (50th percentile) reflected a combination of statistical considerations and ad hoc business constraints (such as conforming to a known total market opportunity, i.e., the sum over all companies). This estimated wallet was sometimes smaller than the realized 2004 IBM revenue for some companies. The final reported estimates were therefore taken as the maximum of the *k*NN model estimate and last year's revenue (we refer to this operation as *flooring*). These wallet predictions were aggregated into opportunities by sales account according to the IBM internal account structure. The MAP workshops covered a about 1200 important sales accounts. Figure 17.3 presents the expert-validated opportunity for a major IBM software brand as a function of the calculated opportunity estimates from the 2005 workshops.

We can make a number of interesting observations here:

1. Forty-five percent of the opportunity estimates are accepted without alteration. The majority of the accepted opportunities are for smaller accounts. This shows a strong human bias toward accepting the provided numbers and emphasizes the value of supplying estimates where the experts have little knowledge.

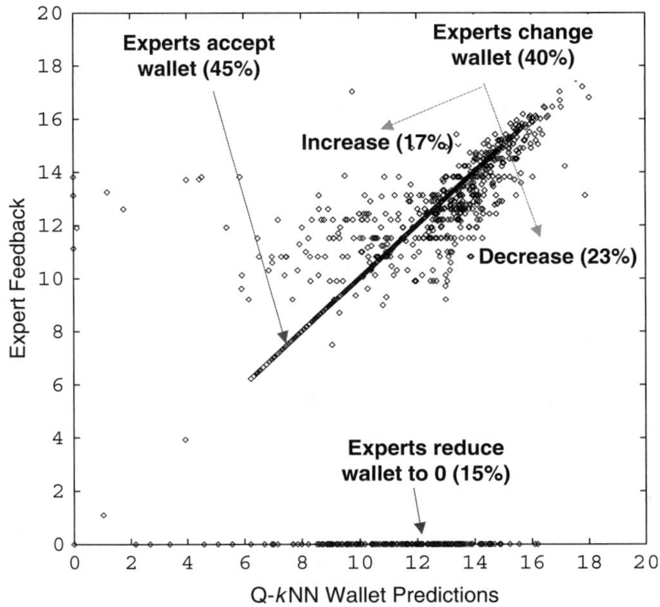

Figure 17.3 Expert feedback versus Q-*k*NN wallet predictions.

2. For 15% of the accounts, the experts concluded that there was *no* opportunity—mostly for competitive reasons.
3. For the remaining 40% of the accounts, opportunity estimates were decreased (23%) slightly less often than they were increased (17%).
4. The horizontal lines reflect the preference for round numbers.
5. The opportunities and the feedback appear almost jointly normal in a log plot. This suggests that the opportunities have an exponential distribution with potentially large outliers and that the sales experts corrected the percentage of opportunities.

17.5.3 Evaluating Wallet Models

While the purpose of the MAP workshop was not primarily to validate our models, as a side effect we now have 1200 true wallets (if you want to believe the expert opinions) for 2006 at the sales-account level that we can use to compare and evaluate our different wallet modeling approaches. We decided to eliminate the 15% of the accounts where the experts reduced the opportunity to zero for competitive reasons. This information is not available to the models, and we did not want to bias our evaluation. It may be an interesting classification problem to identify accounts without opportunity, but currently we want to focus the quality of our estimates if there is an opportunity.

It is a well-established fact that monetary quantities typically have a very long-tailed exponential distribution. The few largest numbers, corresponding to the biggest IBM customers, would typically dominate modeling and evaluation, and homoscedasticity assumptions underlying most modeling and evaluation approaches typically do not hold for monetary quantities. This is clearly shown by the experts' tendency to make wallet adjustments by percentages rather than dollars (corresponding roughly to homoscedasticity on the log scale), as discussed above. On the other hand, the success of models in a business environment is ultimately measured in dollars, not log-dollars. We therefore chose to evaluate model performance on three scales: error on the original (dollar) scale, square root scale, and log scale. In addition to the sum of squared errors for each scale, we considered the absolute error. This provides us with a total of six performance criteria.

We adopted the modeling approaches discussed in the previous section, and to account for our lack of knowledge about what truly defines a customer wallet, we allowed the model parameters—such as the quantile being modeled and the neighborhood size—to vary. In total, we built nearly 100 different models, counting all variations of model parameters, input variables, and different quantiles. We followed the same aggregation process and calculated the resulting opportunities for 2006 at the sales-account level for each model. Finally, we ranked all models according to each of the six performance criteria and compared how often a given model appears within the top 20 models. Table 17.1 shows the relative performance of the best variants, and includes as reference points the performance of the shown Q-*k*NN model and a very naive model that predicts for each account the maximum revenue over the last three years.

The results in Table 17.1 support the following conclusions:

- Q-*k*NN performs very well after flooring but is typically inferior prior to flooring.
- The 80th percentile seems to be the most appropriate quantile to capture experts' definitions across multiple approaches.
- Linear quantile regression performs consistently well (flooring has a minor effect).
- Models without last year's revenue do not perform well.

TABLE 17.1 Model Performance in Terms of Number of Times a Wallet Model was within the Top 20 Models Across the Six Different Performance Metrics

Model	Brand 1	Brand 2	Brand 3
Shown Q-*k*NN	6	5	6
Max revenue,	1	3	4
Linear Q-regression	6	4	5
Q-*k*NN	1	0	2
Q-*k*NN+flooring	3	6	6
Q-tree	1	4	4

Based on this analysis for three major product brands, we concluded that the linear quantile regression model showed the best overall performance when a quantile of 0.8 was used. In other words, this quantile regression model provided the best agreement with the expert feedback collected during the initial 2005 MAP workshops. Hence, this model was selected to provide the revenue opportunity estimates for MAP workshops conducted in late 2006.

17.6 OTHER APPLICATIONS OF HIGH-QUANTILE MODELING

In this chapter, we have considered high-quantile modeling as a tool for estimating customer wallets. We have shown how we can use it to model a customer's REALISTIC wallet as the 0.9 or 0.95 quantile of their conditional spending. This can be interpreted as a highly optimistic (yet still attainable) estimate of what they could spend on IT from IBM. This task of modeling "what we can hope for" rather than "what we should expect" turns out to be of great interest in multiple other business domains, including the following:

- When modeling sales prices of houses, cars, or any other product, sellers may be very interested in the price they may get for their asset if their negotiations are successful. This is clearly different from the average price for this asset and is more in line with a high quantile of the price distribution of equivalent assets. The buyer may also be interested in the symmetric problem of modeling a low quantile.
- In outlier and fraud detection applications, we may have a specific variable (such as the total amount spent on a credit card) whose degree of "outlyingness" we want to examine for each one of a set of customers or observations. This degree can often be well approximated by the quantile of the conditional spending distribution given the customer's attributes. For identifying outliers, we may just want to compare the actual spending to an appropriate high quantile, say 0.95.

Detailed discussion of these applications is beyond the scope of this chapter. Some experimental results on a set of real-life problems, using a large battery of quantile modeling approaches, can be found in our recent paper (Perlich et al. 2007).

17.7 SUMMARY

We have proposed three definitions for the customer wallet—TOTAL, SERVED, and REALISTIC—and argued that the last two are the ones that should be modeled. We have described some of the difficulties in evaluating the performance of wallet models without relying on primary research results, which are expensive, of questionable quality, and usually relevant only for TOTAL wallets. In this context, we have suggested the use of the *quantile regression* loss function for evaluating REALISTIC wallet predictions.

Our main focus is on developing predictive modeling approaches for SERVED and REALISTIC wallet estimation. We proposed and discussed in detail two new methodologies for modeling the REALISTIC wallet: quantile nearest neighbor and quantile regression. We reviewed the statistical considerations involved in designing methods for high-quantile estimation and described some existing quantile modeling methods, as well as our own adaptations of *k*NN and CART to quantile modeling.

Next, we described the MAP application and utilized the output from its first iteration to analyze which of a large candidate set of models is most consistent with IBM sales executives' notion of wallet. One interesting conclusion from our analysis is that the experts relied quite heavily on the numbers we presented to them (which were the output of a simplistic first-approximation model). The other conclusion is that the experts' notion of customer wallet seems most consistent with a high-quantile model for the 0.8 quantile compared to our previously proposed definition of the 0.9 quantile (Rosset et al. 2005). The model which performed best overall was a linear quantile regression model which (naturally) relies heavily on previous-year observed sales revenue to predict the current-year wallet.

ACKNOWLEDGMENTS

We would like to acknowledge the early work of Sholom Weiss on the nearest neighbor approach and of Bianca Zadrozny on quanting. We thank Paulo Costa, Alexey Ershov, and John Waldes of IBM for useful discussions.

REFERENCES

Breiman, L., Friedman, J.H., Olshen, R.A., and Stone, C.J. (1984). *Classification and Regression Trees*. Belmont, CA: Wadsworth International Group.

Chaudhuri, P. and Loh, W.L. (2002). Nonparametric estimation of conditional quantiles using quantile regression trees. *Bernoulli*, 8(5): 561–576.

Du, R. and Kamakura, W. (2005). Imputing customers' share of category requirements. *INFORMS Marketing Sciences Conference*.

Epsilon. (2001). Solving the customer value puzzle. Technical Report, Epsilon Data Management.

Garland, R. (2004). Share of wallet's role in customer profitability. *Journal of Financial Services Marketing*, 8(8): 259–268.

Karalic, A. (1992). Employing linear regression in regression tree leaves. *Proceedings of the European Conference on Artificial Intelligence*. Sage, Thousand Oaks, CA.

Keiningham, T.L., Perkins-Munn, T., and Evans, H. (2003). The impact of customer satisfaction on share-of-wallet in a business-to-business environment. *Journal Of Service Research*, 6(1): 37–50.

Koenker, R. (2005). *Quantile Regression*. Econometric Society Monograph Series. Cambridge University Press.

Langford, J., Oliveira, R., and Zadrozny, B. (2006). Predicting the median and other order statistics via reduction to classification. *Proceedings of the 22nd Conference on Uncertainty in Artificial Intelligence (UAI)*.

Lawrence, R., Perlich, C., Rosset, S., Arroyo, J., Callahan, M., Collins, M., Ershov, A., Feinzig, S., Khabibrakhmanov, I., Mahatma, S., Niemaszyk, M., and Weiss, S. (2007). Analytics-driven solutions for customer targeting and sales force allocation. *IBM Systems Journal*, 46(4): 797–816.

Malthouse, E.C. and Wang, P. (1998). Database segmentation using share of customer. *Journal of Database Marketing*, 6(3): 239–252.

Meinshausen, N. (2006). Quantile regression forests. *Journal of Machine Learning Research*, 7: 983–999.

Morgan, J.N. and Sonquist, J.A. (1963). Problems in the analysis of survey data and a proposal. *Journal of the American Statistical Association*, 58(302): 415–434.

Perlich, C., Rosset, S., Lawrence, R., and Zadrozny, B. (2007). High-quantile modeling for customer wallet estimation and other applications. *Proceedings of the 13th International Conference on Knowledge Discovery and Data Mining (KDD)*.

Pompa, N., Berry, J., Reid, J., and Webber, R. (2000). Adopting share of wallet as a basis for communications and customer relationship management. *Interactive Marketing*, 2(1): 29–40.

Quinlan, R. (1993). Combining instance-based and model-based learning. *Proceedings of the 10th International Conference of Machine Learning (ICML)*.

Rosset, S., Neumann, E., Eick, U., and Vatnik, N. (2003). Lifetime value models for decision support. *Data Mining and Knowledge Discovery Journal*, 7: 321–339.

Rosset, S., Perlich, C., Zadrozny, B., Merugu, S., Weiss, S., and Lawrence, R. (2005). Wallet estimation models. *International Workshop on Customer Relationship Management: Data Mining Meets Marketing*.

Takeuchi, I., Le, Q.V., Sears, T., and Smola, A. (2006). Nonparametric quantile regression. *Journal of Machine Learning Research*, 7: 1231–1264.

Torgo, L. (1997). Functional models for regression tree leaves. *International Conference on Machine Learning*.

Zadrozny, B., Costa, P., and Kamakura, W. (2005). What's in your wallet? estimating customer total category requirements. *INFORMS Marketing Sciences Conference*.

18

APPLICATIONS OF RANDOMIZED RESPONSE METHODOLOGY IN E-COMMERCE

PETER G.M. VAN DER HEIJDEN
Department of Methodology and Statistics, Utrecht, The Netherlands

ULF BÖCKENHOLT
Faculty of Management, McGill University, Montreal, Canada

18.1 INTRODUCTION

Randomized response is a method for intentionally misclassifying or perturbing part of a dataset. This misclassification can be done either by a researcher or by a respondent in a survey or other research study. The motivation is the same in both cases: Misclassification or data perturbation provides a convenient mechanism for controlling the amount of information that can be extracted from a dataset. The Web age has brought us an avalanche of information about individual behavior related to e-commerce, ranging from searches for information about commodities to shopping and postpurchase activities. At a result, there has never been a stronger need to apply and develop statistical methods that enforce individual control of the dissemination of personal information. This chapter reviews and discusses the use of randomized response methodology as a means to provide this privacy protection in the context of two e-commerce problems.

The first problem is collecting e-commerce self-report data on sensitive topics while protecting the privacy of individuals. For example, in online-surveys, individuals may be asked for self-reports on such sensitive topics as their financial situation (e.g., loans, or income), their use of male cosmetics or drugs to overcome impotence,

Statistical Methods in e-Commerce Research. Edited by W. Jank and G. Shmueli

their health problems such as urinary incontinence, or medical prescriptions, but also on such potentially less sensitive issues as shopping expenditures and default rates in credit card usage. It is well known that surveys on topics like these lead to refusals to answer or untruthful answers. Randomized response has been shown to be the currently best method of dealing with such sensitive issues (see Lensvelt-Mulders et al. 2005) and also to work work well in Web-based applications (Lensvelt-Mulders et al. 2006). Therefore, we see much potential for randomized response in e-commerce data collection applications.

The second problem that randomized response can address in the context of e-commerce is statistical disclosure control (SDC). The need for SDC in e-commerce has been reviewed recently by Fienberg (2006). Randomized response can be used to misclassify part of a data source so that the data source can be analyzed by third parties who are not the original owners of the data source. A special application in this area is in the context of data integration of multiple databases. Here multiple databases are linked, with the aim of mining the data while preserving data confidentiality. Privacy-preserving data-mining randomized-response methods misclassify the parts of the data that are *not* used for linking the databases (for a review, see www.csee.umbc.edu/kunliu1/research/privacyreview.html; also see Evfimievski 2002; Ambianis et al. 2003; Du and Zhan 2003). Related work is reported by Trottini et al. (2004), Karr and Fienberg (2005), and Fienberg (2005). Although SDC applications of randomized response are well known and recently were further developed under the name of PRAM (Kooiman et al. 1997; Gouweleeuw et al. 1998), an abbreviation of *postrandomization*, these methods have not yet become mainstream.

Originally, randomized response was proposed by Warner (1965) as a tool for collecting information on sensitive topics. He offered a respondent two complementary statements such as "A. I did use hard drugs last year" and "B. I did not use hard drugs last year." A randomizing device such as a die, with an outcome unknown by the interviewer, indicates to the respondent which statement required a response. If, for example, an outcome of 1, 2, 3, or 4 is linked to answering A and an outcome of 5 or 6 is linked to answering B, one could view the answer to A as a correct answer and the answer to B as a misclassified answer, with a misclassification probability of 2/6. Due to the potential misclassification that a respondent may or may not encounter, respondents may feel safe in responding to the sensitive topic, yet it is possible to estimate the prevalence of the sensitive topic defined in A.

Characteristic of this application of randomized response is that respondents are offered a sensitive question and the misclassification is carried out by the respondent. However, very early in the development of randomized response, it was noticed that the same methodology could be applied to SDC (Warner 1971). That is, after data are collected, partial misclassification is carried out by the owner of the data. Thus, the owner of the data can hand over the data to others without risking privacy violations of individuals. The receiving agency only needs to know the misclassification key for the sample so that statistical analyses can be carried out that take the misclassification into account (cf. Chen 1989; Kuha and Skinner 1997).

Randomized response tools for data collection and for SDC are mathematically identical. For example, the data collection application and the SDC control application use the same equations to obtain univariate estimates and variances. Yet, as was pointed out by van den Hout and van der Heijden (2002), in practice the procedures differ. The differences are as follows:

1. Randomized response as a data collection tool presupposes the cooperation of respondents in the sense that they are allowed to follow the rules prescribed by the randomized response design. There is evidence that some respondents do not do this. This problem has to be handled in the analysis phase. In contrast, randomized response as an SDC tool does not have this problem.
2. Whereas randomized response as a data collection tool is used mainly to randomize response variables (the sensitive topics), randomized response as an SDC tool can be used to randomize both explanatory variables (background characteristics of respondents) and the response variables of a study. When a single source of data is to be protected by SDC, it may be most efficient to use SDC to hide the identity of the respondent by misclassifying his background characteristics so that answers to sensitive questions in a questionnaire cannot be linked directly to individuals. When a data source is released with the aim of integrating it with other data sources, background characteristics should not be misclassified, as these are necessary for linking, but randomized response can be used for misclassifying the response variable. In applications where distinctions between response and explanatory variables are important, this may lead to different statistical analyses for both applications.
3. Related to 2, randomized response as a data collection tool should use a randomization scheme that is perceived as protective by respondents and, at the same time, provide as much information as possible. The need for protection calls for large misclassification probabilities, whereas the demand for information calls for small misclassification probabilities. In randomized response as a tool for SDC, the misclassification probabilities can be chosen in such a way that the outcome is optimal, without the necessity to take the perceptions of respondents into account.

In this chapter, we give an overview of statistical methods for the analysis of randomized responses. We discuss bivariate and multivariate methods and recent developments in this field. In Sections 18.2 and 18.3, we give an overview of the literature on randomized response, with an emphasis on the similarities and differences of the two applications of this method. In regard to the differences, we focus on the statistical analysis issues 1 and 2, introduced above. For issue 3, we refer to van den Hout and Elsayed (2006). In Section 18.4, we summarize a new development that takes into account that respondents may not follow the regulations laid out by the randomized response design. As e-commerce data are often used for data-mining purposes and classification trees are a popular mining tool, in Section 18.5 we provide some new results regarding the use of classification trees in the context of randomized response data.

In this chapter, we focus on the methodology for misclassifying categorical data. Similar ideas are available for perturbing continuous data or mixtures of continuous and discrete data (see Fox and Tracy 1986; Chaudhuri and Mukerjee 1988). Since the same principles apply in these more general settings, we do not consider different data types explicitly.

18.2 UNIVARIATE ANALYSIS

This section introduces our notation and gives a short overview of available tools of analysis for applications of randomized response that are shared by the data collection and SDC approaches.

As an example, consider the following form of the so-called forced classification design (Boruch 1971). Let the outcome of two dice determine whether the answers will be misclassified. For the data collection approach, this would mean that if the sum of the outcomes of the two dice is 2, 3, or 4 (probability $1/6$), then the respondent is asked to give the answer "yes"; if the sum equals 11 or 12 (probability if $1/12$), then the respondent is asked to give the answer "no"; if the sum is in the range 5–10 (probability is $3/4$), the respondent is asked to reveal his true status. Let θ_1 be the probability of "yes" for the true status ("no" has probability $\theta_2 = 1 - \theta_1$) and let θ_1^* be the probability of "yes" for the observed (and partly misclassified) status ("no" has probability $\theta_2^* = 1 - \theta_1^*$). It follows that

$$\theta_1^* = 1/6 + (3/4)\theta_1$$

and

$$\theta_2^* = 1/12 + (3/4)\theta_2. \tag{18.1}$$

These equations can be rewritten in terms of conditional misclassification probabilities. Let the conditional misclassification probabilities be

$$p_{ij} = IP(\text{category } i \text{ is observed} \mid \text{true category is } j). \tag{18.2}$$

Then these probabilities can be collected in a matrix given by

$$P = \begin{pmatrix} p_{11} & p_{12} \\ p_{21} & p_{22} \end{pmatrix} = \begin{pmatrix} 11/12 & 2/12 \\ 1/12 & 10/12 \end{pmatrix}. \tag{18.3}$$

If we collect the probabilities θ_1 and θ_2 in vector θ, and θ_1^* and θ_2^* in vector θ^*, we can write

$$\theta^* = P\theta \tag{18.4}$$

(cf. Chaudhuri and Mukkerjee 1988; van den Hout and van der Heijden 2002). Many randomized response designs can be written this way, and this formulation in 18.4 can also be extended to situations where the number of categories is larger than two. Further background and more complex randomized response designs can be found in Chaudhuri and Mukerjee (1988), Fox and Tracy (1986) and Kuk (1990).

18.2.1 Estimation

Estimation of θ^* and θ is discussed in detail in van den Hout and van der Heijden 2002, and we summarize their results here. If we want to estimate the probabilities for the true status of θ in (18.4), we plug in the observed proportions p^* for θ^* and an estimate for θ is obtained by

$$\hat{\theta} = P^{-1}p^*. \tag{18.5}$$

Thus, $\hat{\theta}$ is a moment estimator. If the elements of $\hat{\theta}$ are between 0 and 1, then $\hat{\theta}$ is also a maximum likelihood estimator. Note that (18.1) shows that if p_1^* is below chance level, the moment estimator $\hat{\theta}_1$ is negative. In SDC applications this can happen only because of sampling fluctuation. In data collection applications, this can also happen when respondents do not follow the instructions for the randomized response design; we come back to this issue in Section 18.4. If a moment estimate is negative, the maximum likelihood estimate will be on the boundary of the parameter space (i.e., the estimate is 0). For binary variables, maximum likelihood estimation is trivial, as the estimate on the boundary of 0 will place the other estimate on the boundary of 1. For variables with more than two levels, an expectation/maximization (EM) algorithm could be used to find the maximum likelihood estimates (cf. van den Hout and van der Heijden, 2002). In later sections, we discuss this boundary problem and solutions in more detail.

18.3 BIVARIATE ANALYSIS AND EXTENSIONS

Bivariate analyses can be carried out by generalizing equation (18.4). Two types of extensions have to be taken care of: First, the matrix P in (18.3) has to be adjusted so that it can handle bivariate problems; second, bivariate problems allow for the definition of restrictive models for θ.

First, we consider three types of extensions of equation (18.3). These are:

1. the situation where both variables are randomized response variables and can be considered as misclassified;
2. the situation where the dependent variable is misclassified, a condition we encounter in the data collection approach of randomized response as well as in applications where multiple data sources are integrated; and
3. the situation where the explanatory variable is misclassified, a result we encounter in the SDC approach of a single data source.

We consider the situation that both variables are binary (but this can be easily generalized to a situation with two polytomous variables). Thus, we can collect the four probabilities referring to the true status into a vector θ and the four probabilities referring to the observed status into a vector θ^*. In all three situations, we generalize (18.3) and (18.4).

We first generalize (18.3). Let A and B be two variables that are possibly misclassified. For these variables, the misclassification probabilities are p_{ij}^A and p_{ij}^B, respectively, defined as in (18.2), and these are collected in misclassification matrices P_A for variable A and P_B for variable B. Thus, variables A and B are misclassified independently.

The joint misclassification probabilities can now be collected in a 4×4 transition matrix P_{AB} that can be constructed using a Kronecker product:

$$P_{AB} = P_A \otimes P_B = \begin{pmatrix} p_{11}^A P_B & p_{12}^A P_B \\ p_{21}^A P_B & p_{22}^A P_B \end{pmatrix}. \tag{18.6}$$

We again consider situations 1, 2, and 3. Situation 1, where both variables are misclassified, yields simply the general form of (18.6). For situations 2 and 3, let variable A be the explanatory variable and variable B be the response variable. In situation 2, the response variable A will be misclassified by randomized response. In (18.6) P_{AB} simplifies since $P_B = I$. In situation 3, the explanatory variable B will be misclassified by randomized response, so P_{AB} simplifies since $P_A = I$.

If we collect the bivariate probabilities for the true status in a four-vector θ and the bivariate probabilities for the observed status in a four-vector θ^*, then θ and θ^* are related by

$$\theta^* = P_{AB}\theta. \tag{18.7}$$

Two interesting statistical models for the bivariate probabilities for the true status θ are, first, the model where variables A and B are dependent and the elements in θ are free and, second, the model where variables A and B are independent:

$$\theta_{ij}^{AB} = \theta_i^A \theta_j^B. \tag{18.8}$$

We note that in both situations when the explanatory variable or the response variable is misclassified by randomized response, (18.8) refers to the situation of no subgroup differences in the probabilities of the response variable (e.g., the prevalence of the sensitive behavior).

18.3.1 Estimation

This section considers the dependence and independence models. We will discuss two methods for estimating these models, one by setting up the likelihood and maximizing it and another one based on analyzing the observed data directly.

Summarizing the literature, Van den Hout and van der Heijden (2004) describe two ways to find maximum likelihood estimates for the dependence and independence models; one uses the EM algorithm, and the other involves maximizing the likelihood directly. The idea of maximizing the likelihood directly is that (i) the likelihood is set up in terms of the observed responses and the probabilities for the observed status θ^*; (ii) for the dependence model, we replace θ^* by $P_{AB}\ \theta$ and maximize the likelihood over θ; and (iii) for the independence model, we replace θ^* by P_{AB} θ with elements of θ equal $\theta_{ij}^{AB} = \theta_i^A\ \theta_j^B$ and maximize the likelihood over θ_i^A and θ_j^B; (iv) by comparing the likelihoods of the two models, a likelihood ratio chi-square test can be carried out testing the null hypothesis of independence against the alternative hypothesis of dependence (see also Maddala 1983).

More conventional methods analyze the counts for the observed status directly, ignoring the fact that one or more variables are misclassified by randomized response. It may perhaps come as a surprise that a conventional analysis of the observed counts cross-classifying variables A and B gives a correct answer about the cross-classified status of true variables. Thus, if a chi-square test is carried out testing for independence in the 2×2 table of observed responses, this test provides the correct answer about independence in the 2×2 table of the (unobserved) true responses. If only one of the two variables is misclassified, then parameter estimates for θ_i^A and θ_j^B can be found using univariate methods. If both variables are misclassified, the joint probabilities can be found in a slightly more complicated but similar way (see Fox and Tracy 1986, p. 52).

We illustrate with an example taken from van den Hout and van der Heijden (2004), where a randomized response design is used that employs red and black cards. In Table 18.1(a), two variables are cross-classified that were collected in a study on compliance with social benefit regulations (Van Gils et al. 2001). The variable G denotes gender. The observed red/black answers to the randomized response question are denoted by F^*. The question is whether or not the respondents earned money by doing some odd jobs without informing the office that provides their social benefit. This is a sensitive question, since not informing the office is against regulations. Let the binary variable F denote the not-observed yes/no answers that we will call the *true answers*.

First, we use the conventional method. Applying the chi-square test to the observed values in Table 18.1(a) yields $X^2 = 3.377$ with 1 degree of freedom and p-value 0.066. When we choose significance level $\alpha = 0.05$, the data do not support rejection of the null hypothesis.

We now show that maximizing the likelihood taking the misclassification into account leads to the same value of X^2. Let $n^* = (n_{11}^*,\ n_{12}^*,\ n_{21}^*,\ n_{22}^*)^t$ denote the

TABLE 18.1 (a) Classification by Gender (G) and Randomized Response Answer (F^*) and (b) Estimated Classification by Gender (G) and True Answer (F)

(a)				(b)			
	F^*				F		
G	Red	Black	Total	G	Red	Black	Total
Male	218	500	718	Male	124.00	594.00	718
Female	152	438	590	Female	56.67	533.33	590
Total	370	938	1308	Total	180.67	1127.33	1308

observed frequencies in Table 18.1(a). We first construct P_{GF}. Since gender (G) is not misclassified, we obtain

$$P_{GF} = \begin{pmatrix} 8/10 & 2/10 & 0 & 0 \\ 2/10 & 8/10 & 0 & 0 \\ 0 & 0 & 8/10 & 2/10 \\ 0 & 0 & 2/10 & 8/10 \end{pmatrix}. \tag{18.9}$$

This matrix is used to estimate frequencies $n = (\hat{n}_{11}, \hat{n}_{12}, \hat{n}_{21}, \hat{n}_{22})^t$ in the classification by G and F by

$$n = P_{GF}^{-1} n^*; \tag{18.10}$$

see Table 18.1(b). (We note that all counts are positive, so the moment estimate is equal to the maximum likelihood estimate in this case. If one or more of the counts were negative, we would have to maximize the likelihood equation through the EM algorithm or directly.) Next, we estimate the expected frequencies in this table, denoted by $\hat{m} = \hat{m}_{11}, \hat{m}_{12}, \hat{m}_{21}, \hat{m}_{22}$, under the model of independence by $\hat{m}_{ij} = \hat{n}_{i+}\hat{n}_{+j}/N$. Since we want to fit the model of independence, we compute the fitted frequencies under this model, denoted by $\hat{m}^*$, by

$$\hat{m}^* = P_{GF}\hat{m} \tag{18.11}$$

and compare them with the observed n^*. Again, we get $X^2 = 3.377$. Thus, this example illustrates that prevalence estimates can be obtained either from Table 18.1(a) by applying univariate estimation methods for males and females separately or directly from Table 18.1(b).

18.3.2 Extensions to Multivariate Analyses

The approach method used to maximize the likelihood directly, presented above, is also useful in other situations. Basically, the framework can be summarized by the following steps:

1. Set up the likelihood in terms of the observed responses and the probabilities for the observed status θ^*.
2. Replace the probabilities for observed status θ^* by the product $P\theta$.
3. Define a (restrictive) model for θ.

For step 3, many models can be chosen and thus the framework is very powerful. We give a few examples from the literature. Van den Hout and van der Heijden (2004) generalize the results for the dependence and independence models to loglinear models in general. In this situation, the P-matrix may accommodate more than two variables. Böckenholt and van der Heijden (2004, 2007, in press) and Fox (2005) use this framework to estimate the Rasch model (Rasch 1980) designed for measuring individual differences with items that form a psychometric test.

An early application of this approach can be found in Maddala (1983) for logistic regression in the data collection approach of randomized response. Here the response variable is misclassified and the framework can be applied in a straightforward way (cf. also Scheers and Dayton 1988; Lensvelt-Mulders et al. 2006). Van den Hout et al. 2007 also apply the model to multivariate logistic regression models where the response variables are all misclassified by using the data collection approach of randomized response. On the other hand, Van den Hout and Kooiman (2005) generalize the linear regression model to the situation where one of the explanatory variables is misclassified using randomized response. This latter class of models may be more useful in the SDC approach of a single data source.

18.4 NONCOMPLIANCE IN THE DATA COLLECTION APPROACH

Since in the SDC approach of randomized response the misclassification process is computerized, we know the misclassification probabilities collected in P. This is not necessarily true in the data collection approach of randomized response, where people may not follow the rules indicated by the randomized response design. If this problem is present in the data but ignored in the analysis, prevalence estimates are biased downward, with the possible consequence that one or more elements of $\hat{\theta}$ fall on the boundary of the parameter space.

Recently, Böckenholt and van der Heijden (2007) and Cruyff et al. (2007) tackled this problem by defining models for multivariate randomized response data that allow for a specific form of noncompliance of respondents. In this form of noncompliance, a respondent either follows the rules indicated by the randomized response design or answers negatively to every sensitive question. Let λ be the probability of following the rules of the randomized response design. Then

$$\theta^* = \lambda P\theta + (1 - \lambda)v, \tag{18.12}$$

where v is the D-dimensional vector $(0, \ldots, 0, 1)^t$. Clearly, if the model specified for θ has as many parameters as θ has elements, then (18.12) is not identified (see Cruyff et al. 2007 for an illustration). Therefore, restrictive models are needed for θ in order to arrive at an identified model. Böckenholt and van der Heijden (2007) use the Rasch model, a restrictive latent variable model well known in psychometrics. Cruyff et al. (2007) use constrained loglinear models.

Interestingly, as was noted by Cruyff et al. (2007), the model can also be rewritten as

$$\theta^* = Q\theta, \tag{18.13}$$

where the transition matrix Q has elements

$$q_{ij} = \begin{cases} (\lambda)p_{ij} & \text{for } i \neq D, j \in \{1, \ldots, D\} \\ (\lambda)p_{ij} + (1 - \lambda) & \text{for } i = D, j \in \{1, \ldots, D\}. \end{cases} \tag{18.14}$$

This shows that the framework presented in Section 18.3, using θ^*, P and θ, can now be replaced by the following steps using θ^*, Q and θ:

1. Set up the likelihood in terms of the observed responses and the probabilities for the observed status θ^*.
2. Replace the probabilities for observed status θ^* by the product $Q\theta$.
3. Define a (restrictive) model for θ.

Even though the form of noncompliance that a respondent is allowed to exhibit is limited in comparison to the types of noncompliance one can theoretically envision, our experience with these models is that in practical situations, models based on (18.12) fit the data rather well. More sophisticated models for noncompliance (i.e., involving the matrix Q) can be found in Böckenholt et al. (in press), where more sophisticated psychometric models for θ also can be found. The models defined in Böckenholt and van der Heijden (2007) also allow for covariates, both for the parameter λ and for the latent variable.

In the framework just presented, estimates of θ yield unbiased estimates of prevalence of the sensitive behavior for those respondents who answer the sensitive questions following the rules laid out by the randomized response design. Univariate estimates for a variable are obtained by adding up elements in θ over the other variables. An example from Böckenholt and van der Heijden (2007) is shown in Table 18.2. In a survey on social benefit fraud, six randomized response questions were posed to the respondents concerning violation of health and work regulations. The table illustrates the importance of taking into account that some respondents do not follow the rules dictated by the randomized response design. The estimated probability of not following the design is 17% for the work items and 13% for the health items. When these percentages are taken into account, the prevalence for the work items, for example, increases from .030 and .110 to .074 and .159, respectively.

Estimating θ also allows for studying the probabilities related to the number of sensitive characteristics displayed (i.e., for three items this can be zero, one, two, or three). This is illustrated in Table 18.3. Here we find that—corrected for the respondents who do not follow the randomized response design—71% of the respondents follows both work and health regulations.

TABLE 18.2 Noncompliance Estimates and 95% Bootstrap Confidence Intervals

Domain	Items	No Bias Correction	Bias Correction
Health	1	.002 (.000, .015)	.033 (.010, .050)
	2	.014 (.000, .034)	.053 (.033, .075)
	3	.048 (.027, .070)	.083 (.055, .112)
	4	.085 (.063, .107)	.130 (.087, .159)
Work	1	.030 (.009, .052)	.074 (.050, .096)
	2	.110 (.086, .133)	.159 (.104, .190)

TABLE 18.3 Estimated Compliance Percentages for Health and Work Domains

Counts Health	Work 0	1	2	Total
0	71	7	3	81
1	7	2	1	11
2	3	1	1	5
3	1	1	1	2
4	0	0	0	1
Total	82	11	6	100

18.5 CLASSIFICATION TREES

Frequently, e-commerce data are analyzed with exploratory data-mining tools such as classification trees. We now discuss classification trees in the context of randomized response data. To our knowledge, this methodology has not yet been considered in this context.

Classification trees provide a breakdown of a population into subpopulations. In our situation, where the response variable is a sensitive characteristic, the aim is to find subpopulations that differ as much as possible in terms of their display of the sensitive behavior. A typical result could be this: Let there be an overall prevalence of .20; if the explanatory variables (in this context also called *splitting variables*) are age and gender, the first split is between males and females (with a respective prevalence of .25 and .11, say), and a second split is found only for the males, namely, males younger than 22 years with a prevalence of .18 and males older than 22 years with a prevalence of .28.

We discuss two applications of randomized response when the response variable is randomized and when the explanatory variables are randomized. We first note that the estimation problem of a classification tree differs from the estimation problems discussed thus far because, for classification trees, there is no overall likelihood to be maximized; instead, a subproblem has to be solved for each separate split in the classification tree. That is, if we consider the example above, the first subproblem is determining the splitting variable used for the first split. Specifically, we need to determine whether there is a split on gender or a split on age. Then, once the first split is determined, for every node separately (here: males separately and females separately), again a split has to be determined.

One possible method to determine the splitting variable is the chi-square test. If the p-value of the chi-square for gender is smaller than the p-value of the chi-square for age, then the variable gender will be used for the first split. Thus, the correct tree can be found as follows when the dependent variable is a randomized response variable by:

1. Determine the classification tree on the observed (i.e., randomized) responses.
2. Transform the prevalence estimates found for each subpopulation using equation (18.5).

We now present an informal proof that this approach leads to the correct classification tree when the dependent variable is randomized. For this result to hold, it is only necessary to show that the order of the splitting variables for the observed data is identical to the order of the splitting variables for the true data. In order to show this, we use the results from estimation Section 18.3.1 for bivariate problems. There we illustrated that there is a one-to-one correspondence between, on the one hand, a chi-square test on the frequencies in an observed 2×2 table where one variable is measured with randomized response and the other variable has two subgroups, and, on the other hand a chi-square test on the frequencies in the corresponding true 2×2 table. It follows immediately that, if the choice between splitting variables in a classification tree is made using the highest possible value of the chi-square test, then the order of chi-squares for the different splitting variables derived for the observed table is identical to the order of chi-squares for the different splitting variables derived for the observed table, which concludes the proof.

Table 18.4 provides an illustration. On the left side of this table we find the true answers and on the right side the observed (i.e., randomized) answers. At the top, we find frequencies for the $2 \times 2 \times 2$ table of Gender by Age by Answer on the sensitive question. In this $2 \times 2 \times 2$ table the question is whether the first splitting

TABLE 18.4 (a) Gender by Age by True Answer and (b) Gender by Age by Randomized Answer

(a)				(b)			
	Answer				Answer		
2 × 2 × 2	Yes	No	Total	2 × 2 × 2	Yes	No	Total
Male, young	2	100	102	Male, young	18.5	83.5	102
Male, old	5	60	65	Male, old	14.6	50.4	65
Female, young	5	30	35	Female, young	9.6	25.4	35
Female, old	2	15	17	Female, old	4.3	12.7	17
Total	14	205	219	Total	47.0	172.0	219
Gender	Answer			Gender	Answer		
	Yes	No	Total		Yes	No	Total
Male	7	160	167	Male	33.1	133.9	167
Female	7	45	52	Female	13.9	38.1	52
Total	14	205	219	Total	47.0	172.0	219
Age	Answer			Age	Answer		
	Yes	No	Total		Yes	No	Total
Young	7	130	137	Young	28.1	108.9	137
Old	7	75	82	Old	18.9	63.1	82
Total	14	205	219	Total	14	205	219

Note: chi-square for the marginal gender table in (a) is 5.694 and in (b) is 1.137; The Pearson chi-square for marginal age table in (a) is 1.007 and in (b) is .201.

variable will be Gender or Age. For the true answers, the chi-square for gender is 5.694 and the chi-square for age is 1.007, showing that the gender chi-square is $5.694/1.007 = 5.656$ times as large. For the observed answers, the chi-square for gender is 1.137 and the chi-square for age is .201, showing that the gender chi-square is also $(1.137/.201 =)$ 5.656 times as large. This illustrates that the relative order of p-values for different splitting variables in the true table will be identical to the relative order of p-values for different splitting variables in the observed table.

This result holds when the response variable is a randomized variable. However, when the explanatory variables are randomized response, the results are more complicated. One of the reasons is that some of the explanatory variables are not necessarily randomized; as a result, the chi-square tests using these explanatory variables will have more power than the explanatory variables that are randomized. Therefore, in this situation, it seems preferable first to transform the observed proportions p^* into p using $p = P^{-1}p^*$ and then to construct a classification tree on the elements of p.

18.6 DISCUSSION

Privacy protection is a crucial objective for both data collection and statistical analyses in e-commerce work (Fienberg 2006). The randomized response methodology can make major contributions to this objective. As we have shown in this chapter, much statistical machinery is already available to address directly problems of privacy protection in analyses of publicly available databases and *privacy-preserving statistical databases*, where the data are altered prior to delivery for data mining. Moreover, randomized response methods can be beneficially employed for collecting information about sensitive topics in e-commerce. Although some arguments could be raised against the use of randomized response for data collection purposes, there are counterarguments that support the use of this methodology:

1. It could be questioned whether the randomized response method is too difficult to be understood by respondents, especially when it is used in an Internet survey. Yet, we have good experience with this data collection method. See Lensvelt-Mulders et al. (2006) for the instruction that we offer respondents to explain randomized response to them; also see our website, www.randomizedresponse.nl. It took considerable fine-tuning to provide these instructions, and we believe that they can be used beneficially as a starting point for anyone who wants to apply randomized response. In general, we emphasize that any randomized response instructions should be tested carefully using cognitive survey lab methods (see Boeije and Lensvelt 2002 for an example).
2. Related to this, it could be questioned whether respondents are willing to follow randomized response instructions. However, recent developments described in this chapter allow researchers to take care of respondents who do not follow instructions, making the randomized response methodology more suited to real-life applications.

3. A drawback of randomized response is that it reduces the power of statistical analyses. Clearly, misclassification has the effect that associations between variables in the populations are harder to detect. Yet, we believe that, because Web-based data collection is becoming increasingly inexpensive, this is not a serious drawback for the effective application of randomized response.

In conclusion, the randomized response framework has much to offer in protecting the privacy rights of individuals both during and after the collection of data. Although the framework scores high on usability and transparency criteria, much work remains to be done to make this methodology a mainstream and routine component of statistical work. The practical importance and urgency of this work cannot be underestimated, especially in view of the mounting tensions between confidentiality and the ever-increasing availability of e-commerce data.

REFERENCES

Ambainis, A., Jakobsson, M., and Lipmaa, H. (2003). Cryptographic randomized response techniques. *Cryptology ePrint Archive, Report 2003/027*. Available at http://eprint.iacr.org.

Böckenholt, U., Barlas, S., and van der Heijden, P.G.M. (in press). Do randomized-response designs eliminate response biases? An empirical study of non-compliance. *Journal of Applied Econometrics*,

Böckenholt, U. and van der Heijden, P.G.M. (2004). Measuring noncompliance in insurance benefit regulations with randomized response methods for multiple items. Proceedings of 19th International Workshop on Statistical Modelling.

Böckenholt, U. and van der Heijden, P.G.M. (2007). Item randomized–response models for measuring noncompliance: Risk–return perceptions, social influences, and self-protective responses. *Psychometrika*, 72: 245–262.

Boeije, H. and Lensvelt-Mulders, G. (2002). Honest by chance: A qualitative interview study to clarify respondents' (non)-compliance with computer-assisted randomized response. *Bulletin de Methodologie Sociologique*, 75: 24–39.

Boruch, R.F. (1971). Assuring confidentiality of responses in social research: A note on strategies. *The American Sociologist*, 6: 308–311.

Chaudhuri, A. and Mukerjee, R. (1988). *Randomized Response: Theory and Techniques*. New York: Marcel Dekker.

Chen, T.T. (1989). A review of methods for misclassified categorical data in epidemiology. *Statistics in Medicine*, 8: 1095–1106.

Cruyff, M., van den Hout, A., van der Heijden, P.G.M., and Böckenholt, U. (2007). Loglinear randomized response models taking cheating into account. *Sociological Methods and Research*, 36: 266–283.

Du, W. and Zhan, Z. (2003). Using randomized response techniques for privacy-preserving data mining. Meeting of Association for Computing Machinery, Special Interest Group Knowledge Discovery and Data Mining, Washington DC.

Evfimievski, A. (2002). Randomization in privacy-preserving data mining. *SIGKDD Explorations: Newsletter of the ACM Special Interest Group on Knowledge Discovery and Data Mining*, 4(2): 43–48.

Fienberg, S.E. (2005). Confidentiality and disclosure limitation. *Encyclopedia of Social Measurement*, 463–469. Amsterdam: North Holland.

Fienberg, S.E. (2006). Privacy and confidentiality in an e-commerce world: Data mining, data warehousing, matching and disclosure limitation. *Statistical Science*, 21: 143–154.

Fox, J.A. and Tracy, P.E. (1986). *Randomized Response: A Method for Sensitive Surveys*. Newbury Park, CA: Sage.

Fox, J.-P. (2005). Randomized item response theory models. *Journal of Educational and Behavioral Statistics*, 30: 1–24.

Gouweleeuw, J.M., Kooiman, P., Willenborg, L.C.R.J., and de Wolf, P.-P. (1998). Postrandomisation for statistical disclosure control: Theory and implementation. *Journal of Official Statistics*, 14: 463–478.

Karr, A.F. and Fienberg, S.E. (2005). Data confidentiality, data quality and data integration for federal databases: Foundations to software prototypes. Available at the Digital government II home page: http://www.niss.org/dgii/index.html.

Kooiman, P., Willenborg, L.C.R.J., and Gouweleeuw, J.M. (1997). PRAM: A method for dislosure limitation of microdata. Research Paper No. 9705, Voorburg/Heerlen: Statistics Netherlands.

Kuha, J. and Skinner, C. (1997). Categorical data analysis and misclassification. In *Survey Measurement and Process Quality* (in L. Lyberg et al., eds.). New York: Wiley.

Kuk, A.Y.C. (1990). Asking sensitive questions indirectly. *Biometrika* 77: 436–438.

Lensvelt-Mulders, G.J.L.M., Hox, J.J., van der Heijden, P.G.M., and Maas, C. (2005). Meta-analysis of randomized response research: 35 years of validation. *Sociological Methods and Research*, 33: 319–348.

Lensvelt-Mulders, G.J.L.M., van der Heijden, P.G.M., Laudy, O., and van Gils, G. (2006). A validation of a computer assisted randomized response survey for measuring fraud in social security. *Journal of the Royal Statistical Society, Series A*, 169: 305–318

Maddala, G.S. (1983). *Limited Dependent and Qualitative Variables in Econometrics*. New York: Cambridge University Press.

Rasch, G. (1980). *Probabilistic Models for Some Intelligence and Attainment Tests*. Chicago: University of Chicago Press. (Original published 1960, Copenhagen: Danish Institute of Educational Research.)

Rosenberg, M.J. (1979). Multivariate analysis by a randomized response technique for statistical disclosure control. Ph.D. dissertation, University of Michigan.

Scheers, N.J. and Dayton, C.M. (1988). Covariate randomized response models. *Journal of the American Statistical Association*, 83: 969–974.

Trottini, M., Fienberg, S.E., Makov, U.E., and Meyer, M.M. (2004). Additive noise and multiplicative bias as disclosure limitation techniques for continuous microdata: A simulation study. *Journal of Computational Methods in Sciences and Engineering*, 4: 5–16.

Van den Hout, A. and Elsayed, E.A.H. (2006). Statistical disclosure control using post randomisation: Variants and measures for disclosure risk. *Journal Of Official Statistics*, 22: 711–731.

Van den Hout, A. and Kooiman, P. (2005). Estimating the linear regression model with categorical covariates subject to randomized response. *Computational Statistics and Data Analysis*, 50: 3311–3323.

Van den Hout, A. and van der Heijden, P.G.M. (2002). Randomized response, statistical disclosure control and misclassification: A review. *International Statistical Review*, 70: 269–288.

Van den Hout, A. and van der Heijden, P.G.M. (2004). The analysis of multivariate misclassified data with special attention to randomized response data. *Sociological Methods and Research*, 32: 310–336.

Van den Hout, A., van der Heijden, P.G.M., and Gilchrist, R. (2007). The logistic regression model with response variables subject to randomized response. *Computational Statistics and Data Analysis*, 51: 6060–6069.

Van Gils, G., van der Heijden, P.G.M., and Rosebeek, A. (2001). *Onderzoek naar regel-overtreding, Resultaten ABW, WAO en WW.* Amsterdam: NIPO. (In Dutch.)

Warner, S.L. (1965). Randomized response: A survey technique for eliminating answer bias, *Journal of the American Statistical Association*, 60: 63–69.

Warner, S.L. (1971). The linear randomized response model. *Journal of the American Statistical Association*, 66: 884–888.

INDEX

Statistical Methods in e-Commerce Research. Edited by W. Jank and G. Shmueli

STATISTICS IN PRACTICE

Human and Biological Sciences

Brown and Prescott · Applied Mixed Models in Medicine
Ellenberg, Fleming and Demets · Data Monitoring Committees in Clinical Trials: A Practical Perspective
Lawson, Browne and Vidal Rodeiro · Disease Mapping with WinBUGS and MLwiN
Lui · Statistical Estimation of Epidemiological Risk
*Marubini and Valsecchi · Analysing Survival Data from Clinical Trials and Observation Studies
Parmigiani · Modeling in Medical Decision Making: A Bayesian Approach
Senn · Cross-over Trials in Clinical Research, *Second Edition*
Senn · Statistical Issues in Drug Development
Spiegelhalter, Abrams and Myles · Bayesian Approaches to Clinical Trials and Health-Care Evaluation
Whitehead · Design and Analysis of Sequential Clinical Trials, *Revised Second Edition*
Whitehead · Meta-Analysis of Controlled Clinical Trials

Earth and Environmental Sciences

Buck, Cavanagh and Litton · Bayesian Approach to Interpreting Archaeological Data
Glasbey and Horgan · Image Analysis in the Biological Sciences
Helsel · Nondetects and Data Analysis: Statistics for Censored Environmental Data
McBride · Using Statistical Methods for Water Quality Management: Issues, Problems and Solutions
Webster and Oliver · Geostatistics for Environmental Scientists

Industry, Commerce and Finance

Aitken and Taroni · Statistics and the Evaluation of Evidence for Forensic Scientists, *Second Edition*
Brandimarte · Numerical Methods in Finance and Economics: A MATLAB-Based Introduction, *Second Edition*
Chan and Wong · Simulation Techniques in Financial Risk Management
Lehtonen and Pahkinen · Practical Methods for Design and Analysis of Complex Surveys, *Second Edition*
Ohser and Mücklich · Statistical Analysis of Microstructures in Materials Science

*Now available in paperback.